T0138284

Darwin's Evolving Identity

Darwin's Evolving Identity

Adventure, Ambition,
and the Sin of Speculation

Alistair Sponsel

The University of Chicago Press
Chicago and London

The University of Chicago Press, Chicago 60637
The University of Chicago Press, Ltd., London
© 2018 by The University of Chicago
All rights reserved. No part of this book may be used or reproduced in any manner
whatsoever without written permission, except in the case of brief quotations in
critical articles and reviews. For more information, contact the University of
Chicago Press, 1427 E. 60th St., Chicago, IL 60637.
Published 2018
Printed in the United States of America

26 25 24 23 22 21 20 19 18   1 2 3 4 5

ISBN-13: 978-0-226-52311-8 (cloth)
ISBN-13: 978-0-226-52325-5 (e-book)
DOI: 10.7208/chicago/9780226523255.001.1

Library of Congress Cataloging-in-Publication Data

Names: Sponsel, Alistair William, 1978– author.
Title: Darwin's evolving identity : adventure, ambition, and the sin of speculation /
    Alistair Sponsel.
Description: Chicago ; London : The University of Chicago Press, 2018. | Includes
    bibliographical references and index.
Identifiers: LCCN 2017031816 | ISBN 9780226523118 (cloth : alk. paper) | ISBN
    9780226523255 (e-book)
Subjects: LCSH: Darwin, Charles, 1809-1882. | Darwin, Charles, 1809–1882—
    Authorship. | Darwin, Charles, 1809-1882—Knowledge—Geology. | Lyell,
    Charles, Sir, 1797–1875. | Geology—Great Britain—History—19th century.
Classification: LCC QH31.D2 S66 2018 | DDC 576.8/2092—dc23
LC record available at https://lccn.loc.gov/2017031816

♾ This paper meets the requirements of ANSI/NISO Z39.48-1992 (Permanence of
Paper).

Contents

Preface   *ix*

Introduction   1
   Themes   5
   Plans   7

PART I   THEORIZING ON THE MOVE   13

   1   **Darwin's Opportunity**   *15*

      Coral Reefs as Objects of Fascination and Terror   19
      Studying Reef Formation as an Objective of the
         Beagle Voyage   24
      Darwin's Training in the Sciences   27
      Enthusiasm for the South Sea Islands   30

   2   **An Amphibious Being**   *33*

      Darwin's Approach to Scientific Work at the
         Beginning of the Voyage   36
      Hydrography Becomes a Resource for the Naturalist   37
      An Ambitious Plan for Studying Zoophytes   41

   3   **Studying Dry Land with a Maritime Perspective**   *47*

      Applying the Lessons of Hydrography to the
         Interpretation of Geology   50
      Elevation and Subsidence   53

4   The Making of a Eureka Moment   *63*

The Dangerous Reefs of the Low Archipelago   65
The View from Tahiti   68
Theorizing Like Humboldt in a Floating Library   73

5   The Surveyor-Naturalist   *81*

Darwin's Sea-Level Study of the South Keeling Reef   82
Seeing Underwater: The Hydrographic Survey at
    South Keeling   91
Darwin's Hydrographic Initiative at Mauritius   98

PART II   TRAINING IN THEORY   103

6   Lyell Claims Darwin as a Student   *105*

Homeward Bound as an Aspiring Geologist   108
Lyell as an Author   112
Master and Student   114
The Primacy of Geology in Darwin's Private, as Well as
    Public, Activities   121

7   Darwin's Audacity, Lyell's Choreography   *125*

Going Public   126
Putting the Coral Theory to Work   131
Species   136
An Astonished Response from the Geological Elite   140
Darwin's Emergence as a Practitioner of Lyellian
    Geological Speculation   144

8   Burned by Success   *149*

Darwin's New Persona   151
The Obligations of a Student to His Master   156
The Beginnings of Darwin's Anxiety about Speculation   160

PART III   A DIFFERENT APPROACH TO AUTHORSHIP   167

9   The Life of a Tormented Geologist (and Enthusiastic
    Evolutionist)   *169*

Darwin's Turn toward Empiricism and the Ideal of
    Comprehensiveness   170
The Pressure of Public Expectations   173
Lyell's Appropriation of the Coral Reef Theory   178
Studying Species as a Diversion from the Task at Hand   182

10   A Finished Task: Darwin's Treatise on Coral Reefs   *185*

The Space between Lyell and Darwin   186
A Mountain of Facts   188

The Theory Emerges   192
The Immediate Reaction to *Coral Reefs*   203
A Theory in Use and in Memory   209

PART IV   WRITING THE *ORIGIN* WITH HIS "FINGERS BURNED"   213

11   **Atoning for the Sin of Speculation**   *215*

Balancing Speculation with Facts   217
Rejecting Lyell's Suggestion to Publish a "Sketch"   224
Lyell Choreographs Another Debut   230
Publishing an "Abstract" After All: *On the Origin of Species*   240
Dealing with Darwin's "Recollections"   248

Conclusion   *259*

Lyell, Darwin, and Authorship   259
Studying Practices, Learning about Theories   264

Acknowledgments   *269*

Notes   *277*

Bibliography   *329*

Index   *349*

Gallery of color plates follows page 150.

Preface

I wrote this book with several audiences in mind. I want people who have a personal or scholarly interest in the history of science to find here an illustration of how our understanding of scientific *theories*, in particular, may be enhanced by attention to the local contexts of scientific practice. While I hope this book will be read for its broader significance to the social studies of science, my focus on the genesis and development of Charles Darwin's theories led me to adopt a new perspective on his motivations and accomplishments. Therefore my second goal is that Darwin enthusiasts of all types, including (but not limited to!) practicing scientists and practicing historians, will find it rewarding to encounter new perspectives on his youthful immersion in maritime culture, his early zeal as an ambitious geological theorist, his anxieties about publishing and about meeting the expectations of various intended audiences (not least his scientific "master," Charles Lyell), and his reasons for keeping the idea of natural selection out of public view for more than twenty years. Finally, and happily, because chronology is important for my argument about various developments over time, it is also possible for all sorts of readers to enjoy this book as a narrative.

Historians seeking a speedy overview of my arguments might proceed by reading the introduction, the condensed narrative offered by the chapter and part titles, and the conclusion, which addresses broader themes in science studies. The endnotes contain several brief essays relating my argument to other scholarship on science studies in general and Darwin in particular.

Finally, a note on my transcription of manuscripts: I have retained the occasional (or, in the case of Darwin's *Beagle* manuscripts, frequent) misspellings and, as far as possible, punctuation. Such elements have proved useful to scholars in determining the date and manner of composition of various passages. For example, a proliferation of full stops (or pen rests) may suggest halting thinking. Deletions by the writer of the manuscript are indicated by text ~~that has been struck out~~, and the writer's insertions are contained within <<pairs of angle brackets>>. Words in manuscripts that were underlined in the original are <u>underlined</u> in this text. Words or phrases I want to emphasize within quotations are in *italics* (with a note in the corresponding citation to confirm that this is my own emphasis).

Introduction

Charles Darwin's career as a scientific author has a great contradiction at its center. His most consequential book, *On the Origin of Species*, which he published in 1859 at age fifty, was in one respect the culmination of more than twenty years' work. Yet the published text was the product of just a few months of frantic writing. The book was a substantial volume of nearly five hundred pages of dense text laying out a broad and erudite argument for Darwin's theory of evolution by natural selection. Yet, as Darwin was at pains to point out to his readers, *Origin* was a mere "abstract" of the full-scale work he had originally intended to write.

The reason for *Origin*'s rushed composition is well known. Another naturalist had unwittingly jolted Darwin into action by sending him an essay that contained an idea almost identical to natural selection. And it might seem obvious why Darwin had originally planned to keep his theory unpublished until he could reveal it in an enormous, exhaustive, multivolume work. Evolution was a controversial topic in nineteenth-century Britain, and Darwin might very reasonably have been reluctant to expose himself and his family to the repercussions of admitting he believed in descent with modification.

In this book I offer a new explanation for Darwin's caution. I show that caution was not his *original* attitude toward publishing at all, and I suggest that Victorian squeamishness about evolution played a smaller role in shaping his plans than did his anxiety about a rather different social convention. My argument is that Darwin's restrained approach to publishing the species theory was a conscious attempt to avoid repeating missteps he felt he had taken as a rash young geological author. Those missteps were specific to the challenge of publishing *theories*.

Once upon a time, as a young naturalist traveling around the world, he had exulted that hypothesizing on geology was akin to "the pleasure of gambling."[1] After returning to Britain, being groomed for a partisan role in the geological debates of the day, embracing a strategy of speedy publication, and consequently absorbing criticism for his theoretical boldness, Darwin became convinced that he had committed an authorial "sin of speculation."[2] Thus his apparent reluctance to publish his species theory was not a product of innate caution, or even of being taught to proceed deliberately, but instead was a counterreaction to the way he *had* been taught to operate as a scientific author.

It is rather unconventional for me to argue that Darwin's early career as a brilliant, ambitious geologist taught him how *not* to become the author of his work on species. Many historians have justly pointed out that geology was his main preoccupation during a five-year voyage around the world on HMS *Beagle* (1831–36). Upon his return to England he rapidly achieved a high profile in the geological community, and his chief public and private identity was for many years that of a geologist.[3] Scholars who take an interest in Darwin's geological work tend to argue for its significance to his later career by pointing out the ways it served as an affirmative model for his work on species, for example, by noting that it was as a geologist that he first theorized about change over long time scales. I wholly endorse such statements. Indeed, Darwin's work as a geologist was as innovative in field research methods as it was far-reaching in scope.

My asserting that Darwin's geological career had important negative consequences is therefore not to say it was unsuccessful in some abstract sense, but rather to claim that it did not yield the type of success Darwin had expected. His experience as a geological author exhilarated and then chastened him, raising his ambitions before reshaping his understanding of how scientific ideas succeeded and failed. He did not, on the whole, conclude that his geological ideas had been wrong. On the contrary, he retained a strong sense of their accuracy and utility. This made it all the more distressing to think he had undermined those ideas by letting himself be too bold, too soon in public.

Darwin found himself so stung by the outcome of his geological publishing strategy precisely because his stated ambitions, and his colleagues' consequent expectations, were so high. He originally claimed he would follow a set of short, provocative publications after the voyage by writing a substantial book. This would provide the evidence for a grand theory of the earth, which he had gestured toward in the short papers, claiming that a single cause, or mechanism, linked volcanoes, earthquakes, the appearance and disappearance of islands and continents, and perhaps even the origin of species. Those ambitious early papers earned him considerable attention but created demands that his geological book be not just comprehensive and rigorous, but also completed in short order. Darwin found himself unable to endure these combined pressures. Gradually his plans changed: he would publish not a great geological treatise but instead a series of more narrowly defined books. The longer he took to complete even the first of those books, the more premature and speculative his original publications appeared in retrospect.

Finally in 1842, after laboring for many more years than he had originally expected, he published not a grand theory of the earth but instead a slim treatise about one element of the planned synthesis, his theory of coral reef formation.[4] This was followed by a second volume on volcanic islands in 1844, and then a hefty tome on the geology of South America in 1846. Each of these books was less theoretical than the last. His initial willingness to make theories public in short articles rather than couching them in a great mass of facts had, ironically, weakened his standing—and his appetite—to speculate, when those facts were finally marshaled.

These missteps unfailingly entered his thoughts in later moments when he was trying to decide how, and how soon, to publish his evolutionary theory. As I show in chapter 11, there were several occasions between 1842 and 1858 when he was forced by circumstance to wrestle with this decision. It is striking how little his deliberations hinged on the *topic* of that theory and how much they focused on the broader question of how theories should be presented. While he did acknowledge with some trepidation the likelihood that some of his readers would focus on the topic and brand him an atheist, what he really agonized over was the prospect that the elite naturalists he most wanted to convince would dismiss his whole approach to science by branding him as reckless, speculative, and "unphilosophical." His certainty that they would do so was drawn from those earlier experiences, and it mattered little that his soberest geological books had eventually secured his reputation in that science. The best chance to have *any* theory accepted, he concluded, was to be sure it would face public scrutiny for the first time only when it could be shielded in a thick armor of facts.[5]

The idea that Darwin's deliberate approach to publishing on species had less to do with the topic of that theory than with his general views on authorship ultimately led me to reassess much of what I previously thought about how Darwin viewed his species theory. It is well known that he grew increasingly anxious and ill from 1837 to 1842, the years when he was first compiling his private notebooks on species, and many scholars have drawn a direct connection between his fretfulness and his decision to reject the doctrine of special creation. I continue to share the view (with scholars such as Ralph Colp, Adrian Desmond, and James Moore)[6] that Darwin became distressed during these years, but I have come to a different conclusion about the primary cause of his strain. Darwin's private woes were very often linked to his perception of mounting *external* pressure to complete the credible, empirically rigorous geological publication(s) he had promised would follow soon after his startling, speculative ones.[7] It might initially sound absurd to think of the "tormented evolutionist" as a "tormented geologist" or—stranger still—a "tormented reef morphologist," but he was all these things at one time or another. The common theme that links all three is that he was a *tormented theorist*: tormented by the difficulty of presenting his ambitious theories to a scientific community that often condemned the act of theorizing.[8]

In these years Darwin kept a diary to record his progress on a range of scientific undertakings. The language he used in that journal reveals that he considered the geological publishing project—no longer the grand theory of the earth but the more narrowly defined coral reef theory—to be his chief task between 1837 and 1842 (the exact years he spent making his private species notebooks). Hard as it might be to believe in retrospect, Darwin felt that his reputation hung on whether he could finish a book about coral reefs. This is the only project for which he recorded episodes when he had *failed* to make progress; otherwise he simply recorded what he had actually done. When those activities instead involved species, he often registered his "idleness." For example, on 14 September 1838 he wrote that he had "frittered these foregoing days away in working on Transmutation theories." That evening, in a letter to his taskmaster Charles Lyell, he admitted, "I wish with all my heart that my Geological book was out. . . . I have every motive to work hard. . . . I have lately been sadly tempted to be idle, that is as far as pure geology is concerned, by the delightful number of new views, which have been coming in . . . bearing on the question of

species—note book after note book has been filled."[9] "Idleness" was synonymous not with inactivity but with dereliction of whichever task ought
to have taken precedence. Far from being the cause of his anxiety, Darwin
found making notes on species to be an exhilarating diversion from the
high-pressure work of writing his geological book.

Themes

This book may be about Charles Darwin, but its purpose is to examine a set
of fundamental questions about science as a vocation and a body of knowledge. How does an individual become a member of a knowledge-making
community? What are the grounds for making a credible scientific claim?
In what ways do rhetorical skills and interpersonal skills, so to speak, operate alongside skills like collecting and experimenting in the production of
scientific knowledge? Are those different skills acquired in distinct ways?
What is the relation between theorizing as an abstract cognitive process (if
indeed it can be isolated as such) and theorizing as an act of authorship?
And who, finally, can "own" a theory—or even a theoretical tradition—that
has become commonly acknowledged intellectual property?

Many of these questions were originally posed by sociologists of science
who sought their answers by studying the workings of contemporary laboratory groups or scientific communities.[10] There are several advantages to
studying these themes from a historical perspective instead, and in particular with respect to Charles Darwin and the scientific community in early- to
mid-nineteenth-century Britain.[11] A general virtue of historical distance
is that I am able to write about developments over the course of several
decades. This allows me to study long-term change in individuals' priori-
ties and in community norms and to treat authorial identity, for example,
as something contingent on specific moments in a career, a scientific discipline, and a broader social context. And, as we shall see, a sustained historical case creates the conditions of possibility for discovering the role
retrospection and memory played in reshaping theories, and vice versa.[12]

There is a lot at stake in choosing Darwin's entry into the scientific
community and the attendant development of his theories as a case for
understanding the production of scientific knowledge. Perceptions about
Darwin's life and his theories have played a disproportionately large role in
shaping many people's ideas about the history of science as a whole. I acknowledge that one consequence of devoting my energy to this case study
is to perpetuate a cycle that keeps a handful of major figures at the fore-

front of our histories of science.[13] I come not to praise Darwin, however, but to embed him within a type of study usually reserved for less renowned figures.

My method has been to "follow" Darwin through his immediate interactions with a series of collaborators, interlocutors, patrons, and audiences.[14] I am, as I explain in the conclusion, studying everyday practices in an effort to gain a new understanding of theories in natural history. Darwin is the focal point of this book, but my interest is in understanding how theories came to be shared (and credit attributed) within whole communities. In imagining Darwin's life, though, it can often seem that he operated outside the social world of science, either because he was on a voyage or because he was cloistered owing to the secrecy of his work and the severity of his physical ailments. And indeed, many familiar episodes from his life would have us believe that he simply did not need face-to-face contact of the kind that is often invoked to explain how other scientists developed their ideas and their reputations.[15] There are good reasons to think that Darwin's scientific career should have come more easily than it would have for almost anyone else. He had the fortunate combination of affluence, male gender, and respectable social class that opened doors to him in England and around the world. He had a family name that gave him a reputation in the sciences before he had contributed anything himself. He had the opportunity as a young traveler to amass a spectacular wealth of experience and evidence without having to surrender his time or his specimens to the dictates of a patron or an employer. Given all this, not to mention his formidable intellect and acquired diligence, we might well expect that very few obstacles ever stood in his way—that Darwin's scientific legitimacy and his evolutionary theory's value would simply have been self-evident.

So, did Darwin of all people require the kind of private mentorship that historians and sociologists of science have shown was necessary for the acquisition of specialist laboratory skills and for admittance to rarefied intellectual communities? Did even the privileged Darwin face challenges in establishing his credibility within a scientific community whose every social convention already favored him? And did the man who believed he had solved the scientific puzzle so perplexing and important that it was known as the "mystery of mysteries" have to work under the assumption that his ideas would never be judged simply on their merit?

In what follows I will argue that the answer to all these questions is yes. The moral is not simply that we should admire Darwin even more. The moral is that insights from the social studies of science have a great deal to offer to our understanding of "exceptional" figures. A now familiar dictum in science studies is that examining successful *and* unsuccessful science symmetrically, and from an impartial perspective, sheds a great deal of light on the processes by which "matters of fact" get established in the first place.[16] I set out to study a successful figure *as though* he would not inevitably succeed. As a result I have gained a new perspective on Darwin's goals, motivations, and insecurities, a revised understanding of the very "success" I took him to have had, and an added appreciation for the ways this success was fostered by people who surrounded Darwin in everyday life.

Plans

The *question* this book addresses is, "How did Darwin transcend the public identity of a traveling naturalist to become a theoretical author and, ultimately, the type of theoretical author he became?" My *answer* is that developing and publishing his coral reef theory, within the context of his broader geological ambitions, was formative in making Darwin the type of author we now remember him to be. Therefore the *topic* of this book is, in the main, the history of Darwin's theory of coral reef formation. This is of necessity a book about how the young Darwin found himself on coral reefs of the Pacific and Indian Oceans, why he thought it was worth trying to explain their origin, and whether (as he much later claimed) his coral reef theory was a simple and successful idea that he never needed to revise.

In part I, "Theorizing on the Move," I focus on the *Beagle* voyage, during which Darwin developed a bold theory about the marine origins of the continent of South America in addition to a new explanation for the formation of coral reefs. My aim is to understand when and why Darwin originally came to believe that he was developing *theories* rather than merely making observations and collecting specimens. By revealing how Darwin learned specific field practices from his shipmates' everyday work of hydrography (maritime surveying), I argue that he gained a new type of knowledge about the ocean that distinguished him from other naturalists of his day. His subsequent decision to incorporate the surveyors' methods into natural history research, it turns out, predisposed him to attend closely to the interrelations and geographical distribution of organisms and rocks, indeed to adopt what might be called a proto-ecological perspective.[17] His hydrographic experience also allowed him to make novel comparisons between sedimentary rocks and fossils on land (the relics of former seabeds) and the present-day seafloor. He thereby realized geologist Charles Lyell's fanciful yearning for the superior form of geology that might be practiced by an "amphibious being."[18] In other words, the ways of seeing the world that seem so distinctively Darwinian—the focus on interactions in the environment and the concern with change over time—were fostered from the first months of the voyage by the lessons he learned from his shipmates.

In reassessing the *Beagle* voyage I also argue against the conventional understanding of how Darwin created and developed his theory of coral reef formation. He himself wrote, late in life, that it had been a product of pure deduction, that he had come up with the theory in South America before he had ever actually seen the coral reefs of the Pacific and Indian Oceans. (See the chart in figure 1, which is designed to illustrate the sequence in which the *Beagle* visited these places.) Over the years, scholars have affirmed that story and have treated the coral reef theory as being a fait accompli by the late stages of the voyage, albeit one that showed Darwin's early genius for theorizing. I have pieced together a much different story, one that contradicts almost every aspect of this defining accomplishment of the *Beagle* years. He did not hit on an explanation for the formation of coral reefs in South America before ever seeing a coral reef; he had this eureka moment in Tahiti while looking directly at one. He had not set out to study coral reefs from the perspective of a geologist; in the voyage's first four years he expected studying them to be the climax of his project to become the world's leading authority on the *zoology* of corals. And the coral reef theory was not just a single good idea, formulated during the voyage and expounded in publication: it was not a single idea at all. It began as a surprise consequence of Darwin's attention to hydrography and became a new organizing principle for his work in the field and in the library, as well as a loosely defined piece of intellectual property that helped establish his place in the world of science. In a sense, he had precisely as many distinct coral reef theories as he had distinct audiences to cultivate.[19] This was not Darwin's sense, however, for as I illustrate in the closing pages of this book, he understood himself to have a single theory that had developed over time.

All this raises a challenging question: If Darwin's coral reef theory was so many things at so many different times, when and how did it become the theory we have always thought it was? In part II of this book, "Training in Theory," I answer that question by arguing that Darwin's ideas about coral reefs were molded into a new and explicitly geological theory—and

Voyage of H.M.S. *Beagle*, 1831-1836

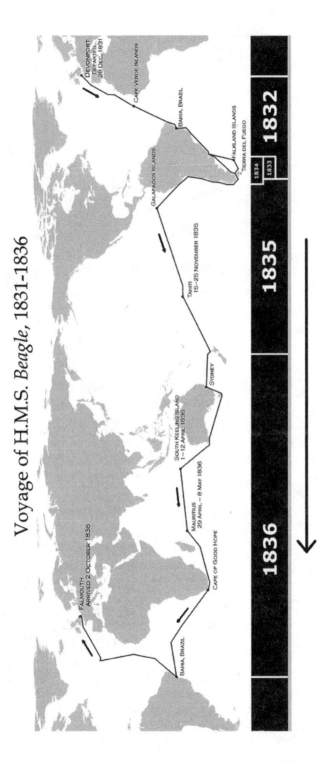

Figure 1. Chart showing the *Beagle's* track from 1831 to 1836 as a continuous line that can be read from right to left. This map illustrates the sequence in which Darwin encountered the Atlantic Ocean, South America, the Pacific Ocean (including Tahiti), the Indian Ocean (including the coral island of South Keeling and the reefs of Mauritius), and finally the Atlantic Ocean again. Map by the author.

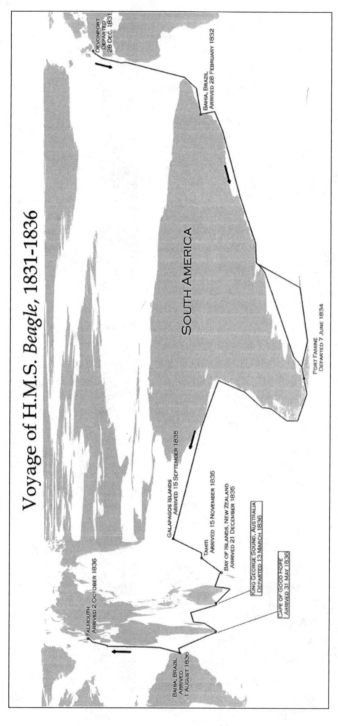

Figure 2. Track chart of the *Beagle* voyage with base map skewed to illustrate the rate of travel. The widths of the map segments between each named location are scaled in proportion to the fraction of the voyage that elapsed between the dates provided. South America's exaggerated size reflects the fact that the main task of the voyage was to carry out surveys that would form the basis for new charts of the continent's southeast and southwest coasts. Map by the author.

Darwin was simultaneously molded into a particular kind of theoretical author—through the active intervention of a new geological mentor, Charles Lyell. Among other sources, I draw on newly accessible private notebooks that remain in the Lyell family home in Scotland in order to show how intensively Lyell and Darwin were working in tandem during this critical period in the careers of both authors.

Lyell is most famous today for the role his *Principles of Geology* is taken to have played in shaping Darwin's worldview *before* he had returned from his five-year voyage around the world.[20] I emphasize instead the face-to-face interactions between Lyell and Darwin during the five years *after* the voyage, arguing that the pair became more than mere friends or allies. They established a relationship with a very particular structure: master and student. Each man recognized the mutual obligations this entailed, and they took those responsibilities seriously. Together they created for Darwin a dazzling debut, in which he presented a series of audaciously theoretical papers to the Geological Society of London while Lyell worked behind the scenes, and occasionally at center stage, to bolster the credibility of Darwin's ideas. This caused a surge of enthusiasm among geologists for Darwin's forthcoming treatise. Having prepared Darwin for success by guiding his writing and cultivating his audiences, Lyell in return assumed the privilege of incorporating Darwin's fresh ideas into his own publications and accepted acknowledgment as (in effect) the senior author of the entire body of work to which Darwin's publications contributed. Like many such relationships, however, this one became tense when Lyell found Darwin's productivity (on geological work) slipping and when Darwin sensed that his own interests and personality might be better served by forging a more independent authorial persona.

Thus in part III, "A Different Approach to Authorship," I demonstrate that Darwin came to believe his bold geological debut had made it more difficult to ensure the lasting success of his theories. He struggled over several years to deliver a book that could live up to his audiences' expectations and his own ambitions. Even as he was recalibrating his goals for a geological treatise that he could not quite complete, Darwin was also resolving never again to let a good theory be compromised by a rash publishing strategy. When he finally published his first geological treatise, on coral reefs rather than on the fuller theory of the earth, his curbed—but still significant—speculations were supported by a resolutely empirical catalog of every coral reef in the world.[21]

My parts 1–3 might have constituted a book by themselves, making an overt point by decentering Darwin's species work right through to the end. Sure enough, Darwin's life, and his place within a scientific community, look very different when we follow where he spent his time and energy instead of according special interest to the small subset of his activities that bore directly on his theory of evolution by natural selection. It should be possible, after all, to write a book about Darwin that resists the urge to make everything turn out to be about the *Origin*.

In fact, that is exactly what I had in mind through most of the years I spent on this project. When I had finished drafting the book that formed the basis for the first three parts of this text, though, I could not help noticing that the familiar story of why Darwin adopted his deliberate approach to publishing on species no longer rang quite true. The final, fourth part, "Writing the *Origin* with His 'Fingers Burned,'" is the result of my effort to see what impact Darwin's formative years as a theoretical author had on the way he decided to publish the species theory. I have already foreshadowed some of the results of that exercise. Darwin remained obsessed with fashioning a reputation for meticulous caution. His decisions about when to publish continued to be shaped by Lyell's advice, though not always in a positive sense. The master's and student's shared experience of cultivating audiences and managing the attribution of credit took on a new urgency when, in 1858, a surprising letter arrived from the naturalist Alfred Russel Wallace. Just as cannily as Lyell had once cultivated and co-opted Darwin's work in support of his own geological principles, he now acted to position Darwin as the master of Wallace's oeuvre. When the *Origin* finally appeared it was not the book Darwin had intended to write, but it was the book he ended up having to defend. From the opening lines of that book he urged his readers to believe he had "not been hasty" in developing his theory and deciding to publish. I finish by tracing Darwin's strategies from the *Origin* to the words of the autobiographical "Recollections" he wrote late in life, in which he reinvented his coral reef theory one last time as a foil for the species theory that had proved so controversial after all.

PART I

Theorizing on the Move

1

Darwin's Opportunity

Nineteenth-century readers were introduced to the idea of a fanciful "amphibious being" by the barrister-turned-geologist Charles Lyell in his *Principles of Geology* (vol. 1, 1830). He subtitled his treatise *An Attempt to Explain the Former Changes of the Earth's Surface by Reference to Causes Now in Operation* and intended the book as a methodological object lesson. Lyell wanted to demonstrate that it was possible to provide a convincing historical explanation for the current state of the globe without referring to physical processes of any type or intensity other than those demonstrably acting in the present. The book is remembered as one of the most important in the history of geology primarily for convincing some readers, notably Charles Darwin, that if the history of the earth was long enough, then the types of organic and inorganic processes at work in the modern world, operating at their observed intensities, could eventually produce monumental changes in the earth's landscape and its inhabitants. In Lyell's view there was no necessary direction to these changes. It was clear that some spots on the surface of the globe that had once been dry land had become sea, and that some of those had again become land.

Lyell evoked this cyclical movement of the earth's crust with his volume's striking frontispiece, an engraving of three ancient columns in Pozzuoli, Italy, that had belonged to a building known as the Temple of Serapis (see fig. 3). Though the columns stood upright, each one had been pockmarked in the section now between twelve and twenty-four feet above ground by *Lithodomus*, a mollusk that still inhabited the nearby Bay of Naples. By 1830 the question of how these columns could have been submerged to a depth of twenty-four feet, subjected to the ravages of stone-eating mussels, then exposed to the air again was already a well-known geological puzzle. Whereas many observers believed that a change in sea level, whether local or global, was the only explanation compatible with the fact that the columns had not been toppled, Lyell drew on analogies with the observed effects of certain recent earthquakes to contend that the level of the land could have oscillated without necessarily disturbing the pillars.

In support of his methodological precept of reasoning from observed causes, Lyell dedicated the bulk of the *Principles*, which eventually comprised three volumes, to a catalog of physical and organic forces now operating on the globe. This required acknowledging the limitations of human perception, and it led to the passage that seems so noteworthy in retrospect. Lyell explained that as natives of the earth's surface, our inability to dwell underground or in the water had constrained the progress of geology.

"Although the reader may, perhaps, smile at the bare suggestion," he continued, a geologist who could dwell beneath the sea would gain a better understanding of the processes that were shaping, and had shaped, the physical world. This "amphibious being" would, Lyell declared,

> more easily arrive at sound theoretical opinions in geology, since he might
> behold, on the one hand, the decomposition of rocks in the atmosphere, and
> the transportation of matter by running water; and, on the other, examine
> the deposition of sediment in the sea, and the imbedding of animal remains in
> new strata . . . and might mark, on the one hand, the growth of a forest, and
> on the other that of the coral reef.[1]

Over the next few chapters I will explain how a young man named Charles Darwin transformed himself into an amphibious being and, remarkably, fulfilled each element of Lyell's prophecy. Darwin's metamorphosis began just a few months into his voyage on HMS *Beagle*, when he recognized the scientific opportunity presented by his shipmates' everyday activity of maritime surveying, or hydrography. He gradually came to apply the tools and practices of hydrography to his zoological work and then to his geological work as well. He did so not by emulating Lyell, as

Figure 3. "Present State of the Temple of Serapis at Pozzuoli." The frontispiece to the
first volume of Charles Lyell's *Principles of Geology* (1830) depicted the well-known
marble columns near Naples. Pockmarks in the middle part of the columns (the dark
bands about halfway up) revealed that they had once been partially submerged. The
fact that they remained standing was an obstacle to any explanation predicated on
a violent change in the level of the land or the sea. Image courtesy of the History
of Science Collections, University of Oklahoma Libraries; copyright the Board of
Regents of the University of Oklahoma.

might be imagined, but by drawing on practical experiences from his education in Edinburgh and Cambridge and by emulating the fact-gathering approaches of Alexander von Humboldt. It was only much later, when Darwin was back in London, that the full significance of these approaches for *Lyell's* geological views would be amplified by the personal interactions between Lyell and Darwin.

One of the pleasures of encountering the young Darwin through his notes and letters is witnessing the exuberance with which he studied the natural world during the *Beagle* voyage. His vigor was fueled by increasingly grand ambitions for what he might achieve in science when the voyage was finished. Over the years as the *Beagle* proceeded around South America from the Atlantic to the Pacific and eventually onward into the Indian Ocean, Darwin self-assuredly recorded a series of objectives for his future career: he aimed to rewrite the geological history of South America and the natural history of zoophytes and to advance a theory of coral reef formation he had developed at Tahiti. Several questions must be answered in order to comprehend the scope of Darwin's opportunity and his ambitions. Why was the *Beagle* passing through these oceans in the first place, and what was his role on board? How did he know these were promising topics on which to establish a scientific career? And what was the foundation for the confidence he developed in his own capacities as a geologist and marine zoologist?

In this chapter I examine three major contexts for Darwin's eventual ambitions. The first was the history of previous efforts to explain the origin and shape of reefs. To understand a theory in historical terms, we must uncover the questions to which it offered an answer. In order to grasp the particular features of reefs Darwin would feel most compelled to explain, I trace inquiries during the previous two generations of European imperial activity in the Pacific, culminating in the twin puzzles of reefs' annular shape and their apparent need for a shallow foundation to build on. This lively and ongoing European discourse about reef formation had a significant impact on the stated objectives of the *Beagle* voyage itself when it was commissioned by the British Admiralty in 1831. Indeed, one of the Admiralty's directions to Robert FitzRoy was to investigate the origin of coral reefs. The larger maritime and imperial purposes intended for the voyage are thus the second context I examine here. The story of how Darwin came to be aboard the *Beagle* has been told many times; I pay special attention to a little-noticed but elaborate discussion between Darwin, FitzRoy, and the Admiralty before the voyage over whether they would be directed to the "South Sea islands." Finally, I examine the intellectual and practical knowledge Darwin brought to the voyage. Here I call attention to his previous experience alongside experts in the sciences of marine zoology and terrestrial geology and to his early exposure to the work of Alexander von Humboldt. These three factors shaped his interpretation of (and original attention to) several of the key phenomena that proved relevant to the theories he eventually developed.

Coral Reefs as Objects of Fascination and Terror

The theory of reef formation that Darwin developed during and after the *Beagle* voyage offered an elegant new answer to a vexing and surprisingly widely asked question. Given that corals can live only in shallow water, how do they build reefs in the deepest parts of the ocean? The mystery was intensified by mid-ocean reefs' strange and distinctive shapes. They were narrow, and curved to form rings surrounding placid lagoons of shallow water. By the time Darwin first witnessed one in 1835, Europeans had been trying to explain the origin of such reefs for more than sixty years, dating back to the late-eighteenth-century wave of Pacific exploration that made household names of Cook and Bougainville, of William Bligh and Fletcher Christian.

The issue of coral reef formation became particularly compelling for several kinds of Europeans between about 1770 and 1830. There were those navigators and naval administrators for whom the threat reefs posed to intertropical navigation was a matter of serious practical concern. They were joined by natural philosophers who took a scientific interest in these structures because of the special position they appeared to hold in the so-called economy of nature. The idea (first proposed in the late eighteenth century) that reefs had been created by living organisms was a prominent topic of discussion among those who sought to catalog and explain the physical and organic causes of change in the natural world. Finally, writers and theologians who drew on the accounts of navigators and voyaging naturalists helped turn corals and reefs into powerful cultural symbols as they explored the contradiction between reefs' own vitality and the threat they posed to seafaring humans.

Among the marvels of the Pacific and Indian Oceans, none inspired a greater combination of fear and puzzlement in European navigators than those ring-shaped reefs we now call atolls. For the French circumnavigator Louis-Antoine de Bougainville, the vast expanse of these reefs east of

Tahiti was a seascape of contradictions. He called the group *l'Archipel Dangereux*, not out of concern for the inhabitants of "[reef-top] strips of land that a hurricane could bury at any moment beneath the water," but because those inconspicuous shoals posed a terrific threat to any unsuspecting ship that might be skimming across the Pacific in low light or poor weather. Yet the reefs and their inhabitants seemed so vulnerable too. Bougainville, who would return home reporting that the permissive Tahitians had escaped the fall of man on their mountainous Eden, wondered of the "almost drowned" low islands nearby, "Is this extraordinary land being born, or is it in ruins?"[2]

On 11 June 1770 a coral reef nearly claimed the man who was to become the most famous Pacific explorer before he had finished his first voyage to the South Sea. At a few minutes before 11:00 pm, as James Cook's ship *Endeavour* sailed northward off the eastern shore of the land he had dubbed New South Wales, the crewman who was taking soundings with a plumb line called out a depth of 17 fathoms, or 102 feet. "Before the Man at the [sounding] lead could heave another cast," Cook reported, "the Ship Struck and stuck fast . . . upon the SE edge of a reef of Coral rocks."[3] They had become the first Europeans to encounter the Great Barrier Reef, a labyrinth from which the *Endeavour*'s crew needed two months to sail clear.[4]

Cook and Bougainville returned from their Pacific voyages at the dawn of the 1770s with tales and charts of these tropical mysteries, fueling the imagination of philosophers and the terror of fellow navigators and the public. Corals sat atop submarine walls springing from depths that were often literally unfathomable with the lengths of rope carried by eighteenth-century explorers.[5] Their submarine slope was always steepest on the windward side, as if the reefs had been specially designed to deprive mariners who rode the winds and currents of even the slightest warning of a shoaling sea.

The first person to propose that these terrifying structures had been created by the growth of corals was Johann Reinhold Forster, the naturalist who along with his son Georg zigzagged the Pacific on Cook's second voyage (1772–75). Forster created a taxonomy of tropical islands, distinguishing ring-shaped "low islands" from the high islands like Tahiti and others whose origin he attributed to the volcanic agency of "subterraneous fire."[6] In explaining the origin of the low isles, Forster offered the then novel claim that corals built the entire structure of what we now call "coral" reefs. While Europeans had long been familiar with fossil corals in geological formations, and though there had been debates through the eighteenth century about whether coral organisms, which produced small stony forms in the Mediterranean and elsewhere, were plants or animals, nobody had previously suggested that they could build massive structures such as those in the Pacific and Indian Oceans. Forster proclaimed that the ringlike shape of low islands was entirely due to the "instinct" of the "animalcules forming these reefs." Corals grew upward and outward from a small base on the seafloor in order to "secure in their middle a calm and sheltered place" where the "impetuosity of the winds, and the power and rage of the ocean" could not encroach.[7] The communal structure therefore emulated the radial shape of an individual polyp. The British navigator Matthew Flinders finished surveying the Australian coast in 1803 full of support for Forster's views, explaining in new detail how "future races of these animalcules erect their habitations upon the rising bank, and die in their turn, to . . . elevate, this monument of their wonderful labours."[8] Meanwhile, thanks to the specimens and illustrations produced on the 1800–1803 French expedition under Nicolas Baudin, Pacific corals also began to colonize the pages of taxonomic works being produced by land-bound naturalists such as the Frenchman J. V. F. Lamouroux.[9]

Forster's explanation held tantalizing geological implications, for it suggested that corals were among the few agents with the capacity to remodel the surface of the globe on a large scale. Lamouroux pointed out that "reefs . . . may eventually block communication between the temperate zones of the two hemispheres."[10] As the sovereigns of Europe raced to master the Southern Ocean, these concerns stimulated navigators and their philosophically minded shipmates into a wider range of reef studies. Johann Friedrich Eschscholtz, a physician who sailed across the northern Pacific on Otto von Kotzebue's Russian-sponsored voyage (1815–18), challenged Forster with an argument based on the distribution of coral islands.[11] They were entirely absent across tracts of tropical sea that, by Forster's theory, would have offered ideal conditions for reef-building corals. Just like high islands, they were usually found in "rows . . . and large groups," suggesting they had a common underlying cause. Therefore, Eschscholtz argued, "the corals have founded their buildings on shoals in the sea; or to speak more correctly, on the tops of mountains lying under water."[12] His views were widely noticed but, because they were published in an unsigned appendix to the "Remarks and Opinions of the Naturalist of the Expedition," they were widely attributed to the voyage's botanist, Adelbert von Chamisso.[13]

By the time the *Beagle* voyage was being conceived, there was a new explanation for the formation of ring-shaped reefs that superseded the coral theories of Forster, Eschscholtz, and Chamisso. It was originally generated as something of an afterthought by the French naturalists Jean René Constant Quoy and Joseph Paul Gaimard, the naval surgeons on Louis-Claude de Freycinet's 1817–20 voyage of exploration.[14] Quoy and Gaimard had a specific zoological argument—that stony corals could live only in shallow water—that had broader geological implications. Through close attention to the coral fragments brought up by anchors and sounding leads during the voyage, they determined that the robust reef-building species "do not begin to build at depths greater than twenty-five or thirty feet."[15] Their larger argument was that that reefs could not therefore have been built up by corals from the deep seafloor, or even from deeply submerged mountains. The Frenchmen charged Forster with vastly exaggerating the geological significance of coral reefs, arguing that what he had believed were great coral formations were simply thin veneers of coral growing atop "the same minerals that form islands and all the known continents."

This led, in an inconspicuous footnote, to what would become an influential new explanation for ring-shaped reefs. "Couldn't this [circular] arrangement be due," Quoy and Gaimard reasoned, "to submarine craters, on whose rims the lithophytes have built?"[16] Thus, whereas Eschscholtz and Forster had both believed that reefs' annular shape was due to some general property of coral growth, Quoy and Gaimard were suggesting that a ring-shaped reef was evidence of nothing more than the existence of a ring-shaped submarine mountaintop. The question, then, was whether such submarine volcanoes existed, with enormous craters up to thirty miles in diameter sitting within a few feet of sea level.[17]

When Darwin offered a theory of reef formation a decade later it was pitted against various expressions of this explanation, which (following Lyell's usage) I will call the *crater theory*.[18] Quoy and Gaimard's claim that stony corals were limited to very shallow depths was widely accepted, and their passing suggestion that ring-shaped reefs were founded on ring-shaped foundations was taken up and rapidly expanded by other authors. Notable among them were two men who became close acquaintances in London in the late 1820s: a just-returned seafarer and a land-bound geologist. The navigator was Frederick William Beechey, who had commanded a British voyage to the Pacific from 1825 to 1828, and the geologist was Lyell, who was then at work on his *Principles of Geology*. Published in 1831, Beechey's *Narrative of a Voyage to the Pacific and Beering's Strait* offered the most detailed observations yet made on the structure of coral islands.[19] On his way through the Pacific to meet up (he hoped) with John Franklin's westbound Arctic expedition, Beechey surveyed no fewer than thirty-two coral islands. In addition to charting them for the Admiralty, Beechey used his *Narrative* to offer descriptions of what he took to be typical features of reefs and lagoons. He wrote as though he had surveyed with competing theories of coral reef formation in mind and emphasized findings relevant to establishing the reefs' origin. Beechey sought to decide between two alternative explanations that collectively represented the "general opinion" on the development of ring-shaped reefs: "that [coral islands] have their foundations upon submarine mountains, or upon extinguished volcanoes."[20] While Beechey's primary responsibility had been to document threats to safe navigation, he averred that he was "not inattentive to the subject [of reef formation], and when opportunity offered, soundings were tried for at great depths, and the descent of the islands was repeatedly ascertained as far as the common lines would extend." He offered the results of these "experiments" in a plate showing "a section of a coral island from actual measurement."[21] Ultimately, though, he declined to offer a personal judgment on whether coral islands owed their shape to "the propensity of the coral animals" or to "the shape of [a] crater alone."[22]

Lyell, on the other hand, made his opinion clear, wholeheartedly embracing the crater theory in the second volume of the *Principles*. (The volume appeared in December 1831, just a few days after the *Beagle*'s departure from England. Darwin received a copy by mail from his brother early in the voyage.) Lyell and Beechey had discussed coral reefs even before the 1830 publication of the first volume of the *Principles*, and in the second volume Lyell now felt able to speculate where Beechey had been unwilling to do so. Lyell explicitly built on the results of Beechey's survey, and he included a reproduction of Beechey's reef cross section while approvingly citing Quoy and Gaimard's figures on the depth limits of coral growth. The crater theory happened to correspond helpfully with Lyell's other ideas about volcanic action. While disputing Leopold von Buch's paroxysmal theory of "elevation craters" in the first volume of the *Principles*, Lyell had already explained a mechanism that would cause large craters to be formed beneath the sea in areas likely to be inhabited by reef-building corals.[23]

This was not the only way Lyell integrated his discussion of coral reefs with the lessons of his earlier volume. He had argued that igneous forces modify the earth's crust "by depressing one portion, and forcing out another."[24] These compensatory movements occurred frequently, and over the long term a seabed might be raised into a continent, and vice versa. If a coral reef subsided, its surface would be restored to sea level by the growth of new corals, but the next elevation would turn the new part of the reef into a high island composed of coral. The scarcity of upraised coral islands that Beechey reported indicated to Lyell that in the Pacific "the amount of subsidence by earthquakes exceeds . . . the elevation due to the same cause."[25] Were it possible to study them, the coral formations of the Pacific would probably be stratified by "the arrangement of different species of testacea and zoophytes, which inhabit water of various depths, and which succeed each other as the sea deepens by the fall of the land during earthquakes, or grows shallower by elevation due to the same cause."[26] Thus, for Lyell, coral reefs were a record of the vertical movements of the Pacific Ocean floor.

By 1831, therefore, offering a theory of coral reef formation meant attempting to answer a well-defined pair of questions. How did they originate in unfathomably deep water, and what caused them to assume a ring-like shape? These problems were recognized to have implications reaching into many branches of science. Their solutions appeared to lie in a synthesis drawn from navigators' reports, inquiries into the conditions of coral growth, and the geological study of the earth's topographic relief.

Studying Reef Formation as an Objective of the *Beagle* Voyage

It should not be surprising, given the interest in coral reefs among naturalists, geologists, and hydrographers in 1831, to learn that a survey vessel dispatched to the tropics that year by the British Admiralty carried instructions to study coral islands. What was noteworthy was that these directions envisioned the survey as a direct test of the latest theory of reef formation. They were part of the "Memoranda for Commander Fitzroy's orders" written by Francis Beaufort, a scientifically minded surveyor who had recently ascended to the administrative position of hydrographer of the Admiralty. The first aim of this voyage, which would be undertaken in the brig *Beagle*, was to carry out a new survey of the southern coasts of South America, eliminating the "motley appearance of alternate error and accuracy" that bedeviled older Spanish charts.[27] The shorelines of that continent had become newly interesting now that revolutions against Spanish control in the 1810s and 1820s had opened new markets to British trade.[28] After completing the South American survey, the twenty-six-year-old Robert Fitz-Roy was to continue taking meridians "at some judicious chronometer stages" across the Pacific, making "the intervening islands . . . standard points to which future casual voyagers will be able to refer their discoveries or correct their chronometers."[29] The making of standard points was one of the most important jobs of the official surveys. It meant using all available scientific means to fix the geographical location of a port or an island. Other navigators who arrived there could use the surveyor's result to reestablish their own latitude and longitude, about which a great deal of uncertainty might have accumulated.

Because this westward run of point making away from the coast of South America would put the *Beagle*'s course directly among "circularly formed Coral Islands in the Pacific," Beaufort instructed FitzRoy to investigate them: "While [your astronomical observations] are quietly proceeding, and the chronometers rating," he wrote, "a very interesting inquiry might be instituted respecting the formation of these coral reefs."[30]

Beaufort's directions to FitzRoy were aimed precisely at the point of contention between recent explanations for the form of coral islands. "A modern and very plausible theory has been put forward, that these wonderful formations instead of ascending from the [bottom of the] sea, have been raised from the summits of extinct volcanoes; and therefore the nature of the bottom at each of these soundings should be noted, and every means be exerted that ingenuity can devise of discovering at what depth the coral formation begins, and of what materials the substratum on which it rests is composed." In addition to the usual nautical charts that FitzRoy would produce, Beaufort demanded "an exact Geological map of the whole island [showing] its form, the greatest height to which the solid coral has risen, as well as that to which the fragments appear to have been forced. The slope of its sides should be carefully measured in different places, and particularly on the external face, by a series of soundings, at very short distances from each other, and carried out to the greatest possible depths, at times when no tide or current can affect the perpendicularity of the line."[31]

By instructing FitzRoy in the features of a reef that might be relevant to settling theoretical questions, Beaufort set in motion a long-range plan for the systematic study of coral islands that built on the achievements of Beechey's recent survey.[32] As Beechey had implied, it was taken for granted that general knowledge would issue from the accumulation of specific facts from one survey to the next. Beechey's 1825–28 voyage that was the source of his general observations on coral reefs had been commissioned before Beaufort became the Navy's chief hydrographer. The instructions Beechey had received are in striking contrast to those of the following decade. He was directed to particular coral islands (e.g., to "ascertain . . . whether

Ducie's and Elizabeth Islands be not one and the same") and was told that "your visits to the numerous islands of the Pacific will afford the means of collecting rare and curious specimens in the several departments of [natural history]."[33] Beechey's special attention to the *form* of coral islands had not been mandated by his orders from the Admiralty.

When it came to FitzRoy's orders Beaufort went so far as to mail him "a couple of [tracings of] Beechey's Coral Isl[an]ds not with any view of your visiting these particular islands . . . but that you might see the humour of these formations."[34] In a public notice of the voyage the week it departed, the weekly literary magazine *Athenaeum* suggested that "the most interesting part of the *Beagle*'s survey will be among the coral islands of the Pacific Ocean . . . and the hypothesis of their being formed on submarine volcanoes will be put to the test. . . . [T]he surveys of [the lagoons] will form, with those of Captain Beechey in his late voyage, the basis of comparison with others at a future period, by which the progress of the islands will be readily detected."[35] Sure enough, the next survey sent to the Pacific after the *Beagle* carried even more elaborate instructions. Beaufort requisitioned a well-boring apparatus for the 1835 voyage of the *Sulphur* so that the deep structure of a reef could be studied for clues to its formation. Indeed, for the rest of Beaufort's tenure the wording of FitzRoy's instruction served as the template for a similar paragraph written into the orders of all surveyors dispatched to the coral seas.[36]

It seems likely that the specific wording of the instructions FitzRoy and his successors received was in part a consequence of Lyell's particular interest in coral reefs. Beaufort's agenda for reform was carried out with advice and cooperation from the metropolitan scientific elite at every step. The coral reef investigation was no exception. Lyell had become a frequent visitor to the Hydrographic Department, and he delighted in telling fellow geologist Gideon Mantell that "our new hydrographer, Beaufort, is very liberal to all geologists, and you may get what unpublished information you like from the Admiralty, and there is an immense deal there."[37] Surveyors' reports were an important source of material for the *Principles*, and Lyell had one of Beaufort's officers vet the book in "all the *nautical* or hydrographical parts."[38] This exchange was of mutual benefit, as Lyell wrote to his fiancée just after the *Beagle*'s departure, "for I find every day the hydrographers are coming to me for instructions. I have just drawn up some for Captain Fitzroy, who has my book, and is surveying in South America. Captains Hewett, Beaufort, King, Vidal, and others, are in continual communication."[39] Lyell himself believed the seeds he planted in the surveyors'

instructions would yield an abundant crop of new facts to support his treatise, all the better to stifle the views of his geological opponents.[40] He must
have been pleased to know that Beaufort had added the status of official
policy to a test of the very coral reef theory that Lyell was sending to press.

There must be no doubt, however, that Beaufort also perceived that
the navy had much to gain from incorporating the latest theory of coral
reefs into FitzRoy's directions. Documenting the location of reefs was, of
course, fundamental to the mission of the Hydrographical Department (as
it was then known) to combat the dangers of navigation. However, it was
acknowledged at the Admiralty that the variable nature of living reefs presented a special obstacle to surveyors. When, for example, in 1842 Francis Price Blackwood was sent to survey the Torres Strait at the northern
end of the Great Barrier Reef because so many vessels had been lost after
becoming "entangled within the reefs," Beaufort exhorted him, "Do not
hurry over the hidden dangers which lurk *and even grow* in that part of
the world."[41] Beaufort made the reason for such substantial investments
of naval time and money clear in his 1846 directions to Owen Stanley on
dispatching him to make a further survey of the Great Barrier Reef. It
was not merely that reefs were dangerous, but that coral growth potentially made surveying marine hazards a fruitless endeavor. "There would
be much of discouragement attached to such surveys," Beaufort wrote, "if
changes should be constantly & rapidly at work in [coral] seas." Stanley was
encouraged, therefore, to "direct [his officers'] attention more particularly
to the formation and growth of coral reefs."[42] If it was possible to account
for these processes, one could suggest where reefs were likely to spring
up and indicate how long existing charts might be relied on before new
growth rendered them obsolete. Beaufort lived by the principle that facts
taken from the periphery of the known world were to be systematized by
savants in the metropolis, who in turn would direct the next wave of collecting. As in his projects to bring the collective observations of surveyors
to bear on magnetic variation and the tides, Beaufort married theorists'
questions with Admiralty interests to pursue a philosophical solution to
navigators' practical problem of coral reefs.[43]

Darwin's Training in the Sciences

Charles Darwin was twenty-two in the summer of 1831 when Beaufort invited him to accompany FitzRoy on the *Beagle* voyage and to take advantage of the scientific opportunities it would present. Beaufort had learned

of Darwin through inquiries with the Cambridge mathematician George Peacock, who in turn received Darwin's name from colleagues. Although Darwin's gentlemanly status was at least as important as his scholarly achievements in earning him the place with FitzRoy, his idiosyncratic education had made him particularly familiar with two distinct approaches to the study of the natural world: geology and the zoology of marine invertebrates. As historians including James Secord, Phillip Sloan, and Jonathan Hodge have shown, there is little truth to the former perception that Darwin departed on the *Beagle* as an inexperienced naturalist.[44] In two years as a medical student at Edinburgh and three more at Cambridge, where he passed his examination for the bachelor of arts degree in January 1831, Darwin had learned zoology, botany, chemistry, and geology from men with diverse, and sometimes competing, perspectives. In lecture halls, museums, and field excursions, Darwin was exposed to the extreme viewpoints of Wernerian and Huttonian geology.[45] He saw models of the scientific lifestyle that ranged from the pious curiosity of the parson naturalist to the subtle ruthlessness of a junior academic. Most important, he collaborated in genuine research with adepts in marine zoology and field geology.

As a diversion from his medical studies in Edinburgh, Darwin devoted himself to spending time with zoologist Robert Grant, who had studied in Paris with the great naturalists Cuvier and Geoffroy.[46] Grant was fascinated by the marine "zoophytes," plantlike colonial creatures—including corals—that he considered to straddle the boundary of the animal and vegetable kingdoms.[47] Grant extended recent work by Lamarck and Lamouroux in an effort to discern whether the organization of these animal-plants represented truly intermediate forms.[48]

At just this moment Edinburgh was becoming the focal point of a new British science of marine zoology (embodied most famously during the next decade in the work of Edward Forbes), in which naturalists co-opted the tools and skills of fishermen in order to gather specimens from deeper and more distant waters.[49] Darwin and a fellow student, John Coldstream, joined Grant in collecting along the shore of the Firth of Forth estuary and in the company of oyster dredgers and fishing trawlers who sailed from the harbor at Newhaven. He enthusiastically followed Grant's research program into zoophytes, striking out on an independent study of their reproduction with a microscope borrowed from his mentor. He was a keen enough observer to gain insights worthy of a short scientific paper, which he read to the student-run Plinian Society in 1827.[50] Indeed, his conclusions on the "ova" of the zoophyte genus *Flustra* were sufficiently interesting that Grant preempted him by mentioning them in a presentation of his own to the Wernerian Society three days earlier.[51] Darwin later soured on his teacher, but he retained Grant's lessons that these marine organisms rewarded close study and could be a potent source of broader insights into generation and development.[52] Darwin carried this informed curiosity about the zoophytes with him on the voyage, along with a wealth of experience in collecting and studying the marine fauna of Edinburgh.

Darwin's second university career, at Cambridge, also provided the opportunity to undertake original fieldwork with a companion much more experienced than himself. In the summer of 1831, shortly after finishing his degree, he accompanied Adam Sedgwick, the professor of geology, on his first study of North Wales. The two had been introduced by John Stevens Henslow, the professor of botany, who had taken Darwin under his wing and welcomed him into an extracurricular world of scientific rambles and intellectual dinners with the "learned men" of the university.[53] Henslow encouraged Darwin to get a clinometer for measuring the direction and inclination of geological strata, but it was Sedgwick who taught him, over a week or more together, how to study stratification in the field.[54] It was an opportunity to see how terminology and concepts that Darwin had learned at Edinburgh and Cambridge would be applied by a competent geologist when he made his first visit to a new landscape.[55] From Sedgwick, as from Grant, Darwin learned the practical skills and unwritten rules of pursuing new knowledge in a discipline with established methods and conventions. His most advanced practical and theoretical work during the voyage came in geology and marine zoology, the two sciences he had learned as an assistant to his mentors' labor.[56] His study of corals and reefs proved to be just one of the myriad ways Darwin worked to merge those pursuits.

Henslow and Darwin shared an enormous enthusiasm for the writings of the scientific traveler Alexander von Humboldt, whose footsteps they aspired to follow over the volcanic rocks of Tenerife. In fact, their ambition to make that journey together was abandoned only when Henslow nominated Darwin for the *Beagle* voyage. With their original objective in mind, however, Darwin had remained in Cambridge in the spring of 1831 after completing the requirements for his degree (and before the trip he eventually took with Sedgwick), studying Spanish while Henslow "crammed" him in geology.[57]

Their mutual hero, Humboldt, was a Prussian who had trained with Abraham Gottlob Werner, Europe's leading practitioner of mineralogy and geognosy. After working for half a decade as a mining engineer, Humboldt used the wealth from his parents' estate to finance a five-year trip to the Spanish Americas, where he and his companion Aimé Bonpland explored the Orinoco River, scaled the Andes, and accumulated an enormous mass of quantitative data produced through the enthusiastic use of the staggering load of scientific instruments their porters carried. On returning to Europe in 1804, Humboldt settled in Paris and generated an endless stream of publications on the results of the expedition, compiling, representing, and analyzing data in tables and distribution maps as well as in prose. Among these publications was his *Personal Narrative of Travels to the Equinoctial Regions of the New Continent*, which Darwin and Henslow read with such zeal.[58]

The documents of the exciting autumn of 1831, when Darwin was preparing to join the *Beagle* voyage, prove that virtually everyone in his circle considered Humboldt the ideal role model of a gentleman traveler. Adam Sedgwick, who was no enthusiast of the Wernerian school of geology in which Humboldt had trained, nevertheless advised Darwin that "Humboldts personal narrative you will of course get—He will at least show the right spirit with wh[ich] a man should set to work." A week later Darwin got his own copy of the English translation of the *Personal Narrative*, a gift from Henslow inscribed "to his friend C. Darwin on his departure from England. upon a voyage round the World." The book was a guide that no philosophically inclined traveler could afford to be without. As FitzRoy told Darwin in the romantic days before the voyage, "You are of course welcome to take your Humboldt . . . but, I cannot consent to leaving mine behind."

Enthusiasm for the South Sea Islands

Although Darwin was eager to see areas of the tropics that Humboldt had described, he proved equally eager to see islands of the tropical Pacific that Humboldt had never visited. Indeed, he used what little leverage he had with the Admiralty to advocate for including the Pacific in FitzRoy's itinerary. Darwin arrived in London in early September after gaining his father's permission to accept the *Beagle* invitation. At the time it was not yet confirmed whether FitzRoy would be ordered to complete a circumnavigation after finishing his work on the west coast of South America or whether he would be ordered to return directly home by recrossing the Atlantic. Unaware of Beaufort's own eagerness to send FitzRoy via the Pacific and Indian Oceans, Darwin made repeated visits to the Admiralty to lobby for the circumnavigation and its tropical island itinerary. Indeed, as long as the point remained in question Darwin withheld his final acceptance of a place on the ship. "The only thing that now prevents me finally making up my mind," he reported to his family after almost a week in London, "is the want of *certainty* about [the] S[outh] S[ea] Islands, although morally I have no doubt we should go there whether or no it is put in the instructions: Cap. Fitz says I do good by plaguing Cap Beaufort: it stirs him up with a long pole."[59]

Darwin was undoubtedly already aware of the coral reefs of the Pacific and even of the efforts of previous voyagers to explain their origin. Indeed, one of his favorite childhood books, *Wonders of the World*, contained a chapter on reefs and their possible modes of formation.[60] Robert Jameson, the professor of natural history at Edinburgh, likely brought Darwin up to date on debates over the origin of coral islands in 1826–27, the same time Grant was stoking his enthusiasm for zoophytes. Jameson's five-month lecture course, though the subject of derisory comments in Darwin's autobiographical recollections, had a remarkable scope unparalleled at any British university.[61] I have identified lecture notes that appear to date from the 1810s and 1820s in which Jameson refers to the "coral riffs that surround the islands in the south sea" as contributors to the "unequalities of the bottom of the sea, or that part of the globe which is still covered with water."[62] Another lecture analyzed reefs' process of formation by "the myriads of calcareous zoophytes," which "effect a chemical change on one of the mineral products carried into the sea by every river that flows through a limestone district. . . . The most important productions by the apparently insignificant race of the Polypi are the accumulations of these calcareous skeletons of the Anthozoa, which form the coral islands and reefs, the dread of the navigator."[63]

Jameson was also responsible for perhaps the most comprehensive review of coral reef knowledge available in the English language until Darwin's *Structure and Distribution of Coral Reefs* appeared in 1842. To the fifth British edition of Cuvier's *Essay on the Theory of the Earth*, Jameson added a twenty-page appendix on coral islands that contained long extracts from Forster and Flinders, along with nearly complete translations of Eschscholtz's report (credited to Chamisso) and Quoy and Gaimard's 1825 article.[64] Darwin bought the book when it came out in 1827, the same year he was taking Jameson's course.[65] Jameson kept proof sheets of the coral islands appendix among his lecture notes, which suggests that Darwin also heard the material delivered aloud in Jameson's dry style.[66] In his course at Cambridge, Sedgwick too described coral reefs as one of the "great agents by which the earth's surface is modified," though the extent of Darwin's attendance at these lectures is not clear.[67] On the other hand, if he had not already heard Sedgwick expound on the topic before their excursion together in Wales, Darwin could have received an impromptu lecture during the trip as an explanation for the fossil corals to be found in formations there.[68]

As the *Beagle*'s departure neared, Darwin gathered advice for his trip from "several great guns in the Scientific World," including useful hints from Grant (who had since moved from Edinburgh to the recently founded London University) on keeping zoophytes in good condition for dissection.[69] Grant's other former acolyte John Coldstream, who was still in Edinburgh, responded to a request from Darwin by sending illustrations on the construction and use of a dredge for collecting marine specimens, along with a ringing endorsement of Beechey's *Narrative* and other hints on the latest Scottish methods of "obtain[ing] a rich supply" of "the rarest . . . zoophytes."[70]

By the time Beaufort's official instructions for the voyage were entered into the record, Darwin had accepted his place and traveled to Devonport to await the *Beagle*'s departure alongside the crew. The hydrographer's knowledge that FitzRoy was to be accompanied by this young savant with a special enthusiasm for investigating the South Sea Islands could only have encouraged him to include the coral reef instruction.

2

An Amphibious Being

During the *Beagle* voyage Darwin gained a familiarity with the seafloor that was unprecedented among naturalists of his day. The hydrographers furnished him with techniques for visualizing underwater topography and for taking samples from the bottom of the sea. This in turn made his forays onto dry land opportunities for undertaking just the types of comparisons between terrestrial and submarine processes that Lyell envisioned for his amphibious being. Indeed, this method of comparison, when combined with the precise geographical knowledge yielded by the surveyors' work, became for Darwin a fruitful source of what Lyell had called "sound theoretical opinions." In time this amphibious approach to natural history played a central role in the process by which Darwin developed his theory of reciprocal geological uplift and subsidence, his theory of coral reef formation, and his theories of descent with modification.

It was significant not simply that hydrographers' sounding methods yielded samples from the seafloor that provided Darwin insight into submarine zoology, botany, mineralogy, and geology, but that the implement they used for sounding the depths predisposed him to study submarine animals, plants, rocks, and terrain *together*, and to pay attention to their interrelations. The sounding lead captured samples in a fashion that led Darwin against the grain of ordinary natural history practice. Collecting via sounding lead meant gathering whatever was to be found in a particular location on the bed of the sea rather than setting out to collect organisms belonging to a particular group. Decades before inventorying a patch of ground became the standard practice of a new science called ecology, Darwin was doing this on the spots of seafloor that had been touched by the sounding lead. Because this sampling was done in the course of a geographical survey, moreover, the specimens he gathered were enhanced from the beginning by data about the location and depth they occupied. The features of the underwater realm that Darwin was able to study through hydrography became his preoccupations on land as well: the interrelations between animals, plants, and minerals, and their geographical distribution across space and altitude. And because most of the dry land he studied showed evidence of having formerly been underwater, the ability to envision landscapes in their past submarine state turned into Darwin's most reliable method of thinking about change over time.

The matrix of practices and concepts I have just described as being a product of Darwin's exposure to the hydrographers' work resembles a phenomenon that historian Susan Faye Cannon named "Humboldtian science."[1] Alexander von Humboldt was indeed a hero of Darwin's, and the geographical sensibility that underlay so much of Darwin's work is often seen as a consequence of his admiration for Humboldt's *Personal Narrative*, which described the Prussian's 1799–1804 travels in the Americas.[2] As is well known, Humboldt focused on the geographical distribution of plants and other phenomena—by altitude as well as across horizontal space—in several works that Darwin devoured before and during the voyage. I aim to show, however, that Darwin's orientation toward distribution (in three dimensions, no less) need not have emerged exclusively from reading Humboldt's books.

As scholars in the history and sociology of science have argued in emphasizing the importance of face-to-face interactions in the production of knowledge, it was difficult if not impossible to learn everything one needed to know about replicating an experiment or constructing an instrument merely by reading instruction manuals.[3] Likewise, I illustrate how the face-to-face practical training available on the ship provided an impetus for Darwin's geographical thinking that might otherwise be seen as the mere consequence of his reading. Daily life aboard a survey ship, with the

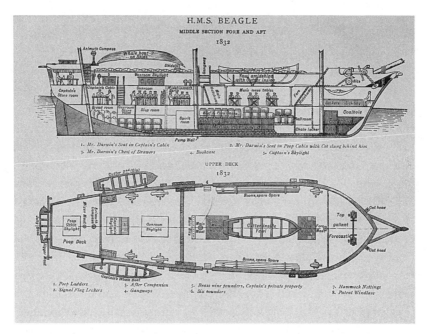

Figure 4. HMS *Beagle* in longitudinal section and in overhead view, as illustrated in the 1890 edition of Darwin's *Journal of Researches*, published by John Murray. Darwin hung his hammock above the map table in the poop cabin in the aft part of the ship. Image courtesy of Biodiversity Heritage Library. http://www.biodiversity library.org.

attendant opportunities that Darwin could have seized only from working hydrographers, made it possible for him to think and behave in ways that have retrospectively been dubbed Humboldtian.

Given that the first years of the voyage were spent surveying southerly shores far too cold to be inhabited by reef-building corals, it is remarkable how often Darwin mentioned coral reefs in letters and notes from this period. He expected that when he eventually did visit coral reefs in the Pacific Ocean, he would study them as a *zoologist*, continuing work on Atlantic Ocean zoophytes that had been aided so much by the hydrographers' activities. As we will see, up until the moment when his ambitions were reshaped in late 1835 by a new theory of reef formation that he developed at Tahiti, Darwin fully intended that he would one day integrate the "fine Coralls" of the Pacific into an ambitious program of zoological research.

Darwin's Approach to Scientific Work at the Beginning of the Voyage

"I do not think," Darwin wrote in his diary, "the impression this day has made will ever leave me." Such was the intensity of his reaction to standing on tropical soil for the first time.[4] After three awful weeks spent battling seasickness while under sail and lamenting a quarantine that denied him a chance to land on Humboldt's hallowed ground at Tenerife, Darwin found his zeal for the sciences that he had learned from Grant and Sedgwick quickened by the *Beagle*'s first landfall. Ashore, finally, at the island of St. Jago (Santiago) in the Cape Verde Islands, he wrote, "The first examining of Volcanic rocks must to a Geologist be a memorable epoch, & little less so to the naturalist is the first burst of admiration at seeing Corals growing on their native rock." As he was to do throughout the voyage, he associated tropical corals with lesser members of the zoophyte group he had begun studying in his university days. "Often whilst at Edinburgh, have I gazed at the little pools of water left by the tide: & from the minute corals of our own shore pictured to myself those of larger growth: little did I think how exquisite their beauty is & still less did I expect my hopes of seeing them would ever be realized."[5] He was inspired to take a boat out to dredge for corals in the manner he had learned at Edinburgh.[6] Over three weeks at the islands he adopted a "usual occupation of collecting marine animals in the middle of the day & examining them in the evening," when he would sketch them as they appeared under his microscope.[7] He collected and observed marine fauna, in other words, in the way he had learned from Grant and examined them as to their form, behavior, and "irritability."[8]

Meanwhile, the geological skills he had acquired from Sedgwick led him eventually to interpret features of the islands in ways that lent credence to claims Darwin encountered in the first volume of Lyell's *Principles*, which he had received as a present from FitzRoy.[9] Darwin believed he saw evidence of a gradual, cyclical change in the relative levels of land and sea. In particular he considered that a layer of white marine rock that remained perfectly level despite having been elevated well above the sea taught the same lesson of steady change that Lyell had illustrated using the still-upright columns of the Temple of Serapis.[10] Darwin's first letter back to Henslow emphasized "how much I am indebted to [Sedgwick] for the Welch expedition," which had made him capable of discerning that "the geology [of St. Jago] was preeminently interesting & I believe quite new [with] facts on a large scale . . . that would interest Mr Lyell."[11]

This letter to Henslow reveals a great deal about how Darwin conceived of himself and the practice of various sciences at the outset of the voyage. In a moment of humility he confessed to being anxious "whether I note the right facts & whether they are of sufficient importance to interest others." Yet, as we have seen in the reference to Lyell, he already possessed a tacit sense of the kinds of facts that would be relevant to particular individuals' theories. He also already had ample confidence in the novelty and relevance of his observations in the science he knew best. He trusted his zoological training enough to draw enthusiasm, rather than unease, from anomalous findings. Of a stony coral he wrote, "I examined pretty accurately a Caryophyllea & if my eyes were not bewitched former descriptions have not the slightest resemblance to the animal."[12] Between the thrill of reconstructing the geological history of a new place and the pride of knowing that he could identify marine organisms and produce original insights into their microscopic structure, the beginning of the journey only reinforced his two strongest scientific affiliations: "Geology & the invertebrate animals will be my chief object of pursuit through the whole voyage."[13] This early letter to Henslow reveals not just that Darwin was anxious about whether he noted the right facts, but that his training mattered a great deal. He happened to find those features of St. Jago most noteworthy that could be studied using the sciences he knew best, and it was his training with Grant and Sedgwick that allowed him to feel confident that in marine zoology and geology he was seeing interesting and novel things.

Hydrography Becomes a Resource for the Naturalist

The next leg of the voyage included a moment that was crucial in Darwin's development as a practicing naturalist and that I argue was to have massive implications for the theories he later developed. He began for the first time to pay attention to the everyday activities of hydrographic surveying. This gave him access to specimens he could otherwise not have acquired, but it also gave him access to a type of three-dimensional spatial thinking that he had not previously considered. As many scholars have pointed out, some of Darwin's greatest insights were to depend on his attention to the way organisms and rocks were distributed. I argue that the surveyors' work in which Darwin immersed himself did more than just stimulate a general interest in thinking geographically; it supplied a specific means of identifying specimens in the context of the place and depth where they had been found. During this early stage of the voyage Darwin's focus was on marine zoology. The hydrographers' work fed Darwin's ambitious pursuit of the kind of zoophyte studies he had learned from Grant in Edinburgh. His first documented thoughts about coral reefs were expressed in terms of these zoological ambitions. As Darwin began to draw value from the surveyors' work and to imagine incorporating his insights into single-authored scientific publications he was also, perhaps unconsciously, exemplifying Humboldt's relation to the collaborative gathering of zoogeographical and phytogeographical facts.

Even before reaching St. Jago in the first weeks of the voyage, Darwin had begun to experiment with ways to resume the marine zoological studies he had conducted at Edinburgh. "I proved to day," he wrote in his diary on 10 January 1832, "the utility of a contrivance which will afford me many hours of amusement & work.—it is a bag four feet deep, made of bunting, & attached to a semicircular bow [that] is by lines kept upright, & dragged behind the vessel." The contraption yielded an abundant harvest of organisms "exquisite in their forms & rich colours," leaving Darwin to "wonder that so much beauty should be apparently created for so little purpose."[14] This is one of the first known descriptions of a plankton net in use, and it reveals that Darwin boarded the *Beagle* already primed to develop novel methods for collecting marine specimens. Life aboard a working survey vessel would offer another prospect for collecting, however, and Darwin began to exploit this opportunity just a few weeks later.

Given his maritime experiences in Edinburgh, it should be no surprise that Darwin was fascinated to see the officers begin carrying out their survey in Brazilian waters between Bahia (present-day Salvador) and Rio de Janeiro. In March 1832 Darwin recorded in his diary that "the labours of the expedition have commenced.—We have laid down the soundings on parts of the Abrolhos, which were left undone by Baron Roussin." Determining the extent of these shoals was among the first of Beaufort's official instructions to FitzRoy, and the systematic accumulation of data caught Darwin's attention.[15] "The scene being quite new to me was very interesting.—Everything in such a state of preparation; Sails all shortened & snug: anchor ready to let fall: no voice or noise to be heard, excepting the alternate cry of the leadsmen in the chains."[16]

Darwin was curious to analyze the surveyors' results, so on the back of a sheet of zoology notes he began compiling a "table of thermometrical changes during crossing & recrossing the bank" (see fig. 5).[17] He recorded the time of day when each sounding was taken, the water temperature (to the quarter of a degree Fahrenheit), and the depth measured in fathoms. He had already learned to follow the hydrographers' convention for

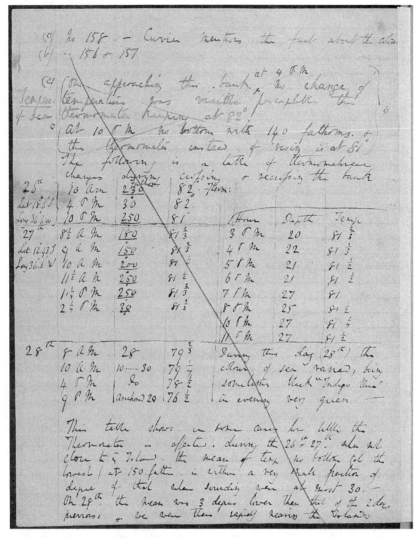

Figure 5. Darwin's notes about the soundings taken on 26–28 March 1832: an example of quantitative data collection based on the *Beagle* survey. This page of his zoological notes contains a "table of thermometrical changes during crossing & recrossing the bank" at the Abrolhos Shoals. This is among the first manifestations of Darwin's attention to the surveyors' work, showing that he was interested in how seawater temperature was affected by depth. The table juxtaposes latitude and longitude with time of day, depth in fathoms, and water temperature in degrees Fahrenheit. The vertical or diagonal pencil lines across this and other manuscripts indicate that Darwin subsequently made use of the material in later notes or writings. DAR 31.1:31v. Reproduced by kind permission of the Syndics of Cambridge University Library.

indicating that no bottom had been found at a given depth by noting the length of the sounding line in fathoms and placing a dot beneath it. By his third day paying attention to the surveyors' work, he discovered something that made him begin to record hydrographic findings on the front of the sheet, among his formal zoological notes.

Before the sounding line was cast, the concave bottom of the sinker, or lead, (pronounced *led*) was "armed" with tallow that would either record an impression of a hard bottom or capture a sample of any loose material on the seafloor[18] (see plate 1). The arming allowed chartmakers to include enough detail about the type of bottom that navigators could place their location, if necessary, by sounding in the same waters.[19] It also presented a wealth of information to the opportunistic naturalist.

On 28 March 1832, this time *within* his zoological notes, Darwin recorded that "10 miles West of Abrolhos; there came up with the lead (17 Fathoms) a piece of Fucus.—on which were growing numerous minute tufts of a Conferva." Under his microscope, the harvest of this one sounding was considerable. The filamentous alga he called conferva had "stems simple cylindrical white transparent jointed; end truncate; length 1/10 of inch, diameter 2/3000." Looking closer, he saw that "on this minute plant & on a small coralline were crowded together a forest of numerous species of Bacillareès & Anthrodieès"[20] This experience showed him that one cast of the lead might produce multiple specimens *and* reveal their interrelations, as in this case diatoms were attached to a plant that was itself growing on another plant. What is more, they would be presented complete with knowledge of the depth of the water they inhabited and the geographical position of the spot where they were found. His hero, Humboldt, had championed attention to just such details of an organism's location and vertical distance from sea level in his study of terrestrial flora and fauna.[21] Darwin had therefore learned that laws about the habitats and interrelations of organisms were to be derived from analogous practices of exact collecting. Lying offshore from the continent that had made Humboldt famous, this one cast of the lead offered Darwin everything a philosophical naturalist could desire. At this moment, ninety days into the voyage, the sounding lead joined the dredge and the microscope among the tools with which Darwin's zoological knowledge was built.

There were two distinctive features about the sounding lead as a collecting device. The first was that it sampled rocks, animals, and plants together while preserving their relation to one another, which helped establish the links between Darwin's geology, zoology, and botany that were later solidified by his more famous land-based study of fossils. Second, the sounding lead drew samples from an identifiable fixed point. Not only that, when hydrographers wielded the lead the geographical location of that fixed point was itself a matter of careful attention. This was much different from collecting marine specimens with an oyster dredge, as Darwin and other Edinburgh naturalists had recently begun to do. A dredge disturbed and jumbled material as it was dragged across the seafloor, rendering each specimen's exact point of origin unknowable and destroying clues to the relations between specimens. Specimens from the lead, by contrast, could be attributed to a specific place, geographically and vertically, and to a set of conditions in which plants, animals, and rocks existed in exquisite inter-relation.

In other words, collecting by sounding lead meant sampling indiscriminately in a precise location. Such an approach became standard practice for ecologists later in the century, but it contrasted sharply with other tool-aided methods for collecting in early nineteenth-century natural history, in which a zoologist armed with a butterfly net, or a geologist with a hammer, might wander widely while sampling only insects or only minerals. Here aboard the *Beagle* a new collecting tool was helping Darwin to link geology, botany, and zoology. Thanks to the surveyors' expertise, he was considering these links geographically as well. I have noted that Darwin's hero, Humboldt, had championed recording just such details of location and vertical distance from sea level in his study of above-water vegetation, and Humboldt had himself previously been a surveyor of sorts. Before going to South America he studied the distribution of vegetation in mines near Freiberg, Saxony, while studying with mineralogist and geognostic theorist Abraham Gottlob Werner.[22] But it was the *Beagle*'s surveyors who empowered Darwin to collect data as Humboldt did rather than Humboldt's writing somehow dictating that Darwin would collect like a surveyor.

An Ambitious Plan for Studying Zoophytes

Marine invertebrates increasingly became a fixation for Darwin, in part because they could be studied both on the ship and on shore. During a long leave from the ship at Rio de Janeiro, he was "busily employed with various animals; chiefly however corallines," in contrast to which he considered the local "Geology [to be] uninteresting, [and the] Botany and Ornithology too well known."[23] But his researches continued apace because he was able to develop unexpectedly rich collections even when he could not leave the ship. In late 1832 he begged Henslow to "recollect how great a proportion of time is spent at sea" as he regaled him with descriptions of the "new & curious genera" of pelagic animals caught in the trawl and the "interesting" zoophytes hauled up by the lead. "As for one Flustra," he raved, "if I had not the specimen to back me up, nobody would believe in its most anomalous structure." He was proud to say of his time on the *Beagle*, "It has been a splendid cruize for me in Nat[ural] History."[24] He remained on or near the ship as FitzRoy surveyed Tierra del Fuego and the Falkland Islands in early 1833, "during [which] time," he noted in his diary, "I have been very busy with the Zoology of the Sea." Thanks in large part to his decktop collecting methods, he was struck with the opinion that "the treasures of the deep to a naturalist are indeed inexhaustible."[25]

Among these marine treasures, Darwin puzzled to understand a variety of organisms that he counted under the heading of zoophytes. In applying this general name he followed Grant's use of a term applied by Cuvier and derived from the name Zoophyta, by which Linnaeus designated a group of animated beings intermediate between the animals and plants.[26] Among the creatures Darwin considered zoophytes were a number of colonial invertebrates, including those he called corals (sometimes "coralls") and the smaller "corallines." At various times in the voyage he referred to reef-building corals (what are today called hermatypic corals)[27] as "lamelliform" (platelike) or "corall forming" corals, or "lithophytes."[28] "Corallina," on the other hand, was his name for a group of encrusting organisms that he came to believe, in early 1834, had no "connection with the family of Zoophites" and were probably better placed in "the grand division of plants."[29] Nevertheless, he collected these algae, and studied their physiology and means of propagation, in the same ways he investigated the colonial invertebrates.

That his invertebrate work built on his training from Grant is explicit in his discussion of zoological specimen 983 (in spirits of wine), a mosslike encrusting zoophyte collected in mid-1834.[30] Among the notes of his dissection, which included five illustrations, he wrote, "I examined the Polypus of this very simple Flustra, so that I might errect at some future day, my imperfect notions concerning the organization of the whole family of Dr Grants paper."[31] This refers to the paper in which Grant preempted by three days Darwin's first Plinian Society publication in 1827, suggesting that the young naturalist recognized his *Beagle* studies as being intellectually continuous with those he began in Edinburgh. In reengaging with his erstwhile mentor's analyses, he considered himself capable of far more than simple description of specimens.

Over several weeks in the summer of 1834, Darwin made explicit his ambitious plan for the study of zoophytes. Relevant as they were to many of the questions that interested him most, he reported to his family, "Amongst Animals, on principle I have lately determined to work chiefly amongst the Zoophites or Coralls: it is an enormous branch of the organized world; very little known or arranged & abounding with most curious, yet simple, forms of structures."[32] His manuscripts of the period, flush with new discoveries, explain much of his optimism about this work among the colonial marine invertebrates.

This letter also hints that his enthusiasm was based on vocational opportunism, fed by his perception that this group of organisms had yet to be mastered. It was surely telling that the two most famous naturalists of the previous generation, Cuvier and Lamarck, disagreed over the propriety of the term zoophyte and the unity of the group. Lamouroux had addressed the uncertainty of the field as stimulus for his own contribution to it: "What we know [of the natural history of Polypes] pales in comparison to what we do not know."[33] A passage in the second volume of Lyell's *Principles* (which contained the chapter on coral reefs, and which Darwin had received by mail in Montevideo in November 1832) must have seemed even more inspirational. "The ocean teems with life—the class of *polyps* alone are conjectured by Lamarck to be as strong in individuals as insects. Every tropical reef is described as bristling with corals, budding with sponges, and swarming with crustacea, echini, and testacea; while almost every tide-washed rock is carpeted with fuci and studded with corallines, actiniae, and mollusca. There are innumerable forms in the seas of the warmer zones, which have scarcely begun to attract the attention of the naturalist."[34] It is not difficult to imagine Darwin believing that this passage, written shortly before the *Beagle* sailed, had been intended for him personally.

In Valparaiso, Chile, the day after the *Beagle*'s first landfall on the west coast of mainland South America (and just after announcing his concentration on zoophytes to his family), Darwin boldly laid out his findings in a long letter to Henslow. He believed he had evidence for major taxonomic revisions, having examined two species of the genus *Sertularia* "taken in its most restricted form as by Lamouroux" and found that "the Polypi quite & essentially differed, in all their most important & evident parts of structure." With this compelling discovery, Darwin had "already seen enough to be convinced that the present families of Corallines, as arranged by Lamarck, Cuvier &c are highly artificial.—It appears they are in the same

state which shells were when Linnaeus left them for Cuvier to rearrange."
His audacious disagreement with the highest authorities went beyond
morphology. During one of his dissections he had managed to stimulate a
collective reaction by multiple polyps of a "little stony Cellaria." He took
this as a remarkable indication of coordination from one polyp to the next.
"This fact, as far as I see, is quite isolated in the history ... of Zoophites.—it
points out a much more intimate relation between the polypi, than Lamarck is willing to allow."[35]

Darwin already expected to continue this line of study on the reef-
building corals of the South Sea once FitzRoy had finished surveying the
shores of the Americas. Lyell had indicated that the tropical oceans were
virtually calling out to be studied by such a zoophyte expert as Darwin aspired to be, and he had specifically noted that few sites offered more promise than the "reef[s] bristling with corals." As Darwin devoted himself to
studying the marine zoology of the Atlantic, there can be no doubt that
coral reefs were indeed never far from his thoughts. Writing to Henslow
while "sea-sick & miserable" on a bleak run from the Falkland islands to the
mainland in April 1833, he affirmed, "I trust that the Corall reefs & various
animals of the Pacific may keep up my resolution."[36] Three months later, in
the dead of the southern winter and facing at least one more season of surveying before they might cross to the "glorious Pacific," he reiterated the
spell that coral reefs cast on his imagination. On 18 July 1833, a day when
the *Beagle*'s log shows that the weather at noon was gloomy and thick with
fog, Darwin burst forth, "I am ready to bound for joy at the thoughts of leaving this stupid, unpicturesque side of America. When Tierra del F[uego] is
over, it will all be holidays. And then the very thoughts of the fine Coralls,
the warm glowing weather, the blue sky of the Tropics is enough to make
one wild with delight."[37]

The coral reefs of the Pacific enlivened more than Darwin's idle yearn-
ings, however, for he envisioned them to be the culmination of his voyage-
long program of marine invertebrate zoology. As he explained to Henslow,
from whom he had just received the *Report* of the second ever meeting of
the British Association for the Advancement of Science (BAAS), "For my
second *section* Zoology.—I have chiefly been employed in preparing myself for the South sea, by examining the Polypi of the smaller Corallines in
these latitudes."[38] What could this mean? To say he was "preparing himself" was to imply that he had some responsibility to uphold in the tropical
Pacific. The most obvious task would be Beaufort's instruction to study
coral reef formation. As his earlier correspondence with Henslow indicates

so strongly, coral reefs were the main feature Darwin associated with the South Sea.

However, this letter makes it clear that Darwin did not view the study of reefs as a "geological" assignment.[39] Rather, he was especially pointed in classifying this work as "zoology," a distinction he happened to emphasize on this occasion because he was mimicking the BAAS's organizational division of the sciences into different "sections."[40] Given that it would be Fitz-Roy, not Darwin, whose soundings were expected to determine whether the Pacific was dotted with submarine volcanoes, it makes sense that in March 1834 Darwin saw his future contribution to the question of reef formation as lying in zoology. The crater theory was predicated on Quoy and Gaimard's novel proposition about the limits on coral growth, and it appears that from Darwin's perspective the study of coral reefs would hinge on understanding this organic process. Thus, studying solitary Atlantic corals, notwithstanding their inability to form reefs, would prepare him for his expected inquiry into the origin of the coral islands of the Pacific. Because of the retrospective shadow cast by Darwin's subsidence theory, it is difficult to imagine his coral reef work as a venture modeled more on the scientific practices of Grant than on those of Lyell.[41] In mid-1834, though, that is how Darwin saw it.

3

Studying Dry Land
with a Maritime Perspective

The *Beagle* voyage would be almost four years old by the time Fitz-Roy concluded the crew's labors on the coast of South America in September 1835 and set a course for the Galápagos Islands. Darwin had spent a great deal of time studying soundings and even more time rambling inland on a series of geological excursions across the Pampas and into the Andes. Along the way he came up with an idea that proved to be a crucial stimulus for his eventual theory of reef formation and his larger theory of the earth—and thus for his rapid ascent in the geological community after the voyage and his private conversion to transmutation. This hypothesis, which he definitely had established by the time he left South America, was that the floor of the Pacific Ocean must be sinking. It grew from his efforts to answer the principal geological question to which he had devoted himself in South America: when and how that continent had emerged from the ocean. The question was provoked by his recurring experience of discovering marine fossils well above sea level during a series of inland excursions on the east coast. Many historians have followed Darwin's lead in emphasizing the significant role these fossils played in shaping his interpretations of the continent's geological history.[1] In this chapter I argue that the way Darwin interpreted these fossils—and, more important, decoded the history of the geological formations in which they were found—was made possible by his attention to the ongoing hydrographic work being carried out on the *Beagle*.

Because Darwin's familiarity with hydrography has previously been underestimated (by historians) or unstated (by Darwin in his published writings), his perception of various South American landscapes as former seafloors has been described as though it were an exercise in pure conjecture. In fact, he based these analyses on his direct knowledge of the real seafloor. Thanks to the hydrographers, he was able to compare the physical features of the South American landscape with detailed descriptions of submarine topography and to compare the sedimentary rocks and fossils he found on dry land with the present-day sediments and organisms collected from the seafloor by the sounding lead. Darwin interpreted South America's geological past from the perspective of an amphibious being.

Darwin's geological work convinced him that the continent had been elevated very steadily in small increments, and he began to believe that the floor of the Pacific was sinking in the same manner. I have identified evidence that he was considering this possibility as early as August 1834, just over a year before the *Beagle* left the coast of South America for the last time and sailed into the Pacific. In notes written during that final year in South America, many devoted to thinking through evidence for various kinds of geological uplift, he occasionally mused on the issue of subsidence, which later became so central to the coral reef theory he eventually published. Some of these notes have been described by historians as evidence that Darwin had in fact conceived his coral reef theory while he was still in South America.

In the last part of this chapter I offer a new interpretation of these notes, arguing that they were stimulated by the immediate problem of how to find evidence for past elevation and subsidence in the stratigraphic record— presented by rock formations of the sort he was surrounded by during his inland excursions in South America. There is no evidence that Darwin thought his solution to this problem was relevant to understanding the origin and shape of coral reefs until a moment of insight at Tahiti suddenly endowed these thoughts with added significance. Darwin himself subsequently implied that this connection had all along been inherent in his work in South America, but this was a retrospective reassessment. The earlier ideas turned out to be necessary for him to eventually invent the coral reef theory, but it was only once he had done so that this connection between the South American work and the understanding of coral reefs seemed inevitable.

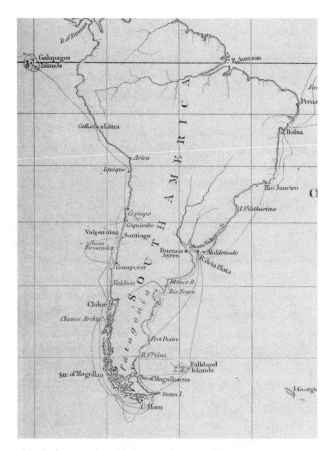

Figure 6. A detail of FitzRoy's published track chart of the *Beagle* voyage illustrating the key destinations visited in South America. Image from FitzRoy's 1839 *Narrative*, appendix to vol. 2. Wellcome Library, London.

Understanding the process by which Darwin worked and developed the coral reef theory depends on recognizing that these connections were not inevitable. It requires hindsight to identify certain moments, ideas, specimens, or landscapes as factors in developing the theory, but having done so, the only way for historians to understand what these experiences meant to Darwin *at the time they happened* is to accept that their future significance was as yet unknowable and to explain his actions in terms of the motivations and frames of reference that were available in the moment.

Applying the Lessons of Hydrography to the Interpretation of Geology

Darwin's curiosity about changing levels of sea and land, first piqued at the Cape Verde islands, was redoubled by the geology of South America. As FitzRoy surveyed southward, Darwin began to learn the geology of eastern South America during frequent excursions on shore. The northern part of Patagonia consisted of great terraces of land, level to the naked eye, that stretched over hundreds of miles between the Atlantic and the Andes. Most of these terraces, which now stood tens or hundreds of feet above sea level, were characterized by distinctive marine remains such as the shells of the "great oyster bed" of Patagonia. Clearly, those oysters had lived in the sea; the question was in what fashion the sedimentary rocks in which the oysters were embedded had become dry land. It seemed more likely to Darwin that the land had been elevated than that the sea had receded to such an extent, but he wondered how such large tracts of the earth could have been raised without any apparent deformation.

Another puzzle was posed by a vast bed of gravel consisting of distinctive porphyry pebbles that seemed to have originated somewhere to the northwest, in the Andes[2] (see fig. 7). What agency, Darwin wondered, could have transported a layer of pebbles so evenly across an area that he had himself "traced for more than 700 miles"?[3] In an essay titled "Reflection on Reading my Geological Notes," written about March 1834, he considered the possibility that after a "vigorous elevation" of the seafloor these pebbles had been carried "by the retreating waters" from the "West foot of the Cordilleras [Andes]" to "a deeper sea." Whatever the exact cause, Darwin felt sure they had been distributed in a "short period." Why did he conclude that they had moved rapidly? Because they were not "encrusted by stony small corallines.—(Which I always have noticed to be the case in these seas)."[4] This statement is clear evidence that Darwin was reasoning by comparing knowledge gained from the sounding lead with that learned through his terrestrial geology.

The specific point of comparison in this case was with pebbles that emerged on the armed sounding lead. Darwin had become very familiar with the pebbles of the seafloor and with the interesting organisms that were often attached to them. In this way the survey not only provided Darwin with submarine geological and zoological specimens that might be of isolated interest, but also gave him a more synthetic view of the seafloor's physical conditions, its flora, and its fauna.

Thus Darwin was able to write of the Patagonian porphyry pebbles,

Figure 7. Rodados Patagónicos in Chubut, Argentina. These are the Patagonian "pebbles" to which Darwin devoted so much of his geological research. I am extremely grateful to Oscar Martinez, Universidad Nacional de la Patagonia San Juan Bosco, for sharing this photo.

"Whatever their origin, they mark a great change in the inhabitants of the ocean[, for] during a succession of elevations [subsequent to the elevation of the gravel bed, and each producing another, lower terrace,] such shells as now exist—flourished on the successive lines of beach & were scattered over the bottom."[5] In other words, a series of elevations had converted new parts of the seafloor into dry land in such a recent geological period that the same organisms that had been alive then could now be found in the waters beneath the *Beagle*. Indeed, Darwin told Henslow, "the most curious fact is that the whole of the East coast of [the] South part of S. America has been elevated from the ocean, since a period during which Muscles [mussel shells] have not lost their blue color."[6] The remains of sea creatures, some identical to those yet living, became his index of successive elevations.

Darwin's technique of comparing terrestrial and submarine topography became a way of thinking about change over time, a macroscopic analogue to comparing fossil and living vertebrates. South American *landscapes* were fossilized *seafloors*.[7] No wonder the familiarity with the undersea world that Darwin had gained from the *Beagle*'s sounding operations gave him an advantage over land-bound geologists. He was becoming Lyell's amphibious being because he had lived and worked aboard a surveying vessel. Such a merging of geological training with intensive maritime experience was all but unprecedented. As he exulted to his older sister Caroline later in the voyage, Darwin had been fortunate that "few or rather no geologists" had undertaken their field excursions "in ships."[8]

The first half of 1834 was also a crucial time in shaping Darwin's eventual view that the elevation of South America must have been offset by subsidence elsewhere. In "Reflection on Reading my Geological Notes," he posited three possible types of elevatory force: "It becomes a problem. how much the Andes owes its height. to Volcanic matter pouring out?.— how much to horizontal strata tilted up.? how much to these horizontal elevations of the surface of continents?"[9] Darwin's approach to answering this question was to depend on the hydrographers' work. "The only method" of solving this puzzle, he wrote, "is to compare the increased height of the plains in the interior between any two points, with the probable slope of the oceans bottom in the same distance."[10] Determining the slope of the ocean's bottom was, of course, an everyday task for his companions on the ship. Although only hydrographers could produce such knowledge, Darwin by now took for granted that it was readily available to him.

Two essays written before the end of the southern winter (mid-1834), on the "Valley of S. Cruz"[11] and "Elevations of Patagonia,"[12] reveal that his ideas about elevation were extraordinarily volatile. Time and again after laying out evidence from the soundings or his inland observations and coming to a careful conclusion, Darwin would trail off into a series of challenges, questions, and self-contradictions. By 1835 he gradually became convinced of the importance of what, in the quotation above, he called "horizontal elevation," which he also described as elevation "concentric" with the earth. Both terms were slightly misleading; they referred to bulging of the earth's crust on the order of thousands of square miles[13] (see fig. 8). Unlike localized injections of molten rock beneath the crust, gradual "horizontal" elevation would result in an apparently level uplifting of beds such as he saw in Patagonia. Eventually he conceived of the entire continent having been uplifted in this manner. The Andes themselves, he believed, had been carried upward by this movement. They must have predated the continent and existed formerly as a chain of islands; as they were raised into mountains, the lower surrounding land would have emerged from the sea. At some point he added a declaration in pencil atop a sheet of

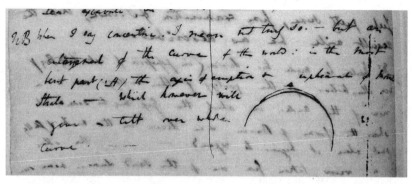

Figure 8. "NB When I say concentric. I mean not truly so." This diagram, accompanying a note on the back of page 8 of Darwin's essay "Valley of S. Cruz," gives insight into the "concentric elevation" that he then believed was the most likely means by which the plains of Patagonia had been elevated. As illustrated by the diagram, concentric elevation described "an enlargement of the curve of the world: in the most bent part (A) the axis of eruption & upheaval of Mountain Strata.—which however will give a tilt over whole curve." DAR 34.2:110v. Reproduced by kind permission of the Syndics of Cambridge University Library.

earlier notes: "The Andes created all S. America."[14] And when, in February 1835, FitzRoy documented that an earthquake had elevated the coast of Chile by several feet in relation to sea level, Darwin gained confidence that this process was continuing by degrees.[15]

Elevation of this sort must have been offset by subsidence of another part of the earth's crust, Darwin believed.[16] Where might this compensatory sinking of the crust be taking place? He surmised that it was concealed from the geologist's gaze beneath the vast Pacific Ocean.

Elevation and Subsidence

The idea that the Pacific floor had been sinking in geologically recent time, and perhaps was still doing so, played a crucial role in the way Darwin interpreted the South Sea islands when he later visited them. He began to juxtapose the high mountains and the bottom of the sea as he ranged into the Andes from the west coast in August 1834. The vista he encountered in the broad valley of the Aconcagua River had, he concluded, once "most clearly [been] marine with Islands"[17] (see plate 2). On 18 August he wrote in his tiny field notebook, "With respect to [the] great valleys . . . Perhaps in Pacific if seen, [our] wonder would be reversed."[18] The words "if seen" could be taken as irony, because there was no obvious way to look directly at the bed of the Pacific Ocean. There is a chance, though, that Darwin had a notion that he might one day "see" the bed of the Pacific via hydrography, a practice that had already proved so useful off the Atlantic coast.

A strenuous mountain journey—all the way over the Andes from Santiago, Chile, to Mendoza, Argentina, and back in March and April 1835—convinced him that vertical oscillations of the earth's crust had occurred on an even grander scale than he previously imagined. "I have certain proof," he wrote in his diary, "that the S[outhern] part of [the] continent of S[outh] America has been elevated from 4 to 5000 feet within the epoch of the existence of such shells as are now found on the coasts." In the high desert between Mendoza and the summits of the Andes Darwin had found the silicified remains of a copse of trees, still standing upright, embedded in sedimentary rock containing marine fossils (see plate 3). He concluded that "at a remote Geological æra . . . this grand chain consisted of Volcanic Islands, covered with luxuriant forests."[19] A euphoric letter to Henslow captures Darwin's astonishment at what he had found. "I hardly expect you to believe me," he gasped, "when it is a consequence [of his discoveries] that Granite which forms peaks of a height probably of 14000 ft has been fluid in the Tertiary period." The Andes were remarkably young in addition to being remarkably tall. He marveled, "The structure, & size of this chain will bear comparison with any in the world. And that this all should have been produced in so very recent a period is indeed wonderful." Darwin nearly trembled at the grandeur of his own hypothesizing, and he felt the need to reassure Henslow that his reasoning was sound. "In my own mind I am quite convinced of the reality of this. I can any how most conscientiously say, that no previously formed conjecture warped my judgement. As I have described, so did I actually observe the facts."[20] Earlier in the voyage he had reveled in making such "previously formed conjecture[s]" before commencing geological fieldwork, telling his cousin William Darwin Fox, "Geology carries the day; it is like the pleasure of gambling, speculating on first arriving what the rocks may be; I often mentally cry out 3 to one Tertiary against primitive; but the latter have hitherto won all the bets."[21]

Sometime before 29 May 1835 Darwin firmly decided to look for evidence that the bottom of the Pacific had subsided. On that day he wrote a letter to Robert Alison, a geological acquaintance in Valparaiso. Darwin's letter is not extant, but Alison's response referred to the *Beagle*'s impending departure from South America by saying, "I wish much to hear of your report respecting the islands in the Pacific, and it will be curious if you find a sinking of the land there, & a rising here."[22] Most likely Darwin was not optimistic that he could prove the action of a process whose effects would be hidden beneath the waves. The seafloor might be subsiding, but did he possess the necessary powers of submarine vision? One week after his letter to Alison, he wrote to his sister Catherine saying, "I have lately been reading about the South Sea—I begin to suspect, there will not be much to see."[23]

However, while he remained for his final months in South America, where uplift had exposed former seabeds to view, Darwin was still contemplating this problem. The previous year when he had been confounded by the distribution of pebbles across Patagonia, he had concluded that their presence across a great geographical distance could be explained only by a gradual horizontal elevation as opposed to any violent upheaval. He then knew from his observations of sounding data, which he had begun compiling in April 1834 as "Observations on the Bottom of the sea between the Falkland Islands & S[anta] Cruz," that the action of the currents at the seabed was minimal[24] (see fig. 10). This was demonstrated by the fact that pebbles just like those in the porphyry gravel came up in the arming still covered by delicate, and completely intact, living corallines. Under existing conditions such matter was never found on the seafloor at points very distant from the coast from which it was eroded. Thus it seemed to him that porphyry pebbles could be distributed over a long distance perpendicular to the coast only during an ongoing period of elevation, when the seabed would, in effect, slide up and out of the water on its way to becoming dry land. The longer this process lasted, the more thinly and widely would a bed of erosional materials be redistributed on the upraised land.

At about this time Darwin wrote a series of notes that I have come to believe were subsequently misinterpreted by other historians of his geological work. These jottings are found in another of Darwin's small pocket notepads, this one known as the Santiago Book. Virtually every scholar who has written about his coral reef theory has mentioned, and accepted, a declaration Darwin made forty years after the *Beagle* voyage. As I discuss in chapter 11, the autobiographical recollections he wrote in his mid-sixties state that he came up with his eventual coral reef theory in South America before ever seeing a coral reef. The reason for Darwin scholars' confidence on this point is a reference to corals in the Santiago Book that at first glance appears to confirm his recollection. There is good reason to revisit this passage, however, for the current interpretation renders it a profound

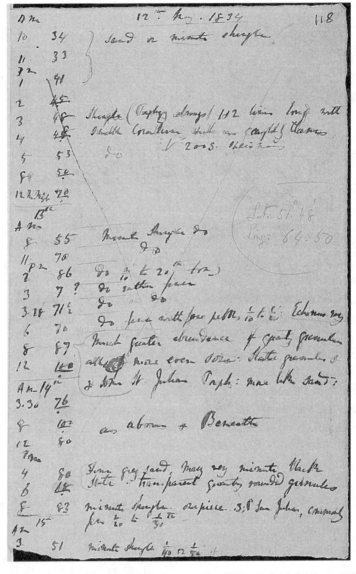

Figure 9. 12 May 1834: "We sailed from S. Cruz. in a SW line to look for the L.Aigle rock.—I attended carefully to the Soundings." The result of Darwin's careful attention was a pair of tables including this one, which shows the depth, type of bottom, size of granules or pebbles, and any organic matter recovered by the arming of the sounding lead. Note the labor that must have been involved for the ship's crew in hauling up the line after soundings of up to one hundred fathoms (six hundred feet) every two or three hours throughout the day. Quotation from DAR 34.1:90; pictured table from DAR 34.2:118. Reproduced by kind permission of the Syndics of Cambridge University Library.

non sequitur in the context of the notes in which it was written. Rather than explaining the location or shape of coral reefs, this note actually contains a theory about analyzing sedimentary rock formations on land. This makes sense, for it was a continuation of Darwin's musings on elevation and, in particular, on the question of how one could determine whether a sedimentary bed had been formed during a period of elevation.

The specific point of departure for the notes in the Santiago Book was the conclusion I mentioned above, that the distribution of pebbles across Patagonia could be accounted for only by sustained, gradual elevation. Now he was trying to answer a different kind of question, however. It was a more abstract and general geological puzzle: Was there a way to tell whether a layer of rock in the stratigraphic record had originally been deposited on a shallow and/or elevating seafloor? He had, in other words, asked whether his interpretation of Patagonian pebbles suggested a general rule for practicing stratigraphy. Must *every* stratum of conglomerate rock containing such pebbles have been formed during a period of uplift? Thus in early-mid 1835 Darwin noted in the Santiago Book, "I believe much conglomerate [in a stratum of sedimentary rock] is an index of [the sea's] bottom coming near the surface." He went on to state this more clearly in the form of a rule of stratigraphic interpretation: "May we not imagine [that] each band of conglomerates [in a succession of strata] marks an epoch when that part of the ocean's bottom was near to a continent or shoal water[?]"[25] As he concluded in a note that he added, after the fact, to his essay on the Rio Santa Cruz (see fig. 10), "I strongly suspect there is a constant relation between the spreading out of gravel & that part of the ocean bottom having been elevated to within a moderate distance of the coast line."[26]

In the same passage of notes from the Santiago Book, Darwin also considered what a sedimentary rock would look like if it had been deposited on a sinking seabed rather than a rising one. He wrote, "As in [the] Pacific a Corall bed. forming as land sunk. would abound with. those genera which live near the surface (mixed with those of deep water) & which would more easily be told the Lamelliform. Corall forming. Coralls.—I should conceive in [the] Pacific. wear & tear of Reefs must form strata of mixed. broken. sorts & perfect deep-water shells (& Milleporae)."

This passage refers not to the constructional part of a coral reef, but to the sediment that would reach the deep seafloor through "wear & tear" on growing reefs. Thus beds would contain a mixture of organisms from shallow and deep water. But the deepwater shells would be embedded in "perfect" condition while the "genera which live near the surface" would

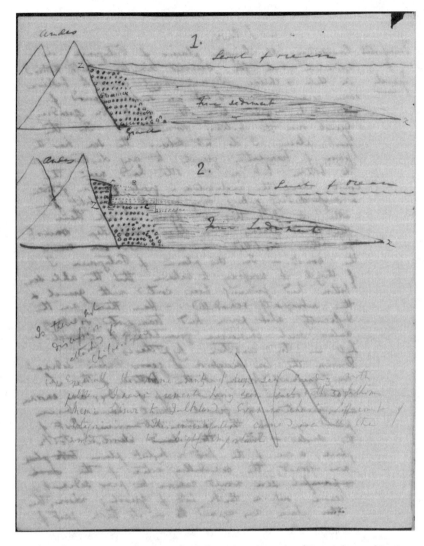

Figure 10. "Let us try another hypothesis." A prime example of Darwin's searching style of theoretical note making. In text that appears to have been added to his essay on the Rio Santa Cruz only after visiting Chiloe on the west coast of South America, Darwin wrote:

> In the plain of Patagonia I thought it necessary to believe that the whole seas bottom had primarily been coated with gravel & this subsequently remodelled.— Here then was the difficulty[:] what power had transported so many miles such immense quantities of gravel. . . . Let us try another hypothesis.—During the long succession of years, when sediment was gradually forming at the bottom of the S. Atlantic ocean a great mass of beds, an enormous accumulation of gravel must

be deposited *only* when they had been removed by "wear & tear" and drifted to the seafloor. Unlike the deepwater shells, they would be found in a "mixed. broken" condition.

Darwin was not certain he had actually seen a layer of sedimentary rock on land that answered to this description. He pondered whether "such appearance correspond[s] to any of the great Calcareous [limestone] formations of Europe." He concluded that if such a bed consisted of organic remains of which a "*large* proportion [was] those Coralls which only live near [the] surface," then "we may suppose [that] the land [was] sinking" when it was deposited.[27] Such subsidence would continually create fresh zones of shallow water where corals could grow, which would in turn mean greater amounts of coral rubble reaching the seabed.

---

have been piled up at the Eastern submarine foot of the Andes.—When ~~the land~~ elevations took place, & one of the first or highest plains ~~took place~~ was formed, the remodelling action of the ~~shoal & powerful~~ sea would remove the finer sediment & leave only a thick bed of gravel; when, this other plain was exposed ~~to~~ in its turn, part of it would be destroyed & the gravel spread out in a thinner layer over another plain (or ~~[illeg]~~ seas bottom).—this process being repeated, the gravel would at length be carried far to sea-ward by the action of the sea near to a beach. but necessarily the thickness or quantity must be greatly diminished <<in>> each ~~time~~ successive plain.—The Diagram will show what I mean. No^r 1 represent what is supposed to have been originally the case; ~~No^r 2. after some elevations~~ Z.X being the bottom of the Atlantic: No^r 2. after some elevations the cliff AE would be formed & the bottom ZX would have a different shape; hence the mass of gravel included between ABCE has been removed & would be remodelled over the bottom as far as tidal power could carry it: Part of this after another elevation, would, as represented, be again spread out in a thinner sheet. & so on an [*sic*] infinitum; the finer particles always being removed to a greater distance.— Now at S. Cruz. at the coast the Gravel is not above 50 feet thick. & at 100 miles inland it is 212 ft thick.—Against this <<hypothesis>> have three objections it leaves unaccounted the transportal of the enormous angular blocks of ancient rocks; the facts, which I have adduced to show how little motion gravel has even in shoal, turbulent seas.—& lastly the appearance of the pebbles having come from the Northward. (It might be conjectured, in answer to this last objection, that an eddy stream swept by the foot of Andes to the South. owing to a. current . . . such. as now exists, sweeping round. C. Horn.)—~~My mind is~~ I cannot make up my mind on this question, but I strongly suspect there is a constant relation between the spreading out of gravel & that part of the ocean bottom having been elevated to within a moderate distance of the coast line. DAR 34.2:151–152v.

After making some reading notes that may indicate the passage of hours or weeks while this train of thought was interrupted, Darwin resumed writing on the topic. At this stage he stopped thinking only of sedimentary rocks that might contain the remains of shallow-water corals that had been sloughed off a reef and began to imagine the possibility of strata actually "containing Corall reefs."[28] He realized that intact fossil reefs might offer irrefutable evidence of the direction of the land's movement, on one condition: "The test of depression <<in strata>> is where [a] great thickness has. shallow. coralls growing in situ: this could only happen when bottom of ocean was subsiding."[29] The "great [vertical] thickness" was the key indication, for under normal conditions the vertical depth of a reef was limited by the shallow range in which reef-building corals could flourish. A sinking ocean floor, on the other hand, would allow corals to grow on top of one another to a "thickness" limited only by the total amount of subsidence.

It is important to recognize that Darwin arrived at these conclusions in answer to his questions about stratigraphy. His notes show clearly that this "test of depression" was meant to guide the interpretation of sedimentary rocks *on land*, and there is no evidence that he saw any way to apply it to his observations at the islands of the Pacific, where depressed strata would be hidden underwater. The test of depression certainly built on Darwin's expectation that widespread gradual subsidence was a genuine phenomenon of the present and the past, which itself was a consequence of his belief in the reality of "horizontal" or "concentric" elevation. Moreover, these notes prove that Darwin was aware before mid-1835 of Quoy and Gaimard's claimed depth limit for the growth of reef-building corals, his assent to their rule perhaps being encouraged by his now wide experience studying zoophytes belonging to various depths.

These notes in the Santiago Book do not, however, prove that Darwin's coral reef theory was thought out on the west coast of South America, notwithstanding the claims made by several historians who have cited these passages, and the letter from Alison, in support of this point from Darwin's autobiographical recollections.[30] Indeed, the notes imply that Darwin was at this time entirely incapable of imagining that he would apply the notion of upward growth of corals on a sinking foundation to his study of a living reef. A few pages later he was still casting about for a subsidence-based prediction that could actually be tested later in the voyage: "If the Pacifick Isl[ands] have subsided there ought to be a peculiar vegetation."[31] The ideas recorded in the Santiago Book did contribute to his eventual coral reef theory, but not until he was actually in the Pacific looking at a living reef.

Taking stock of Darwin's progress in his "chief object[s] of pursuit," studying geology and examining invertebrates, we find that in late June 1834 the two activities were very complementary.[32] I want to emphasize that several of his areas of interest converged because they were tightly meshed with the labor and objectives of the surveyors. Darwin desired to know the elevation of every formation he saw; Beaufort had instructed that "it should be considered an essential branch of a nautical survey, to give the perpendicular height of all remarkable hills and headlands." Darwin was also eager to compare the inclination of the continent above and below sea level. FitzRoy had been commanded not merely to map undersea terrain, but "to note with accuracy the slope, or regularity, of the depths," along with "the quality of their various materials, and the disposition of the coarse or fine parts, as well as of what species of rock in the neighbourhood they seem to be the detritus." To Darwin these were topics of *increasing* fascination and significance because of their relevance to theories of South American uplift that he had arrived at *as a consequence* of things he had learned from the surveyors. His fieldwork and his theorizing evolved together during the voyage, in ways that depended upon *ongoing* access to cooperative hydrographers.

What of FitzRoy's obligation to study coral reef formation? Though Darwin had by this point in the voyage made no explicit reference to Beaufort's instruction in any of his notes, it is clear that the Pacific and its coral reefs were never far from his mind. Recall that Darwin had told Henslow as early as April 1833 that "the Corall reefs & various animals of the Pacific may keep up my resolution" and a year later that he had been "preparing [him]self for the South sea, by examining the Polypi of the smaller Corallines in these latitudes." Writing to his sister Catherine from the Chilean coast "a hundred miles South of Valparaiso," he admitted that he had achieved "so much in Geology & Natural History, that I look back to Tierra del Fuego with grateful & almost kindly feelings." As he told his family, "Amongst Animals, on principle I have lately determined to work chiefly amongst the Zoophites or Coralls: it is an enormous branch of the organized world; very little known or arranged & abounding with most curious, yet simple, forms of structures."[33] Taken together, these letters suggest that Darwin was drawn to coral reefs by more than just a tourist's curiosity to see them. He was making ever more emphatic commitments to an ambitious program of coral research by which he expected to synthesize knowledge of Atlantic and Pacific zoophytes.

4

The Making of a Eureka Moment

There is no evidence that before the *Beagle*'s arrival at Tahiti on 15 November 1835 Darwin had conceived any new answer to the questions of how and why ring-shaped coral reefs were formed, or even that he had actively contemplated these questions during the previous four years of voyaging. By the time the ship departed eleven days later, he had written a coy note in his diary indicating that, in addition to his ambitions in coral zoology, he had gained a sudden confidence that he could overturn the established explanation of reef formation. Then, during the long passage from Tahiti to New Zealand he drafted an essay in which he described a eureka moment at Tahiti and sketched the outlines of an elegant and breathtakingly original theory that claimed to explain the origin of virtually all the reefs in the Pacific.

In this chapter I recreate the circumstances that led to Darwin's moment of insight and argue that it depended on his ability to envision the underwater terrain and its inhabitants with the eyes of a hydrographer and on his suspicion, born of his South American geologizing, that the floor of the Pacific Ocean might be sinking. I identify two further factors specific to Tahiti that, when combined with the knowledge and expectations Darwin brought to the island, made his eureka moment possible. The first of these, recorded explicitly in Darwin's notes, was the vantage on the land-

and seascape of the Society Islands offered by climbing high on Tahiti's mountainous slopes. The second local factor, which Darwin described in his diary but which I am the first to link to his coral reef theory, was the striking succession of different types of plants he witnessed while ascending the mountainside at Tahiti. This was a conspicuous manifestation of the "zonation" of plants according to altitude, a phenomenon Humboldt had depicted in iconic diagrams as distinct rings of vegetation encircling mountains at different elevations. In this chapter I argue that Darwin's new explanation for how rings of shallow-water corals could grow in the deep ocean was a consequence of applying to submarine organisms this law of organic distribution that Humboldt had championed in describing land plants.

I also analyze the twenty-page essay Darwin wrote immediately after leaving Tahiti. This first of Darwin's statements on the formation of reefs is revealing at several levels. It indicates that Darwin's eureka moment caused his attention to shift toward questions he had not been consciously considering, which in turn made a set of previous experiences (the ones that made the eureka moment possible) seem as though they had *always* been inherently directed toward answering these new questions.

Darwin's hydrographic experience helped him to adopt methods and perspectives like Humboldt's, and to transpose them beneath the sea. The coral reef essay in turn provides evidence that he drew on Humboldt's *writings* for concrete data *and* for geognostic theories to a degree that has previously been underappreciated. The essay also illustrates that while his eureka moment had been inspired by a single landscape, his early confidence in the coral reef theory rested on evidence drawn not from his own fieldwork but from using maps to compare the shapes of individual reefs and the distribution of groups of reefs. Therefore the contents of the essay direct our attention to the value Darwin drew from being able to alternate between, or juxtapose, periods of fieldwork and periods of intensive cabinet-style research. I am hardly the first scholar to point out that Darwin made frequent use of the extensive library of books and charts aboard the *Beagle*, but I want now to call particular attention to his high opinion of the knowledge that could be produced by studying facts found in publications and other secondhand accounts (as well as through direct experience). This presaged, as I will later argue, Darwin's ongoing dependence on map-based research on coral reefs *after the voyage* as well.

The Dangerous Reefs of the Low Archipelago

After leaving the shores of South America for the last time, the *Beagle* called at the Galápagos Islands. The five weeks spent there have become the most famous period of the voyage because of the evolutionary implications Darwin later attributed to the distribution of the animals he encountered on the islands.[1] Beyond the Galápagos lay the open Pacific and the islands Darwin had dreamed of seeing when he was "stirring up" Beaufort about the voyage's itinerary. The *Beagle's* next destination was Tahiti, but to get there FitzRoy would have to thread the ship through the ring-shaped coral reefs of the Low Archipelago (now called the Tuamotu Archipelago). Bougainville had christened these islands *l'Archipel Dangereux* in acknowledgment of the threat such inconspicuous landforms posed to safe navigation. Twelve days out from the Galápagos the sight of a black tern flying past the ship indicated that land was near, but it took another week, until 9 November, before one of the slips of coral land was sighted. FitzRoy identified it as Honden Island (now Puka-Puka)[2] from a chart made by Johann von Krusenstern during the first Russian-sponsored circumnavigation. To Darwin the tiny island "bears no proportion & seems an intruder on the domain of the wide all-powerful ocean." Giving the English translation of the Dutch "Honden," he identified it in his diary as "Dog or Doubtful Isd.—The latter name expressing all which was known about it."[3]

The inherent danger of sailing among unfamiliar low islands was amplified by the difficulty of identifying them. Determining a ship's location was tricky enough, especially while in motion, but to make matters worse the many reefs of the Dangerous Archipelago were difficult to tell apart because few had any distinguishing features. Thus it was common for islands to be misidentified or to be "discovered" multiple times by navigators who had miscalculated their own positions (or who were referring to charts made by someone whose position had been miscalculated). In turn it was unsafe to assume that the way ahead was safe even if charts showed open ocean on the course that had been set. To the extent that FitzRoy trusted any existing surveys of the area, he depended on Krusenstern's chart and the narrative it accompanied, which he judged "the only documents of any use to us while traversing the archipelago of the Low Islands." Alluding to Bougainville's account, he continued, "This archipelago is indeed extremely deserving of its appellation, 'Dangerous'; for numerous coral islets, all low, and some extensive, obstruct the navigation, while unknown currents and strong squalls, and a total want of soundings, add to the risk

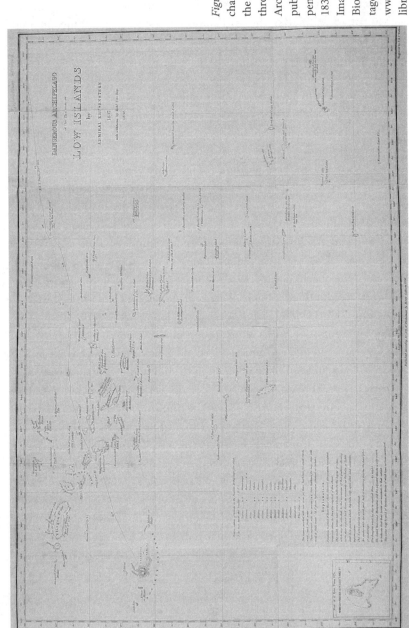

Figure 11. Track chart showing the *Beagle*'s route through the Low Archipelago. It was published in the appendix to FitzRoy's 1839 *Narrative.* Image courtesy of Biodiversity Heritage Library, http://www.biodiversity library.org.

of sailing there at night."[4] The "want of soundings" referred to the abruptness with which the reefs rose from the ocean's depths and the consequent inability to anticipate a reef's proximity. When approaching a continent, by contrast, soundings would reflect a gradually shoaling bottom to indicate that land was nearby. Out here in the open ocean there was no such warning, and it proved possible to sound with a line hundreds of fathoms long and find no bottom despite being nearly within striking distance of a low island.

All these risks weighed on FitzRoy as the next several days brought heavy wind and rain from the west. In a decision unprecedented during their previous ocean crossings, FitzRoy refused to sail in the dark. On a stiflingly hot and humid evening, Darwin put pen to paper. "The air being thick & misty & the night dark, for the first time it has not been thought prudent to run on. So that we are now hove to, wasting the precious time till daylight comes & shows us the dangers of our course."[5] FitzRoy in his narrative recalled "pass[ing] some anxious nights." Daybreak of 13 November revealed an islet that did not appear on Krusenstern's chart. At midday they saw another, larger lagoon-encircling reef to the south, also uncharted. They ranged along the north shore making a running survey. The chart FitzRoy later published shows that they found no bottom with soundings of 142 and 144 fathoms (852 and 864 feet). He reported the larger island's "native name" as Cavahi (now Kauehi), which suggests either that he had some interaction with its inhabitants or that he learned its identity after arriving in Tahiti. Neither he nor Darwin mentioned landing at Cavahi, but FitzRoy described the reef as having "a number of islets covered with cocoa-nut trees, surrounding a lagoon." He lamented that he "could not delay to examine the [islets on the] south side of the lagoon."[6]

Presented at last with his first close-up view of a low island, Darwin climbed the mast in search of a better vantage point. He referred to Cavahi in his diary as Noon Island, it being the one they had spotted at midday. Noon Island seemed in some respects rather underwhelming, for the width of these islets of mere sand atop the reef was "trifling." The "long brilliantly white beach . . . capped by a low bright line of green vegetation" painted a stripe across the ocean's field of blue. In both directions it narrowed into the distance and sank beneath the horizon. From the masthead he was able to see across the smooth surface of the lagoon to the opposite side of the ring-shaped reef. This impressed him, for he judged the "great lake of water" to be ten miles across. (For a similar perspective, see plate 4.) But Darwin became frustrated at seeing the reefs only from a distance.

Even from the top of the mainmast, he concluded, they offered a "very uninteresting appearance."[7] Within days, however, his life would be changed by the opportunity to gaze upon a ring-shaped reef from a considerably greater height.

The View from Tahiti

Tahiti is an island of precipitous mountains encircled by a coral reef. As the *Beagle*'s company caught sight of it for the first time in the dawn light of November 15, FitzRoy was disappointed to see bold clouds obscuring the lofty peaks.[8] It is difficult to overstate the emotional effect that the history of prior voyages to Tahiti and its majestic appearance could combine to produce in Europeans' minds. This was, from the European perspective, the most tantalizing of the South Sea Islands: the place where Bougainville had encountered people whose lifestyle was purported to reflect the "state of nature"—the paradise from which the mutineers of the *Bounty* had been unwilling to depart. Darwin arrived in Polynesia possessing a mass of facts accumulated during the past four years, a set of skills honed during his time in South America, and—as a result of applying both to the study of continental elevation—a specific question to guide his attention in the Pacific: Was the ocean floor sinking? This would be his first chance to look for evidence of the subsidence he felt sure must have occurred to compensate for the emergence of South America. Here in Tahiti, within a matter of days, all these factors coalesced into a novel answer to an entirely different question: the origin and shape of coral reefs.

Arriving at the breathtaking peaks of Tahiti after spending nearly a month as a speck on the seemingly limitless ocean, Darwin again sought perspective on his surroundings. Rather than climbing a mast, he hired guides to lead him up the nearest ridge. Ascending to an elevation of "two or three thousand feet," he realized he had climbed through a series of discrete zones of vegetation. This vertical journey from sea level gave him an uncanny feeling of moving horizontally across climatic regions of the globe, as though he were traveling from the equator back toward his temperate home. Halfway up, after coarse grass had succeeded the dwarf ferns below, Tahiti began to look strangely familiar. "The appearance was not very dissimilar from that of some of the hills in North Wales," he remarked, "and this so close above the orchard of Tropical plants on the coast was very surprising."[9] He pressed onward until "trees again appeared . . . tree ferns having replaced the Cocoa Nut"[10] (see plate 5). This was a remarkable

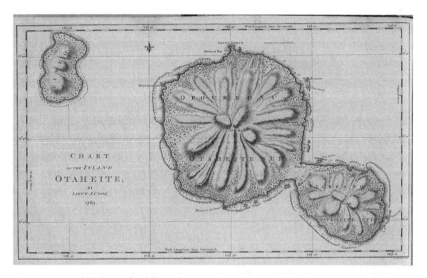

Figure 12. Cook's chart of Tahiti and Eimeo, based on his visit sixty-six years before the *Beagle* arrived there. On Darwin's first weekday at Tahiti he hiked up the mountainside above Matavai Bay, in the northwest part of the island. From a height of three thousand feet, he cast his gaze westward across Tahiti's barrier reef to the neighboring island of Eimeo. David Rumsey Map Collection. www.davidrumsey.com.

experience, but one he might have expected on such a steep climb, for the first pages of Darwin's favorite book asserted that "each group of plants is placed at the height that nature has assigned." Those vertical regions, Humboldt argued in the *Personal Narrative*, "form the natural divisions of the vegetable empire; and in the same manner as the perpetual snows are found in every climate at a determinate height, [plants] have also their fixed limits."[11] Humboldt's plant geography, like his study of snow lines, famously illustrated that ascending in altitude (from sea level to a great height) was equivalent to ascending in latitude (from equator to pole) (see plate 6).

The *Beagle*'s onboard library contained at least one of Humboldt's iconic diagrams of the succession of floras on the flanks of Chimborazo, then believed to be the highest mountain in the world, and Darwin knew of Humboldt's claims to have determined "according to barometrical measurement, in more than 4000 plants of the equinoctial region, the height of each station above the level of the sea."[12] However, he had never seen such a vibrant manifestation of this phenomenon as on the slope of Tahiti, where a succession of characteristic floras ringed the mountain like a series of living contour lines.[13]

Figure 13. The reef-encircled island of Eimeo (Moorea) as viewed from Tahiti in 1839 and depicted by Joseph Pitty Couthouy of the United States Exploring Expedition. Four years earlier Darwin had written, "Viewing Eimeo from the heights of Tahiti I was forcibly struck with this opinion. . . . Remove the central group of mountains. & there remains a Lagoon [within the reef]." Image from Couthouy, "Remarks upon Coral Formations."

High above Matavai Bay, Darwin turned west toward Tahiti's smaller twin, the island of Eimeo (present-day Moorea; see fig. 13). Lying just fifteen miles distant, its jagged peaks were plainly visible even from the *Beagle*'s anchorage. From higher up it became clear that the mountainous Eimeo, just like Tahiti, was encircled by a barrier reef that stood some distance offshore. Darwin likened this perspective to viewing a framed engraving. The landmass of Eimeo was the picture itself, the vivid aquamarine lagoon surrounding it was the mat, and the white line of waves marking the outer edge of the reef was the frame[14] (see plate 7).

This is the moment, gazing at Eimeo from the heights of Tahiti, when it suddenly mattered that Darwin had arrived at the Pacific already believing the seafloor might be sinking. Here, I propose, is where ideas he had previously formed in South America gained new meaning and significance. The notes he wrote about this moment do not bother to explain why he had been pondering the subsidence of the ocean floor. They mention only the consequences of doing so, of imagining that Eimeo was slowly sinking out of sight: "Viewing the Eimeo from the heights of Tahiti I was forcibly struck with this opinion. . . . Remove the central group of mountains, & there remains a Lagoon [within the reef]."[15] If Eimeo were drawn downward below the sea while the reef around it continued to grow, Darwin realized, all that remained would be a circular reef enclosing an empty lagoon. Had he, while staring at a barrier reef, just solved the puzzle of how a different class of reefs was formed? Perhaps ring-shaped reefs did not grow atop submarine volcano craters but were a result of corals growing upward at the circumference of a sinking island.

For the reef to remain visible while Eimeo sank out of sight, corals would have to continue growing up into the shallow layer of water just below sea level. It seems no coincidence, then, that Darwin had this idea immediately after pondering the vertical distribution of plants while climbing to his present vantage point. It seems likely that he recognized these rings of coral as equivalent to the bands of flora encircling Tahiti above the sea. After all, didn't corals grow like a turf of vegetation wherever a suitable foundation lay at the appropriate depth? Just as the land plants would maintain their optimal altitude by migrating up the mountainside as Eimeo sank, living corals would add thickness to the reef whenever subsidence provided new room to grow back up to the ocean's surface. Like the snow line that marked a boundary in Humboldt's mountainside diagrams, the waterline was a fixed limit for corals. Darwin was applying the same principles that constrained the geography of mountainside plants in the Andes to the vegetative growth of zoophytes on the flanks of submarine mountains.

Darwin had been waiting four years to get a look at the zoophytes responsible for building coral reefs, but he continued for several days to explore Tahiti's higher altitudes. It was not until 22 November that he finally fulfilled his wish, in an outrigger canoe paddled by several Tahitians.[16] At the farthest point he reached from shore he learned that a "mound of Coral rock, strikingly resembling an artificial (but low) breakwater," fronted the open ocean.[17] Inside the line of whitecaps marking the reef's highest point was a broad tract filled here and there by patches of coral, with "little narrow twisting channels & holes of deep water, & on the other hand many points. where the Coral reaches to the surface."[18] This inner section of the reef was bathed by the calm waters of the lagoon, which deepened toward the island as the back of the reef fell away, leaving harbors "where a ship can anchor in a fine Sandy bottom."[19] He suspected that the sediment and freshwater that ran off Tahiti into the lagoon prevented any growth of corals immediately near shore.[20]

While examining the lagoon at Tahiti, Darwin was especially eager to discern whether different types of organisms had formed different parts of the reef. The main constituents of the inner reef proved to be "stony & branching generas."[21] In his diary he recalled "admiring the pretty branching Corals."[22] From the boat he collected a specimen of *Fungia* and kept it alive long enough to study its polyps' "considerable powers of contracting & motion" under his microscope.[23] He hoped to compare these specimens with corals living on the other side of the reef, in the water of the open ocean, but the islanders were unable to take him there "owing to the surf . . . breaking violently on the outer margin, continuously pump[ing] over in sheets the water of its waves."[24] Instead he relied on the testimony of the Tahitians themselves. "Showing [lagoon corals] to some intelligent natives, I was assured that such kinds never grow on the outside of the reef or compose solid reefs.—From their descriptions. I imagined the prevalent kinds, so situated are [corals] such as [the genera] Porites. Millepora. & some Meandrina & Astrea. Anyhow, they considered that there is a wide distinction in the two cases."[25]

Very little documentation exists, alas, of this tantalizing interaction between an English naturalist and Polynesian interlocutors for whom reef-building corals were a familiar feature of the local environment. Was Darwin's idea of an "intelligent" Tahitian specific to the questions he was asking? Did he find someone who was especially knowledgeable about corals, or did he assume that Tahitians were uniformly familiar with matters relevant to natural history? And what did the Tahitians who paddled the va'a and led Darwin on inland ascents make of him? The history of contacts between Pacific islanders and visiting scientists is long and complicated. It includes the efforts at mutual cartographic understanding between Captain Cook and the Raiatea-born navigator Tupaia and, two hundred years later, disagreements over whether it was safe to reoccupy atolls in the Marshall Islands after the nuclear weapons tests there.[26] Darwin's interview with the Tahitians was consistent with the practice of many other traveling naturalists who sought aid from locals, and it followed the pattern of his work elsewhere.

Observing—and engaging with—other people was one of Darwin's main preoccupations during the voyage. He published accounts of slave society in Rio de Janeiro, of the indigenous inhabitants of Tierra del Fuego, and of the gauchos in Patagonia. And just as his fellow occupants of the wooden world that was a British surveying vessel helped him to "see" features of the underwater realm that had eluded other geologists and zoologists, locals around the globe taught Darwin to notice and understand features of his surroundings to which he would otherwise have remained blind.[27]

All of his own former experience studying the contents of his dredge and the armings of the sounding lead encouraged him to believe his Tahitian informants when they told him that different types of corals formed the inner and outer parts of the reef. "Analogy. from the habits of all other marine animals," he concluded, "would lead one to suppose that the same species would not flourish in two such different localities, as the foam of furious breakers. & shallow placid [lagoons]."[28] He was convinced that the bulky reef builders could inhabit only the outer margin, meaning they would grow upward, but not inward over the reef flat, if an island they fringed were to sink. This explained why circular reefs continued to have lagoons rather than becoming covered over by a solid cap of coral rock.

Visiting Tahiti had been a revelation. The combination of Darwin's high-altitude eureka moment and his study of the reef from sea level led him to make one of the boldest entries in his entire *Beagle* diary. On determining that the up-close details of the reef offered nothing to contradict his speculations from the mountainside, he wrote, "It is my opinion, that besides the avowed ignorance concerning the tiny architects of each individual species, little is yet known, in spite of the much which has been written, of the structure & origin of the Coral Islands & reefs."[29] All his ambitions are revealed in this brief note. Earlier in the voyage he had decided to study corals—the "tiny architects," as he called them here—precisely because he relished the opportunity to gain expertise in an area of "avowed ignorance." Now, unexpectedly, he could also challenge the "much which [had] been written" on a theoretical problem that was deemed so significant in Lyell's *Principles of Geology* and in Captain FitzRoy's own orders from the Admiralty.

Theorizing Like Humboldt in a Floating Library

Brimming with confidence after leaving the site of his eureka moment, Darwin spent the December 1835 passage from Tahiti to New Zealand writing a detailed statement of his new theory, a twenty-page essay headed with the words "1835 Coral Islands."[30] The *Beagle* was crossing the very waters in which Beaufort had expected FitzRoy to carry out his coral island surveys, Beechey having already charted the islands of the eastern Pacific. Exhausted by the survey of South America, however, FitzRoy chose to sail on without stopping. Because this left Darwin unable to add to his firsthand knowledge of coral reefs, he turned to the books on board for further information.

Along with recent geological texts by Lyell and De la Beche that drew on the work of Quoy and Gaimard, the *Beagle*'s library contained all the "voyages," the travel genre that ranged from Forster's learned *Observations* to Beechey's recent narrative. With ready access to virtually the whole canon of reef science, Darwin felt emboldened to generalize beyond his own narrow experience at Tahiti, laying out the implications of his new theory for the geology of the globe.

"Although I have personally scarcely seen anything of the Coral Islands of the Pacifick Ocean," he began, "I am tempted to make a few observations regarding them."[31] He started his argument not with what he had seen in Tahiti, but with what he had learned while examining the printed material aboard the ship. He used the descriptions provided in the books by Forster and Beechey to synthesize a taxonomy of islands: (1) high islands without coral reefs, such as the Sandwich (Hawaiian) Islands; (2) high islands encircled by coral reefs, such as Tahiti, Eimeo, and other Society Islands;[32] (3) low islands made of coral, such as those "lagoon islands" (atolls) in the Low or Dangerous Archipelago; and (4) islands of dead coral in the annular shape of a lagoon island but uplifted from the water, such as Beechey's Elizabeth Island. Darwin pointed out that archipelagoes tended to be characterized by islands of a given type. Furthermore, many of these island groups shared a general orientation from northwest to southeast, which suggested that some common cause had acted in the original formation of both high and low islands.

He declared that this conventional taxonomy was faulty because it was based on an "artificial" distinction between the rings of coral that encircled high islands and those that encircled empty lagoons. Extending this point, he argued that if you removed the central landmass from within an encircling reef, you would be left with a "lagoon island" identical in "structure & origin" to those now existing in the Low Archipelago.[33] (The term "lagoon island" at this time referred, rather confusingly, not to a high island within a lagoon but rather to a ring-shaped *reef* that encircled an *empty* lagoon.) Pointing out a sequence of intermediate forms from Beechey's *Narrative* and Krusenstern's *Atlas*, including "those extraordinary barriers of Coral, which front for so many leagues the coast of Australia," Darwin demonstrated that no "essential character" remained to distinguish the reefs of the high island class from those constituting the low islands. He finished this line of argument by showing that there was a plausible mechanism by which high islands could be removed from within an encircling reef, namely subsidence of the ocean floor of the kind he had envisioned while viewing Eimeo. "If the proofs of the identity in nature of the two kinds of reefs, are considered as conclusive," he resolved, "there is no <u>necessity</u> that the Lagoon I[slands] should be based on [submarine] Craters."[34]

In the second part of the essay Darwin shifted his focus from the morphology of reefs to the corals that constructed them. He described everything he knew about the factors limiting the growth of different kinds of stony corals and explained their distribution on the circular reef at Tahiti.

He believed Quoy and Gaimard were mistaken in claiming that the same kinds of corals lived within and on the outside of a reef. He then proposed what could be called a thought experiment, imagining "an Island situated in a part of the ocean. which we will suppose at last becomes favourable to the growth of Corall," so that "Corall would immediately commence to grow on the shore & would commence Sea-ward as far as the depth of water would permit its rising from the bottom."[35] He explained that an island fringed by a reef of this kind would "essentially differ from those in the South Sea, in the depth of the water . . . beyond the Wall not suddenly becoming excessive."[36] If there were no subterranean movement, corals could never grow into a reef like that of Tahiti, where soundings beyond the breakwater showed a precipitous change of depth. However, he pointed out, coral growth combined with subsidence of the seafloor would produce structures resembling, in turn, the sequence of real islands he had already described as moving from the taxonomy's second class (reefs encircling high islands) to its third. If, on the other hand, coral growth were interrupted by subterranean elevation (or the equivalent, a drop in the level of the sea), the result would be a fringe of dead coral rock like that of the rare fourth class. He recalled, in agreement, that Lyell believed the paucity of upraised coral in the Pacific implied that subsidence had outweighed elevation in the recent history of the ocean floor. Lyell had not, however, drawn a connection between this observation and the forms taken by Pacific reefs.

Darwin's idea, then, constituted a new answer to the question Beaufort had asked in his instructions to FitzRoy. What was the origin of "circularly formed" coral reefs? Darwin put it succinctly: "The direction of the movement determines the structure of the reef."[37] And the dimensions of a circular reef corresponded not to the extent of an underlying crater but to the former shoreline of a sunken island. The same principle, Darwin argued, was "referrible to those reefs which front a continent"—their depth and distance from shore could be the product of subsidence.[38] Key to it all was what Humboldt might have called the "geography of corals."[39] Their distribution across the globe determined which islands or continents might form the foundation for a reef. Their local distribution in horizontal and vertical space was the basis for a given reef's outline. Thus the shape and breadth of the reef would be determined by the direction and inclination of this foundation where it passed through the several-fathom vertical zone in which reef-building corals could live. Darwin had absorbed the concepts of a three-dimensional plant geography and set them in motion through a conjectured geological past. The insight at Tahiti had, he concluded, removed "much of the difficulty in understanding Coral for-mation" that he believed had confounded Forster, Quoy, Gaimard, and Lyell.[40]

Months earlier on the western slope of the Andes, when he had tried to determine the geological history of South America from the composition and thickness of various layers of sedimentary rock found there, Darwin had supposed that "the Test of depression <<in strata>> is where [a] great thickness has. shallow. coralls growing in situ." Now he had evidence—in the horizontal shape of the world's living coral reefs—of an ongoing pro-cess that must be producing just such massive vertical thicknesses of coral rock. In the "Coral Islands" essay he modified his words from the Santi-ago Book to emphasize how the present could serve as key to the past: "When in any formation there should be found, a great thickness com-posed of Coral & the genera of which resembled those, which now build the reefs, we might also conclude. that during its successive accumulation, the general movement, was one of depression."[41]

In the final pages of the "Coral Islands" essay, Darwin meditated fur-ther on the connection between his new theory and the geological ques-tions that had intrigued him ever since his discovery of geologically recent shells on the high plains of Patagonia. As we know, he had originally imagined Pacific subsidence only because of the particular process he believed had elevated South America. This in turn had led him to posit that island-fringing reefs could be turned, by subsidence, into encircling reefs and eventually into the freestanding reefs known as "lagoon islands." And that supposition appeared, earlier in the essay, to have been confirmed by evi-dence drawn from the charts of the Pacific that FitzRoy carried, which were themselves products of hydrographic work on earlier journeys just like the *Beagle*'s.

But now he reversed his line of thought, using his insights into coral reefs as a basis for reasoning about continents. He took the existence of en-circling reefs and lagoon islands as evidence for subsidence in the Pacific: if he understood the propagation of corals correctly, and if submarine move-ment truly did determine the structure of reefs, then the lessons that could be read from the shape of reefs around the globe would be most "important to Geology." "For then we might assume that groups of Lagoon Is[lands] clearly showed that a chain of Mountains had there subsided."[42] Though he did not draw the link explicitly, this conclusion was supported by the ob-servation with which he opened the essay, that archipelagoes of high and low islands alike seemed to share some connection to an original cause that produced mountains in a northwest-southeast direction. In South America he had seen seabeds elevated into terraces and mused about compensatory subsidence. Now he had a glorious confirmation. "The general horizontal uplifting which I have proved has & is now raising upwards the greater part of S. America . . . would of necessity be compensated by an equal subsidence in some other part of the world.—Does not the great extent of the Northern & Southern Pacifick include this corresponding Area?"

In his last lines Darwin made explicit a connection between his thinking and Humboldt's geognostic theorizing. He quoted his hero as arguing "that 'the epoch of the sinking down of Western Asia coincides with the elevation . . . of all the ancient systems of Mountains, directed from East to West.'" This seemed, in Darwin's view, to lend credence to the idea that elevation and subsidence were *simultaneous*, compensating for one another, and that they were therefore causally linked geological processes.[43] Like other continental theorists, Humboldt thought that parallel mountain chains had been uplifted within the same geological period, meaning that the orientation of the main axis of a mountain chain could be an an indication of its age. Darwin came to reject this line of thinking once he had become Lyell's acolyte after the voyage, and instead he presented his arguments about reciprocal elevation and subsidence as contributions to *Lyell*'s theory of the earth.

For the time being, though, two of Darwin's ideas—about Pacific subsidence and tests for depression in strata—glimmered with new meaning and importance in the light cast by his flash of insight about the shape of coral islands. Now that he had seen their significance for understanding the morphology of reefs, he would never fail to see these two ideas from this new perspective. It was true, as he was later to write in his autobiographical recollections, that those ideas were "thought out on the west coast of S. America before [he] had seen a true coral reef," but it was only the insight at Tahiti that made them part of a theory of coral reef formation. Only then were they incorporated into an explanation of the form and origin of ring-shaped reefs. It is extremely significant that Darwin ended the essay by invoking Humboldt. The implications of Darwin's coral reef theory for understanding the vertical movements of the earth's crust have always been described as though they were fundamentally and indeed inherently Lyellian.[44] In fact, it was only after the voyage when Darwin was working side by side with Lyell himself that the coral reef theory became couched in the code words of Lyell's *Principles*.

Though he did not do it in the "Coral Islands" essay, Darwin felt compelled to give some consideration to evidence he had seen earlier that year that seemed to give credence to the possibility of submarine craters underlying lagoon islands. Not only had the crater theory postulated an enormous number of submarine volcanoes in the Pacific, it required that their cratered summits be of a surprisingly uniform elevation, not quite reaching sea level, but lying in that narrow vertical zone where madrepores could grow. It was little more than a month since Darwin had seen the "vast & almost infinite number of Craters" that gave the Galápagos Islands their "singular & highly characteristic aspect."[45] He cast his mind back to them now on the long passage to New Zealand, adding several new pages to his existing run of geological notes on the Galápagos. Three of the "great Volcanic mounds . . . surmounted by craters" had been found, by angular measurements, to have almost identical elevations, between 3,720 and 3,730 feet. "Inspecting the chart," he admitted, "one is tempted to exclaim; on such foundations, ready placed at an equal height, the Lithophytes, might soon raise to the surface, their circular ridges of Coral rock."[46] Lower craters made of sandstone were also noteworthy, because they shared with many annular reefs the characteristic of being slightly taller and steeper on their windward sides. "I am so much the more bound to point out this coincidence," he noted, "as I am no believer in the theory of Lagoon I[slands] . . . being based on the circular ridges of submarine craters."[47] He reassured himself that because the southwest swell of the Pacific disrupted the relation between the directions of the wind and surf in the coral zones, "the case of the Sandstone craters & that of the Lagoon I[slands] is not entirely similar."[48]

Darwin was also deeply puzzled by "the entire absence of all [coral] reefs" at the Galápagos, despite their being "situated in the Pacifick and under the Equator."[49] Darwin wondered whether it was due to a "deficiency of [the] Calcareous matter" with which corals built their skeletons, but FitzRoy suggested the alternative possibility that reef builders could not withstand the cold water that surged up from great depths against the western shore of South America. Darwin at first thought that testing this "ingenious idea [would] require extended observation."[50] Instead, he got hold of FitzRoy's weather journal and took advantage of the trove of empirical data accumulated on a surveying ship to conduct a retrospective experiment. He assumed that since "the whole ocean, near Tahiti abounds with Coral animals . . . we may presume the temperature of the Sea there [to be] perfectly favourable to their growth."[51] Thus he compared the mean and low values of forty-four water temperature measurements from the

Low and Society islands with ninety-nine such measurements from the Galápagos in order to evaluate whether the water temperatures at the Galápagos could potentially be suitable for coral reefs. He found not only that the Galápagos mean and low temperatures were colder by 9.5°F and 18°F, respectively, but they were also much more variable than those of the coral seas, whose temperature never registered more than one degree below the mean. "It may easily be believed," he concluded, that a marine inhabitant of the South Sea could "never flourish" in temperatures as cold and variable as those of the Galápagos.[52] Therefore apt conditions for coral growth could not be assumed from latitude alone, no doubt confirming a trend that FitzRoy had already noticed in the course of recording the measurements. In his notes Darwin supplemented these conclusions with citations on ocean temperatures from three separate volumes of Humboldt's work.[53] For Darwin, as indeed for FitzRoy and Beaufort, extracting a natural law by organizing precise measurements according to geography was the essence of Humboldtian philosophy. Whereas Humboldt took his own spectacular array of instruments to South America (where a troop of unsung porters performed the indispensable task of transporting them), Darwin became aware of the opportunity to construct theories from collectively generated data because it was being produced all around him by the everyday work of the surveyors.

The evidence I have presented in this chapter indicates that Darwin read Humboldt's work in a manner quite different from the way it tends to be read by historians of science today. Perhaps because of the overwhelming size of Humboldt's published oeuvre, his significance and impact are often described in terms of the *style* of his work: lettered, Romanic, quantitative, geographical, holistic. In contrast, Darwin's notes from during and after the voyage indicate that he pored over Humboldt's work for its *details*. References to Humboldt in the *Beagle* notebooks cover topics as diverse as the migration of birds, the hibernation of crocodiles, the geography and mineralogy of areas across three continents, and, as we have seen, the average slope of volcanic cones and the temperature of seawater. Darwin drew as well on the specific conclusions Humboldt derived from all his data, not the least of these being Humboldt's statements about the relation between altitude and latitude, the zonation of plant distribution and alternating vertical movements of the earth's crust. Darwin may have started reading Humboldt's works for their captivating style and exotic locales, but he returned to them again and again during the voyage for their unparalleled value as compendiums and as theoretical texts. This is not an argument against his having absorbed styles of writing, thinking, and re-searching from Humboldt as well. Rather, I am arguing that the kinds of questions Darwin was driven to ask by his work alongside the hydrographers helped keep Humboldt's work relevant to him.

I have argued in this chapter that the documentary evidence generated during the *Beagle* voyage supports the conclusion that Darwin first perceived himself as having a new answer to the question of coral reefs' shapes and origins while at Tahiti. This is contrary to the claim he made much later in his life, that he had thought up the whole theory on the west coast of South America before ever seeing a coral reef. Instead, my interpretation of his notes suggests that he first conceived of the theory (or more properly, *a* theory) of reef formation precisely at the moment when he first got a sustained look at a coral reef. I believe the most plausible explanation for this later contradiction is that the theory he innovated at Tahiti added considerable value to earlier ideas from western South America.

In demonstrating how the Tahiti theory served to revise the apparent significance of thoughts he had had earlier, I wish to advance the notion that the theory imposed a narrative on Darwin's past and future work. Put more strongly, the theory *was* a narrative of his past and future work as much as a narrative about how and why the location of an island within an encircling reef might hypothetically be transformed into the location of an empty lagoon. Darwin was simultaneously extrapolating the history and future evolution of a reef's form from the single moment at which he was seeing it, but in doing so he also invented a new narrative about himself, retrospectively creating a story that invested certain elements of his experience in South America with relevance to his work at Tahiti and, as I shall describe in the next chapter, defining a future line of inquiry that would now be directed explicitly at understanding the formation of coral reefs. The nature of Darwin's claim about the South American origin of the theory also suggests that the narrative generated at Tahiti was in some respects inescapable for Darwin once he had framed it: he could no longer recall the South American work without taking for granted that it had been directed toward producing a theory of coral reef formation.[54]

5

The Surveyor-Naturalist

At the moment when he wrote his December 1835 essay, Darwin admitted that he had "scarcely seen anything of the Coral islands in the Pacifick Ocean." Nevertheless, he had come up with a theory that would at once answer the Admiralty's coral reef question and perhaps bring him into conversation about the earth's crustal movement with his hero, Humboldt. But with the Pacific behind him after three months at New Zealand and Australia, he had still never set foot on a coral island.[1] Only in the last months of the voyage, in the Indian Ocean, did the chance finally come. With Darwin's encouragement, FitzRoy elected to call at the Keeling Islands—now known as the Cocos (Keeling) Islands—a pair of "low islands" (atolls) seven hundred miles southwest of Java and Sumatra.[2] The decision was evidently made very late, for as Fitz-Roy's lieutenant John Clements Wickham reported in a letter back to Sydney, "Our trip to the Swan [River, site of Perth] is knocked on the head . . . as we go to the Keeling Islands instead, we anticipate a pleasant time there." Elaborating on the tasks awaiting him, Wickham continued, "a plan will be made of [the Keeling Islands], and it is intended to get several soundings outside the reefs, more for the sake of geology than anything else; Darwin is very anxious about it; the account of these I[slan]ds in the Naut[ical] Mag[azine] makes us long to see them, and it will be but little out of our way."[3]

It would be easy to read Wickham's "for the sake of geology" as implying that Darwin's new theory had generated the entire rationale for visiting the Keeling Islands, but it is important to recall that Beaufort's original instructions to FitzRoy had presented it as a hydrographic task to study the structure of a coral island.

In this chapter I examine the only two occasions when Darwin had the opportunity to make a close investigation of coral reefs. The *Beagle* would spend ten days in all at South Keeling in early April 1836. A few weeks later FitzRoy provisioned the ship at Mauritius, which gave Darwin the chance to examine that island's fringing reef. The two visits took place in very different circumstances and offer contrasting case studies of Darwin's fieldwork. On both occasions he was eager to understand the distribution of different kinds of corals on different parts of a reef and to determine whether the submarine structure of each reef had the characteristics predicted by his theory. The critical distinction between the two visits was that FitzRoy carried out a survey at one of these locations but not the other. South Keeling was a remote and little-examined outpost, and he and the officers undertook a full hydrographic survey of the reef. Mauritius, on the other hand, was a well-established depot, and there was no reason for FitzRoy to resurvey the island. Comparing Darwin's activities while studying these two reefs reveals just how far he had come to depend on his shipmates' work. When he learned that FitzRoy would not be surveying the Mauritius reefs, Darwin felt he had no choice but to take to a boat and play the role of hydrographer himself.

Darwin's Sea-Level Study of the South Keeling Reef

Sailing into the lagoon through a channel between reeftop cays, the *Beagle* anchored at South Keeling, the larger of the two Keeling reefs.[4] At first sight Darwin described it as "one of the low circular Coral reefs, on the greater part of which matter has accumulated & formed strips of dry land."[5] These long, narrow cays supported a population of about a hundred indentured workers from the Malay Archipelago and their overseer, John Clunies-Ross, and his family. Clunies-Ross was a Scottish merchant and sea captain who in the previous decade had established a coconut plantation (and ousted a rival colonist) on the formerly uninhabited islands.

From the anchorage, Darwin could see that the land was broken into a chain of islets, each one interrupting his view of the unbroken arc of the reef. Thus the "brilliant expanse [of the lagoon], which is several miles wide, is on all sides divided either from the dark heaving water of the ocean by a line of breakers, or from the blue vault of Heaven by the strip[s] of land crowned . . . by Cocoa nut trees."

With his new theory of reef formation in mind, Darwin turned his attention toward the points of evidence that might determine whether this structure had taken its shape from an underlying submarine crater (as his contemporaries theorized) or from the subsidence of a reef-fringed island.[6] If subsidence had occurred, then the reef's annular shape must have been maintained over many generations of coral growth. Though his 1835 "Coral Islands" essay revealed that Darwin believed Quoy and Gaimard's assertion that reef-building corals could grow only in shallow water, he rejected their claim that tranquil waters favored corals better than heavy surf. Having seen Tahiti's reef even briefly, Darwin had also felt unable to accept the Frenchmen's failure to differentiate the corals growing in the different habitations around the reef. "M. Quoy & Gaimard state, 'that the species, which constantly formed the most extensive banks, belong to the genera. Meandrina, Caryophyllia. & Astrea' & that the Saxigenous [stony] polypi increase most considerably in shallow & quiet water." I am not aware, Darwin continued, "whether they suppose, that these same species form the outer parts of the reefs."[7] On inspecting Tahiti's encircling reef and discussing its composition with the islanders, Darwin had felt strengthened in the conviction, born of his earlier zoophyte studies, that the inner and outer constituents of the reef must be different types of corals. This was critical for Darwin's theory, for he expected to find that differential growth of these distinct corals was responsible for maintaining the circular shape of the reef and keeping the level of the outer "breakwater" higher than the inner flat and the lagoon floor.

In the 1835 essay, however, he had offered little more than a truism based on this conviction that organisms were narrowly adapted to specific environments: "Those species of Lithophytes, which build the outer. solid wall, flourish best, where the sea violently breaks."[8] Having been prevented from approaching the outer reef at Tahiti, he had "not pretend[ed] to conjecture concerning the cause of this prediliction, whether the motion of the fluid, or the quantity of insolved air. is favourable; or whether the light and heat, which must pervade still shoal water is injurious to the growth of their Species."[9] At South Keeling he sought a more detailed explanation of the constraints and inducements to coral growth on different parts of the reef.

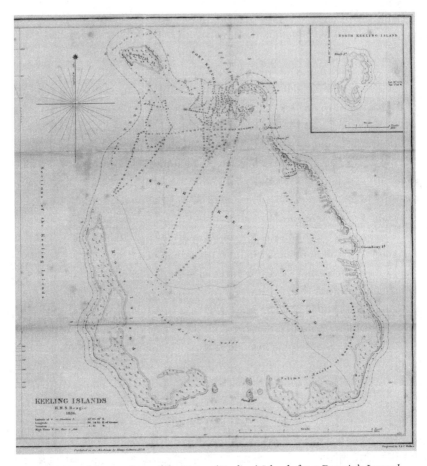

Figure 14. Detail from a chart of the Cocos (Keeling) Islands from Darwin's *Journal and Remarks* (later *Journal of Researches*), which was published as volume 3 of FitzRoy's 1839 *Narrative*. The many small numbers throughout the chart are depths in fathoms, as determined by the hydrographers' work. Reproduced with permission from http://darwin-online.org.uk.

Darwin began investigating South Keeling with traditional, above-water natural history fieldwork, while high winds obliged the hydrographers to postpone beginning to take soundings. Traversing the reef, he was fixated on a particularly vexing puzzle about the link between coral growth and reef shape. Darwin had phrased this most important question in the following way: "Within the lagoon all detritus accumulates, & if as according to M. Quoy and Gaimard. the Coral grows <u>there also</u> most rapidly; how comes it that the Lagoon is not more commonly filled up?"[10] Seeking a specific answer at South Keeling, Darwin transected the reef many times on foot, from the shore of the lagoon, over an islet, and through the ankle-deep water that covered the reef flat at low tide. Approaching the outer margin, Darwin's progress was threatened by pounding waves. He turned to a tool he had learned to use during his many outdoor pursuits in Britain, where he had spent his teenage years absorbed in hunting partridges and collecting beetles. According to his notes, he vaulted "by the aid of a leaping pole . . . very far into the breakers."[11]

Darwin had seen a leaping pole in use during his Cambridge undergraduate days, which he often devoted to country rambles in search of beetles to add to his collection. According to Darwin's son George, in recollections written shortly after his father's death in 1882, "Amongst his Cambridge expeditions I remember his speaking of going down to the fens, then near Cambridge, with a sporting sort of guide who went by the name of Marco Polo, because he carried a leaping pole with a flat board fastened at the bottom for leaping the ditches." A nineteenth-century manual of rural British sports provides a more detailed description of these devices: "The leaping-pole is either of fir or bamboo, about two, three, or even five feet higher than the height of the party using it, and becoming stronger towards the bottom. When used for leaping wide ditches, a pole with a flat disk of several inches diameter at the bottom is of great use in preventing its sinking into the mud, and in peaty bottoms often saves a ducking" (see fig. 15 and plate 8). We may imagine Darwin learning to use the pole as the manual describes, holding it so that "the right hand is placed at the height of the head, and the left on a level with the hips, then grasping it firmly it is dropped into the ditch till it touches the bottom, when making a spring with the left foot, the weight is carried upon the arms, and describes a segment of a circle, the centre of which is at the end of the pole in the ditch. . . . The learner should begin by clearing small ditches, gradually increasing their width, and when expert in these, trying wider ones until he cannot proceed further without a run; then venturing upon a few yards' preparatory run, which will give additional power in clearing space; and finally adopting a good quick run of about six, eight, or ten yards, gradually taking hold of the pole higher and higher as he increases the width of his jump."[12]

Having reached the very limit of the reef, where he stood poised atop "great masses" of living coral with his leaping pole in hand, Darwin studied the composition of the breakwater as the ocean foamed around him (see plate 9). The "chief masses" were of "living Astrea," solid corals "with a curvilinear outline up to 8 ft in diameter."[13] Where they had reached the level of the water, the tops were flattened and no longer growing. Instead, "the As-

Figure 15. "Pole Leaping." An illustration demonstrating the use of the leaping pole, from *Walker's Manly Exercises* (1860). Darwin described attaining the "furthest mound [of coral at Keeling Island], which I was able to reach by the aid of a leaping-pole, and over which the sea broke with some violence." Quotation from *The Structure and Distribution of Coral Reefs*, 1842. Image reproduced courtesy of the George Peabody Library of the Johns Hopkins University.

trea. extend[ed] laterally" in such a way that his "Specimens [would] show. a layer. additional on the sides as compared to the top."[14] Between these great knolls was "an exceedingly strong net work" of intersecting "Millepora . . . in thick vertical plates."[15] Branching corals flourished in the gaps, so that "the interstices [were] soon to be filled up & form solid masses."[16] This was what he had been led to expect at Tahiti. However, Darwin was "most surprised to see. the enormous quantity of matter. which the succesive paper like layers of Corallina [had] accumulated."[17] These stony, encrusting algae appeared capable of enduring "exposure for some time to the air," so that "instantly the surface of the Astrea dies. it is occupied by Corallina [to about] two feet above the level [of] the living solid Corals."[18] The coating layers of rock laid down by coralline algae were evidently integral to the reef's ability to resist the waves, for he found that "3 inches beneath the general level of the Corallina, the breakwater [was] excessively hard." Working "by chisel [and] pixaxe" he "at last attained a fragment & strongly suspect[ed] it [was] Corallina petrified."[19] Enumerating what he took to be "the four Bulwark agents" of the reef, he listed two types of algae along with the coral genera *Astrea* and *Millepora*.[20]

Working through the course of the visit over multiple paths from the breakers to the lagoon, Darwin collected twenty-four zoological and eighteen geological specimens, as recorded in his respective series of notes. The different types of specimens reflected what might be called the "horizontal" composition of the reef from the outer breakers to the waters of the lagoon (see fig. 16). The zoological list included "layers of a pale red encrusting Corallina; from the extreme breakers"; "Astrea [from] the midst of the outer breakers"; "Millepora [from the] Outer reefs in the most exposed places"; "Coral . . . common in holes on the outer reefs"; "Madrepore, in the lagoon"; "Seriatopora. common in the Lagoon"; "White branched Madrepore, exceedingly common in lagoon"; and "one of the commonest Corals in the lagoon: when alive yellow."[21] The description of the geological specimens reiterated Darwin's interest in distribution across the reef, but it also showed his eagerness to document another continuum, from living corals to different kinds of rock. Thus he collected a fragment of "Astrea. the commonest Coral. block on the outer coast"; another of the "next most abundant kind . . . partially petrified"; some "Carb[onate] of Lime. probably Coral petrified"; a specimen that was unambiguously "white petrified Coral"; a "Yellowish white. vesicular stone . . . consisting of particles of shells & Corals. intimately united & blended together"; a similar specimen "which apparently [had] been a lamelliform Coral with the cells. completely filled up"; a breccia that was "very solid [with] fragments [of] a good many branching Coral[s]. from [the] reef"; and an "Astrea converted into <<snow>> white Calcareous rock [with a] glittering Crystall[ine] fracture."[22]

The rocks and organisms Darwin collected were not scattered indiscriminately across the islets and the reef but lay in an order from which he interpreted the origin and development of the sparse arcs of dry land. The reef flat was "composed of a <u>very hard solid</u> rock.—which is petrified Coral & hard Calcareous sandstone."[23] This had been formed when "channels between the living Coral. [had] gradually been filled with detritus [and] . . . petrified & smoothed by the action of the tides."[24] Proceeding in from the reef flat to an islet, one met a beach of "rounded fragments of solid Coral" underlain by a low ledge of breccia (a hard conglomerate rock of calcareous sand and pieces of coral "cemented by the action of atmosphere and tides") that sloped "just perceptibly to seaward."[25] On the windward side of the island, a "succession of beaches" was being laid down over the reef

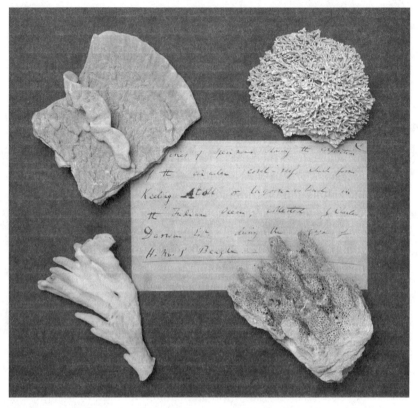

Figure 16. A selection of Darwin's coral specimens from South Keeling, collected in 1836 and now at the Natural History Museum, London. Image © Natural History Museum, London.

flat, "form[ing] the outer parts. of the strips of land."[26] Meanwhile, the land toward the lagoon was made up of ever finer fragments of coral that had been thrown farther inland by waves rushing over the reef flat.[27] The inner beaches of the islets were of a powdery coral sand that became beds of mud below the water level of the lagoon.[28] Darwin agreed with a suggestion from FitzRoy that the finest sand in the lagoon was coral pulverized by strong-jawed fish as they grazed on the reef.[29]

Darwin continued to find order in the way corals were distributed through the "under-water forests of the Keeling islands," whose denizens presented, to FitzRoy's eye, "more difference than between a lily of the valley and a gnarled oak."[30] Of the corals that lived in the lagoon, Darwin found that "the most abundant kinds are the branching sorts."[31] Though he noted with evident surprise that some species within the genus *As-*

trea were to be seen there,[32] their form was recognizably different from the "bulwark species [and] the two other kinds [of *Astrea*] which are found outside."[33] Corallinas, so essential to the breakwater, were likewise "not abundant" in the lagoon.[34] It was true that the conditions in the lagoon cultivated a richer diversity of coral genera, but they almost all grew "brittle & soft."[35] In contrast to the algae-encrusted *Astrea* he had vaulted onto at the outer reef, Darwin found that when merely attempting to stand on the lagoon corals, "a person breaks through them to some depth."[36] Unlike the resilient network being formed under the pounding of the ocean waves, the "dead [lagoon] Coral, showed no signs of adhering & forming as rocks. but rather of wasting."[37] This difference between the inner and outer reef suggested, contrary to the implication drawn from Quoy and Gaimard, that even "independent of repeated depressions," it was "difficult to imagine how [the lagoon] would ever entirely be silted up."[38] "Moreover," Darwin pointed out, again diverging from the Frenchmen's notion that calm, shallow water was most favorable for coral growth, if "the lagoon was nearly filled up the impurity of the water might [further] slow [the] growth of corals."[39] Sand and mud settling on existing corals in the lagoon "must be fatal," he concluded, and new corals would be unable to establish themselves on such a "slippery bottom. of sand or mud."[40] Combined with the lesser bulk of the lagoon's branching corals, these causes "must retard the growth of the Coral in the lagoon as compared to the outer [reef]."[41] Though Darwin found it difficult to distinguish clearly between the effect of any possible subsidence and the effects of the wind and the "comparative growth of corals," it seemed clear to him that present-day processes were helping to keep the lagoon from being filled in altogether.[42]

Based on his observations from sea level, Darwin drew several cross-sections showing the reef's superficial composition. He also sketched a pair of diagrams to illustrate what he imagined to be the nature of the entire submarine structure (figs. 17 and 18). At the bottom, roughly 1,000 fathoms (6,000 feet) deep, he hypothesized a foundation of "Greenstone?", his term for dark volcanic rocks.[43] The image demonstrates that Darwin conjectured that this substrate had sunk in a series of discrete subsidences, each followed by the growth of coral back to low water and the consolidation of a layer of breccia on top of it. Corals could grow only up to the level of low water, while the conglomerate rock, breccia, could be formed from material deposited up to the level of high water. Therefore the thickness of the breccia layer would depend on the difference in elevation between high tide and low tide. In the image Darwin figured this as three feet.[44] Sur-

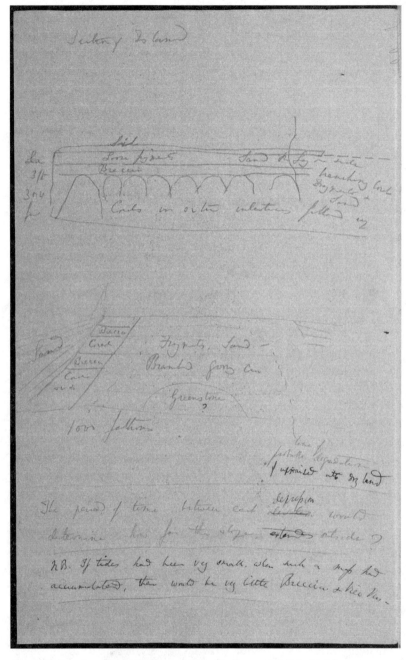

Figure 17. "Section of Island." Drawn before FitzRoy and Sulivan had done any soundings outside the reef, this was Darwin's first guess at the submarine structure of South Keeling. DAR 41:43v. Reproduced by kind permission of the Syndics of Cambridge University Library.

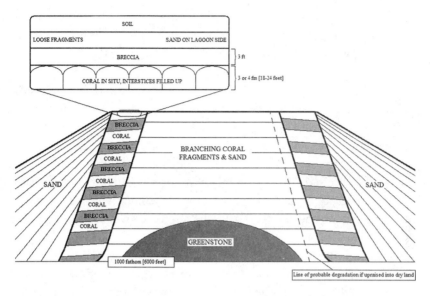

Figure 18. Diagram synthesizing the images in figure 17. In this diagram Darwin's sectional view of the surface structure of Keeling Island (boxed) is combined with his conjectured section of the submarine structure. Note that the results of the soundings taken later in the visit to South Keeling caused Darwin to conclude that the outer profile of the reef did not correspond to that pictured here. Diagram by the author.

mounting the greenstone foundation, and fully enclosed by the ring of corals and breccia in situ, was a massive thickness consisting of the remains of branching corals—fragments and calcareous sand—which he knew to be the main substance of the lagoon floor. Surrounding the entire constructional part of the reef on all sides was a "sand" of pulverized coral that formed an enormous cone of detritus standing on the floor of the ocean.

Seeing Underwater: The Hydrographic Survey at South Keeling

Darwin's leaping pole had propelled him to the very limit of the terrestrial world, allowing him to peer into the breakers at the seaward margin of the reef. However, the evidence that could reveal the island's longer history lay beyond reach of the leaping pole and the pickax, beyond the terrestrial way of seeing. Luckily for Darwin, Beaufort's instructions to FitzRoy had urged him to use "every means . . . that ingenuity can devise of discovering at what depth the coral formation begins, and of what materials the substratum on which it rests is composed." As well, "the slope of its sides" was to be "carefully measured . . . by a series of soundings, at very short distances from each other." The effect of sounding in such a systematic manner in radial lines moving away from the reef would be to extend the transects that Darwin had conducted above water.

Beaufort (and Lyell) intended such coral reef soundings to yield evidence respecting the relative merits of the crater theory and Eschscholtz's earlier claim that reefs grew up from deeper foundations. But the approach could serve as a test of Darwin's theory too. If the foundation of the reef came close to the surface and had the outline of a volcano, or if subaqueous lavas were found at shallow depths, it would offer support for advocates of the crater theory including Lyell and the Frenchmen who had originally posited a depth limit for coral growth. Alternatively, because there appeared (according to Humboldt himself) to be a limit on the inclination at which lava could harden into rock on the side of a volcano, a foundation that sloped more steeply than the sides of known volcanic cones would suggest that the reef had been built by corals growing upward. Because he was convinced that corals could not grow at any significant depth, meanwhile, Darwin seems never to have given serious consideration to older theories (including Eschscholtz's) that postulated corals growing upward from deep in the ocean. Therefore, if the surveyors were to find a steeply inclined reef, the only possible explanation in Darwin's view would be that those corals had grown on a foundation that was subsiding in the manner he envisioned at Tahiti.

Darwin's Keeling field notes reveal a naturalist collaborating intimately with the hydrographers at work.[45] He recorded the detailed results of forty-six individual soundings and summarized the findings of several times that many, and he noted more than a dozen comments about the island from FitzRoy and Lieutenant Bartholomew J. Sulivan (whose name Darwin frequently misspelled in his notes as "Sullivan"). It is difficult to assess the full extent of Sulivan's contribution to Darwin's ideas, but it must have been considerable. Still surviving among Darwin's manuscripts from the voyage are several diagrams of reef structure drawn in Sulivan's fine hand and accompanied by written descriptions of his forays around South Keeling (see fig. 19). As the historian Simon Keynes has emphasized, Darwin joined an intimate circle among the *Beagle*'s "gun-room officers." The traces of their friendships can be found in a book of wagers made among the group during the last year of the voyage when, for example, "Darwin [bet] Sulivan 1 dollar that the Beagle will not anchor at Tahiti by 2pm November 13th."[46]

At Keeling the presence of corals made sounding particularly difficult.

As Darwin remarked in his notes, FitzRoy's mate Peter Benson Stewart "carried away [lost] his anchor in 13 [fathoms] & [his] lead in 16" while "the Capt[ain] when sounding in 10 & 12 fathoms. frequently had the lead jammed. so as not to be without much difficulty to extricate it.—How then rough the bottom must be"[47] (see fig. 20). FitzRoy himself declared, "I was anxious to ascertain if possible, to what depth the living coral extended, but my efforts were almost in vain, on account of a surf always violent, and because the outer wall is so solid that I could not detach pieces from it lower down than five fathoms." The captain spared no effort, however, and "small anchors, hooks, grappling irons, and chains were all tried—and one after another broken by the swell almost as soon as we 'hove a strain' upon them with a 'purchase' in our largest boats."[48]

It devolved on Darwin to diagnose both the type and the present condition of the deeper corals by the impressions they left in the "tallow hardened with lime" that FitzRoy packed into the bottom of his broadest lead. In his notes Darwin gave a special mark of emphasis to one of FitzRoy's early soundings on the outer margin. From a depth of 8 fathoms (48 feet), the tallow came up "beautifully marked with Astrea," the bulwark coral that Quoy and Gaimard had claimed must live within 30 feet of the surface. Seeing that it was "probably alive" (because the tallow was "quite clean," and free of the sand that could accumulate freely on dead coral), this sounding alone demanded a small revision to current zoological knowledge.[49]

Gradually Darwin drew more general conclusions about the extent of the living reef: to a depth of 12 fathoms (72 feet) he found the "armings clean [showing] Millepore [and] Astrea."[50] Beyond the zone of bulky corals he found "a fathom or two of fragments" that occasionally contained smaller bits of animated matter.[51] Below 20 fathoms (120 feet) there was only sand, with "no sign of any thing hard.—in [the] soundings." These observations showed that the reef had a shallow "first inclination" from the breakwater to a depth of about 30 fathoms, with its breadth of 100–200 yards corresponding to a band of "discoloured water" that could be seen beyond the breakers.[52] From the 30-fathom mark, Darwin could see that the bottom "suddenly incline[d]" into darker blue water.[53]

Eager as Darwin was to learn the secrets of the blue deep, it came as a serious disappointment that strong winds "rendered the most important part [of the survey], the deep sea sounding, scarcely practicable."[54] High winds or strong currents made very poor conditions for sounding because an accurate measurement required the lead to drop straight down and the line to remain vertical. To his record of one sounding of 770 fathoms, Dar-

section on the East side between the settlement & ... house.

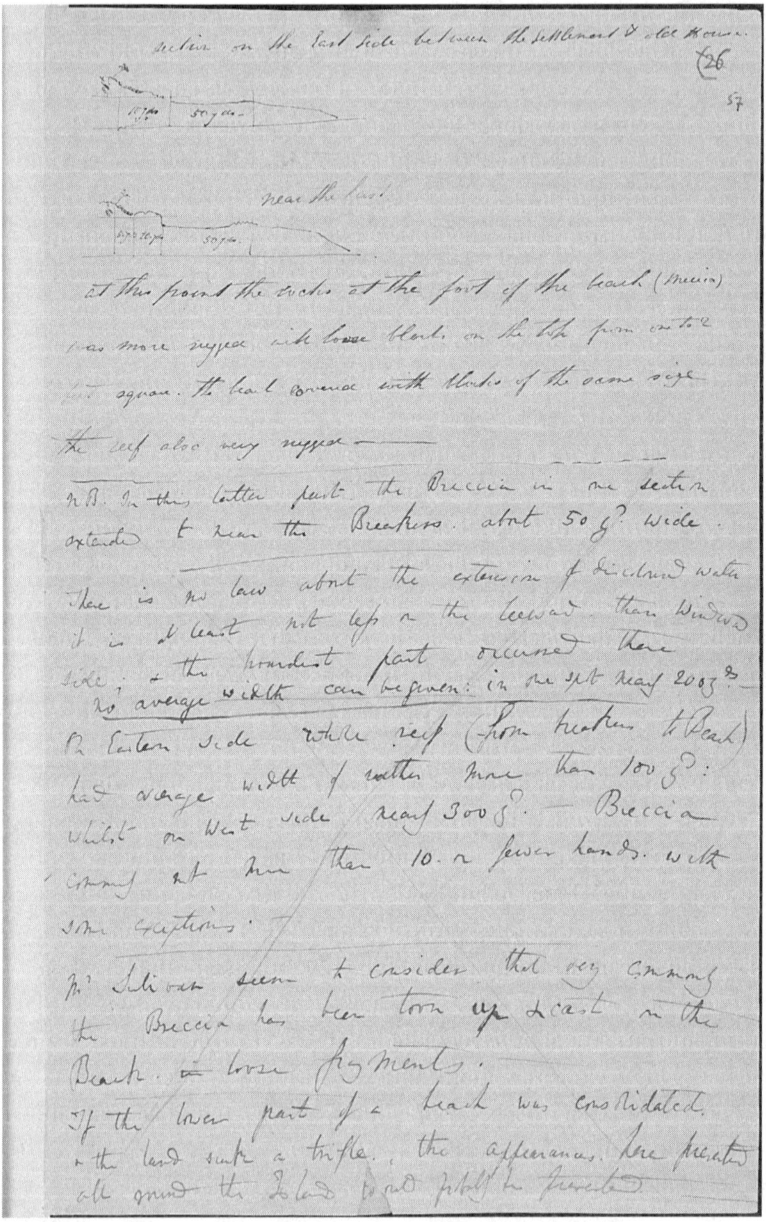

26
57

at this point the rocks at the foot of the beach (...) was more rugged with loose blocks on the top from ... ... square. the ... covered with blocks of the same size the reef also very rugged —

NB. In this latter part, the Breccia in one section extended to near the Breakers. about 50 y.? wide.

There is no law about the extension of disturbed water it is at least not less on the leeward than Windward side; — the hardest part occurred there no average width can be given: in one spot near 200 y.? — On Eastern side whole reef (from breakers to Beach) had average width of rather more than 100 y.? whilst on West side near 300 y.? — Breccia common at more than 10. or fewer hands. with some exceptions.

Mr Sulivan seems to consider that very commonly the Breccia has been torn up at least in the Beach. in loose fragments.

If the lower part of a beach was consolidated, & the land sunk a trifle, the appearance here presented all round the Island would probably be presented

Figure 19. "Mr Sulivan seems to consider . . ." The top third of this page shows sectional diagrams and descriptions of the windward side of South Keeling that one of the *Beagle*'s lieutenants, Bartholomew J. Sulivan, provided to Darwin. Sulivan's fine handwriting sets his notes apart from Darwin's annotations on the rest of the page. DAR 41:57. Reproduced by kind permission of the Syndics of Cambridge University Library.

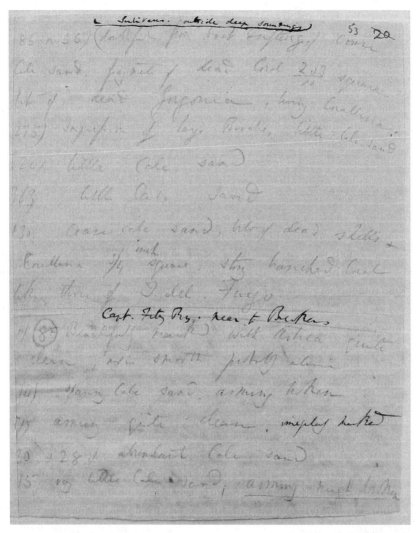

Figure 20. "Sulivans. outside deep soundings" and "Capt. Fitz Roy. near the Breakers." Darwin's knowledge of the outer part of the reef drew heavily on the hydrographers' work. Where the arming of the sounding lead came up "clean" and "beautifully marked," he was confident that the lead had fallen on living coral. DAR 41:53. Reproduced by kind permission of the Syndics of Cambridge University Library.

win added an annotation: "depth doubtful (360 really)."[55] FitzRoy made the best of the conditions, "eagerly tak[ing] advantage" of "two moderate days . . . to go round the whole group in a boat."[56] Sulivan managed to achieve a small number of measurements between 200 and 300 fathoms.[57] More remarkably, at "only a mile from the southern extreme of the South

Keeling" FitzRoy found no bottom at 1,200 fathoms.[58] With nearly one and a half miles of line played out, it stood as one of the deepest soundings yet taken by any navigator. The thought of it boggled Darwin's mind and challenged his mathematical skills. At the bottom of one scrap of Keeling notes are two attempts to multiply 1,200 by 6, with the second yielding 7,200 feet for the depth of FitzRoy's sounding.[59] Working from a list of deep soundings and their "estimated distance" from shore, he preempted the officers by making his own preliminary calculations of the slope.[60] By obscure methods, he found the angle to be 48° at its steepest inclination.[61] This was similar to "Beechey['s] mean slope" given in that navigator's sectional diagram of the reef at Bow Island in the Low Archipelago, which "from the 20 fathom line appears nearly 45[°]."[62]

Darwin sought to understand the factors that shaped a reef by comparing the slopes off different sides of South Keeling, eventually consulting with officers whose facility with trigonometry far exceeded his own. Considering the shallower slopes out to 30 fathoms, he found "no law about the extension of the discoloured water[, which was] at least not less on the leeward. than windward side."[63] Regarding the deeper water, Darwin found it "clear from Mr. Sullivans sections" that there was also "no law with respect to [the] Windward & Leeward . . . shape of [the] lower Mountain."[64] This ambition that the accumulation of geographical measurements would produce knowledge of the physical world's underlying "laws" echoed Humboldt's mode of inquiry.[65] But by incorporating measurements by Sulivan and others into what would become single-author scientific publications, Darwin was also, as I mentioned earlier, emulating Humboldt's authorial relationship to those who had helped the Prussian to gather his own zoo- and phytogeographical facts.

The reef's foundation was indeed exceptionally steep. Sulivan explained that on some attempts the sounding line had been severed between 500 and 600 fathoms, suggesting that it caught the edge of a cliff at that depth. Darwin considered it the "precipice of [an] unfathomable wall" that may have been sculpted by the "Action of [the] sea" at its surface.[66] If the cliff had indeed originally been at sea level, its present situation implied a subsidence of 500 or 600 fathoms. All together the deep soundings plainly contradicted the diagram Darwin had drawn earlier in his visit, which depicted a mound of sand surrounding the island. He now noted that the "very great inclination between the 2 soundings on the SE side [is] so steep that it must be rock."[67] It is not clear that he ever entirely reconciled these contradictory ideas; rather, he abandoned his effort to explain, or explain away, the

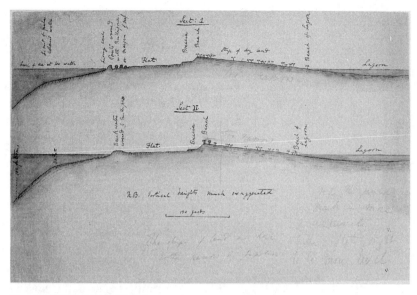

Figure 21. "N.B. Vertical heights much exaggerated." Two watercolored sectional views of islets atop the rim of the South Keeling reef illustrate several of the chief features of the reefs. From right to left, these were the lagoon; the strip of dry land constituting an islet atop the reef, which sloped toward the lagoon; the steeper outer beach and the breccia layer; the reef flat, which was submerged at high tide; the breakwater; the shallow submarine slope of the reef out to a depth of 30 fathoms; and the steep face of the reef plunging down to deep water. DAR 44. Reproduced by kind permission of the Syndics of Cambridge University Library.

imagined talus of debris outside the reef.[68] Perhaps he was willing to do so because the discovery of such a steep foundation was more damaging to the crater theory than to his own. He turned again to Humboldt's published work, where he learned that "cones of Volcano[es] have a medium slope <<from>> 33° to 40° <<Even the steepest parts but little exceeding. these numbers.>>"[69] Because FitzRoy's soundings showed the walls of the reef to be steeper than a volcano's, Darwin emerged more confident in his theory even though the poor surveying conditions had left many of his original questions unanswered.

On departing from South Keeling, Darwin felt emboldened to write about the new coral reef theory in his diary, where he had previously only hinted at its existence. Given that the coral reef theory had to that point appeared among his writings only in the "Coral Islands" essay, Darwin's decision to incorporate it in the permanent record of the voyage was a new sign of self-assurance.[70] This iteration of the theory emphasized the results of FitzRoy's recent deep sounding in arguing, "Hence we must consider this Is[land] as the summit of a lofty mountain."[71] While Darwin acknowledged uncertainty about "how great a depth or thickness the work of the Coral animal extends," he was more convinced than ever that "we must look at a Lagoon Is[land] as a monument raised by myriads of tiny architects, to mark the spot where a former land lies buried in the depths of the ocean."[72]

Darwin's Hydrographic Initiative at Mauritius

The *Beagle*'s next port of call was an island whose bold geography and hold on Darwin's imagination could almost rival Tahiti's. Long a stopping point for Arab, Portuguese, and Dutch sailors, but with no permanent human inhabitants until the end of the sixteenth century, Mauritius (or l'Isle de France) had passed from French to British imperial control just twenty-six years earlier. The lushly vegetated island was familiar to many Europeans as the setting of a 1788 sentimental novel, *Paul et Virginie*, written by the French army officer and botanist Jacques Henri Bernardin de Saint-Pierre. "Imagine what a fine opportunity for writing love letters," Darwin had enthused as his visit to Mauritius approached. "Oh that I had a sweet Virginia to send an inspired Epistle to. – A person not in love will have no right to wander among the glowing bewitching scenes."[73]

Whereas Bernardin's protagonists grew up and fell in love among the island's gardens and mountains, his novel's tragic climax occurred instead on the coral reefs that surround Mauritius. A ship returning Virginie from France was wrecked after trying a new passage through the reef, and she drowned within view of a helpless Paul after being too modest to shed her clothes and swim to safety. Bernardin drew the name of the ship and the details of the wreck from a real 1744 tragedy in which some two hundred people had died when the *Saint-Géran* ran aground on the approach. In his vivid depiction the ship heaved as the waves crashed over it where it had stuck fast on the "belt of reefs that encircle the Isle de France."[74] Bernardin had also written more extensively about those reefs in an anonymously published 1773 book that Darwin cited often in notes made during the Indian Ocean portion of the voyage, but which he credited to an "Officier du Roi," an officer of the king.[75]

Despite the obvious hazard those reefs posed to ships approaching Mauritius, the *Beagle*'s hydrographers had not been ordered to survey the island. Beaufort's instructions did mandate a stop there, but only to determine "the difference in longitude from [the southwest coast of Australia] to the Mauritius." At this late stage in the voyage FitzRoy's highest priority was to get his ship and crew back to England as quickly as possible, and the Mauritius stop proved to be so uneventful for the surveyors that FitzRoy would devote less than a sentence to it in his published narrative.[76]

To Darwin, by contrast, Mauritius offered one final opportunity to study reef-building corals in the field. He still had some very specific queries about the types of organisms responsible for building the various parts of a reef, but he also had a serious problem: only the tools of the hydrographer would let him answer those questions. With FitzRoy and the other officers idle, what was he to do? Darwin's particular solution to this difficulty indirectly reveals the depth of his familiarity with, and reliance on, the work the surveyors had been carrying out at South Keeling and elsewhere. With the help of at least one unidentified accomplice, he took a sounding lead and line and "pull[ed] out to seaward" in one of the ship's boats.[77]

Afloat in the Indian Ocean with surveyors' equipment at his disposal, Darwin took soundings from shallow to deep water the way FitzRoy or Sulivan would have done. But he also innovated by employing a deep-sea lead for shallow-water work so as to exploit the larger sampling area it provided. "I sounded repeatedly with a lead," Darwin wrote, "the face of which was formed like a saucer with a diameter of four inches." This was a considerably broader lead than the one illustrated in plate 1, which has a well of approximately an inch in diameter, and once it was armed with tallow the deep sea-lead would yield vastly more material with which to compare the various parts of the reef.

Darwin was by now intimately familiar with the reef-building corals of the Indian Ocean. With precision unmatched in his previous accounts, he identified four discrete zones as he moved from the beach out to deep water. From the mounds of coral that formed the breakwater out to 8 fathoms (48 feet), the "arming invariably came up deeply cut by the branching Madrepores & marked with the impressions of Astreas; its surface was also, without a single exception perfectly clean. not bringing up a particle of sand."[78] The absence of sand was clear evidence that the corals were alive. "At each cast . . . we pounded the bottom with the lead, & as the sand, if present, would have adhered from any of the blows, I think it is pretty certain, that where coral is most abundant, the bottom is quite clean. . . . This fact would afford a useful aid in ascertaining the depth at which coral flourishes." From 8 to 15 fathoms (48 to 90 feet), the arming was almost entirely clean of sand and "beautifully marked with impressions of Astreas . . . some species of Madrepore, Seriatopora, & fragments of branching Millepora & I think Porites as figured by Lamouroux."[79] The next zone, to 20 fathoms, contained extensive beds of the *Seriatopora* and was free of the massive reef builder *Astrea*. Finally, from 20 to 33 fathoms (120 to 198 feet) most of the soundings showed a sandy bottom.

These notes reveal that Darwin was by now confident in his ability to distinguish several genera of corals by the indentations they left in a gob of tallow four inches in diameter. What is more, these notes illustrate vividly the way a particular method of collecting predisposed Darwin to a particular way of thinking: the sounding lead sampled or recorded whatever was to be found on the bottom, across the spectrum from animal (living corals) to vegetable (calcareous algae) to mineral (sand). As a consequence, what might have been a narrow exercise in identifying coral genera also became something like a proto-ecological investigation of the relation between the deposition of sediment and the growth of corals and reefs.

Armed with the knowledge from his private survey, Darwin gained confidence in his answers to questions on which he had previously been willing only to "conjecture": the depth limit of coral growth, an explanation for the observation that reefs never seemed to form right up to the shoreline of high land, the reason a reef's highest point lay at its outer margin, and the factors that determined the distance of that margin from the shore. Referring to the inhabitants of the zone from 8 to 15 fathoms, he argued "that the limit of 25–30 ft fixed upon by M. Quoy & Gaimard as the extreme depth at which the genus Astrea grows, is three times too little."[80] He went on to enumerate a series of principles of reef growth. No longer was he merely pointing out common features of reefs; he was willing to give an explanation of their immediate causes. He concluded that even at an island like Mauritius, "coral does not usually grow attached to the shores. a fact which probably originate[s] from the want of <<a solid>> foundation & <<the>> injurious tendency of the loose matter washed about by the sea."[81] As to the reason "the highest part of the reef is situated at the outer margin; this must be owing either to the greater motion of the water. or to its greater purity."[82] These factors were so important that, in direct contradiction of Quoy and Gaimard, he claimed that "reefs composed of solid stone are only formed in <<a>> turbulent sea."[83] Finally, he explained that "the distance of the outer margin of the reef from the shore depends on the original inclination of the bottom."[84] "In this island," he remarked, "the thickness of the coral [at the outer margin] need not much exceed the depth at which it is believed coral can spring up from the bottom."[85]

The implication was that subsidence was unnecessary to explain the reef at Mauritius, whereas "the relation of the breadth of the reef to the general angle of inclination is very important with respect to the theoretical origin of the coral reefs of the Pacifick."[86]

It is telling that Darwin listed these principles of reef growth in a geographical sequence from shore to deep water. Sounding had become more than a source of specimens and data. The work at Keeling and Mauritius shows that by his fifth year on the *Beagle*, the practice and logic of hydrographic surveying quite literally ordered Darwin's understanding of coral reefs. Hydrography had initially endowed him with an awareness of the undersea environment that primed him, in many respects, for his insight at Tahiti. It also provided an invaluable way of testing many of his predictions about coral growth and reef structure, to the degree that he considered himself simply unable to study reefs without access to soundings. It was no coincidence that his concept of the "theoretical origin of . . . coral reefs" could be tested by pursuing the hydrographer's obsession with learning the seafloor's "general angle of inclination." Quite the contrary, it was his immersion in hydrographic activity that endowed him with the capacity to think that submarine slopes were of theoretical importance for understanding reef formation.

It makes a telling coda to this episode to examine a set of instructions Darwin wrote more than a decade later as part of his contribution to an 1849 book. The Admiralty's *Manual of Scientific Enquiry: Prepared for the Use of Her Majesty's Navy; and Adapted for Travellers in General* was edited by John Herschel with encouragement from Beaufort. Darwin wrote the chapter on making geological observations. In it he offered instructions for studying coral reefs that attest both to his familiarity with the tools and practices of hydrography and to the central role that hydrographic technique played in his view of reef studies. "The most important point with respect to coral reefs, which can be investigated, is, the depth at which the bottom of the sea, *outside the reef*, ceases to be covered with a continuous bed of living corals. This can be ascertained by repeated soundings with a heavy and very broad bell-shaped lead, armed with tallow, which will break off minute portions of the corals or take an exact impression of them." He continued, ". . . there is reason to suspect that different species of corals grow in different zones of depth; so that in collecting specimens, the depth at which each kind is found, and at which it is most abundant, should be carefully noted. It ought always to be recorded whether the specimen came from the tranquil waters of the lagoon or protected channel, or from the exposed outside of the reef. . . . Whenever it is practicable, soundings ought to be taken at short ascertained distances, from *close* to the breakers in a straight line out to sea, so that a sectional outline might be protracted on paper."[87] Any reader who was familiar with Darwin's 1842 book *The Structure and Distribution of Coral Reefs* would have perceived that he was seeking information uniquely relevant to his own theory of reef formation. But as only the account of his exploits at Mauritius can reveal, Darwin was also instructing future travelers to reenact precisely the hydrographic endeavors that had been the basis for his own knowledge of reefs.

Writing more broadly, he summed up the very essence of the comparative terrestrial and hydrographic view I have called amphibious geology. Under the heading "Nature of the Sea-bottom" he wrote, "As every sedimentary stratum has once existed as the bed of the sea or of a lake, the importance of observations on this head is obvious; and *no one is so favourably circumstanced for making them as a naval officer on a surveying expedition*."[88]

PART II

Training in Theory

6

Lyell Claims Darwin as a Student

Four claims make up the widely agreed-on story of how Darwin developed his coral reef theory. Two of them were stated by Darwin himself more than thirty years after the fact: that he developed the theory in South America, and that it was a product of his geological work on that continent. Contrary to the first claim, I have argued that Darwin conceived of himself as having a coral reef theory only once he reached Tahiti. And, while I showed how significant his geological work indeed was to the origin of his coral reef theory, I also argued that Darwin's interest in coral reefs was shaped by his zoological pursuits during the first years of the *Beagle* voyage. Moreover, I showed that the theory of reef formation (and indeed his geological theories on the origin of South America) depended as much on his knowledge of maritime surveying as on his attention to mainland stratigraphy.

In this chapter and those that follow I turn to the substance of the other two claims. The third claim is implicit in Darwin's late recollections and in the accounts of historians, philosophers, and many scientists who have encountered Darwin's theory: that it was an elegant but also a rather simple theory that Darwin conceived whole and published with no great difficulty. I will present evidence that the coral reef theory was revised and modified several times, in connection with changes in his broader plans as a geological author, and that those successive efforts to publish the theory were all, in their own ways, fraught with anxiety. My main argument through the rest of this book is that by overlooking the unsteady trial-and-error process by which Darwin published his coral reef theory, we have failed to recognize that he had to *learn* how to be the theorist he became. Indeed, we have underestimated how Darwin's attitude toward the act of theorizing itself changed as a result of these experiences, which occurred in the very years when he was privately beginning to construct a theory on the origin of species.

I argue that theorizing, or to be precise *authoring theories*, was a process Darwin learned and refined. He perceived this himself in the closing months of the voyage, as evinced in a description of his forays into theoretical writing on the way to Mauritius that he included in a letter to his sister Caroline. "My occupation consists in rearranging old geological notes: the rearranging generally consists in totally rewriting them. I am just now beginning to discover the difficulty of expressing one's ideas on paper. As long as it consists solely of description it is pretty easy; but where reasoning comes into play, to make a proper connection, a clearness & a moderate fluency, is to me, as I have said, a difficulty of which I had no idea."[1] This quotation indicates that writing theories took practice *and* that theories were refined through the act of writing. Darwin had not simply been born knowing how to express a theoretical idea convincingly. It takes no great leap of imagination to envision how some of his other scientific practices depended on face-to-face training with an experienced teacher, whether it was Grant's lessons in studying zoophytes under a microscope or Sedgwick's field tutorials on using a geological hammer. As I argue in this chapter, Darwin was to receive a similarly intimate training in the authorship of theories.

This brings me to the fourth widely held belief about Darwin's theory, which has been stated independently by all the chief historical interpreters of his coral work: that Darwin's subsidence-based explanation for the origin of ring-shaped reefs was more "Lyellian" than the crater theory that had been advocated by Charles Lyell himself. David Stoddart, for example, has illustrated "how much better Darwin's theory fitted Lyell's general philosophical position than did [Lyell's theory]," insofar as it established the continuity between past and present processes, both organic and inorganic, and showed that "small causes could lead to great consequences." In Sandra Herbert's judgment, Darwin's coral reef theory "adopted Lyell's ideas but transformed them" by showing "Lyell's notions of elevation and subsidence [at work] on a larger scale." Most recently Martin Rudwick has maintained that "Lyell suffered the loss of his own theory quite cheerfully, for he recognized that Darwin had out-Lyelled him."[2] As I will demonstrate, however, Darwin modified the coral reef theory after the voyage to make it conform to, and thus reinforce, Lyell's geological system. This was no mere coincidence. Rather, the coral reef theory acquired its conspicuously "Lyellian" character as the result of a direct collaboration between Lyell and Darwin.

It is well known that Lyell had a profound impact on the development of Darwin's scientific thinking. However, the depth of this impact has largely been attributed to Lyell's indirect influence as author of the book Darwin found so compelling during the *Beagle* voyage. Likewise, the intimate personal friendship that blossomed between the two men after the voyage has been well documented. However, in examining the ways Darwin's coral reef theory changed during his first years back in Britain, I became convinced that an extremely significant aspect of the Lyell-Darwin relationship has been overlooked: that the imbalance of authority between them created a distinct set of obligations for each man.[3]

Darwin occasionally used the words student and master to describe his position with respect to Lyell. Lyell indeed behaved as though his status as Darwin's master endowed him with certain prerogatives, which he unhesitatingly exercised, and also with a set of responsibilities that he took equally seriously. Darwin, for his part, appears to have been conscious of the social and intellectual obligations that Lyell's patronage demanded of him. When in the late 1830s he found himself unable to keep to his promised schedule of writing the geological book (which he had already had to circumscribe so it focused exclusively on coral reefs), he experienced sheepishness, anxiety, and melancholy that seem to have made it even more difficult for him to continue. We should understand Darwin's references to Lyell as his master not simply as a metaphor but as an acknowledgment, if not an assertion, of the compact between them. Their master-student relationship may not have been formalized by an institutional structure, but this does not mean it was not structured by their tacit understanding of what such relationships entailed.

This chapter, then, examines the context in which Lyell began to make himself Darwin's master in geology. He had identified the young man as a potential ally based on Darwin's letters back to Henslow during the voyage. When the *Beagle* returned, Lyell did not simply offer Darwin advice and friendship, he took a collaborative role in Darwin's writing and began to stage manage the young man's nascent career.

Homeward Bound as an Aspiring Geologist

The stakes of Darwin's coral reef work were heightened as the voyage drew to a close by the intensity and specificity of his vocational ambitions. With mere months remaining before he would arrive back in England, Darwin no longer anticipated for himself the life of a country parson. Instead, he planned to establish himself as a man of science in Cambridge or London, where he would be well positioned to distribute specimens from the voyage to experts in zoological, botanical, and mineralogical classification. Darwin expected that the science of geology would most richly reward him for having undertaken the burden of travel. In the letter from Mauritius he told Caroline, "I am in high spirits about my geology. . . . [I]t is a most dangerous task, in these days, to publish accounts of parts of the world, which have so frequently been visited. It is a rare piece of good fortune for me, that of the many errant (in ships) Naturalists, there have been few or rather no geologists. I shall enter the field unopposed."[4] Although elite geologists, like their zoological and botanical counterparts, often studied specimens and drew upon observations that they had not made for themselves, they shared a strong commitment to fieldwork.[5] The Geological Society of London followed a seasonal calendar with meetings held in the winter and spring, freeing the summer and autumn for travel.[6] Each of the leading lights of the society was experienced in the field, and most could be identified by the locations where they had done especially meaningful work, from Sedgwick and Roderick Murchison's sites in Wales to Lyell's destinations in the Auvergne and Sicily.[7] In recognition of this fact, Lyell wrote in the *Principles*, "If it be true that delivery be the first, second, and third requisite in a popular orator, it is no less certain that to travel is of three-fold importance to those who desire to originate just and comprehensive views concerning the structure of our globe."[8] That Darwin too considered geological work to depend on access to the field is evident from notes he made in 1838 when he contemplated settling down to marry: "If *not* marry Travel. . . . If I travel it must be exclusively geological . . . [which] Depend[s] on health & vigour & how far I become Zoological."[9]

Darwin planned to write a book about the geology of the whole world, using his findings in South America to illustrate more general arguments. British geologists of the time were engaged primarily in describing and classifying strata, an undertaking that in the previous two decades had become increasingly based on the identification and correlation of organic fossils.[10] As James Secord has observed, geologists in the early nineteenth century were more concerned with identifying sandstone, for example, than with imagining the history of the primordial sea in which sandstone might have been formed.[11] Darwin, however, wanted to use his interpretation of terrain to decode the history of geological changes. This focus on dynamics was no doubt partly inspired by Lyell's *Principles*, but it surely was also drawn from the example of Humboldt and other theory-oriented European geologists such as Léonce Élie de Beaumont and Leopold von Buch, men Lyell considered his chief competitors.[12]

Over the final two years of the voyage Darwin made increasingly careful, and ambitious, plans for his career as a writer. Included among the theoretical queries and reading notes in his Red Notebook, which he began in the first half of 1836 after filling the pages of his Santiago Book, were instructions to himself about how to organize and present his forthcoming geological treatise.[13] A passage written at Mauritius or shortly after departing indicates that he envisioned a book that was no less than global in scale: "In a preface, it might be well to urge, geologists to compare whole history of Europe, with America." He strategized how he could draw rhetorical strength from the geographical limitations of his work, supposing that "I might add I have drawn all my illustrations from America, purposely to show what facts can be supported from that part of the globe: & when we see conclusions substantiated over S. America & Europe. we may believe them applicable to the world."[14] He believed that the striking evidence of elevation in South America and subsidence in the Pacific illustrated fundamental processes that would make the "Geology of [the] whole world . . . turn out simple."[15]

With the *Beagle* moving homeward in 1836 from the coral formations of the Indian Ocean to the Cape of Good Hope and into the Atlantic, most of Darwin's time was spent organizing his specimen lists and preparing his manuscripts for publication.[16] Along the way he had the opportunity to dine with one of his scientific idols, John Herschel, who was then in Cape Town studying the skies of the Southern Hemisphere.[17] He also began to ask his family and friends to make arrangements for his return. Although he preferred Cambridge's countryside, he imagined that living in London would "in every respect turn out the most convenient," because there he would have ready access to the learned societies where scientific business was transacted among gentlemen.[18] For a man with Darwin's ambitions, the most important of these metropolitan establishments would be the Geological, the Zoological, and the Linnean Societies. He was "very anxious to belong to the Geolog: Society," and wrote to Henslow asking if he would "be good enough to take the proper preparatory steps."[19]

In fact, Henslow had already taken steps to place Darwin's name and his work in front of potential audiences, and by late 1835 many members of the English scientific community were stirred by the thought of his return. On 16 November 1835, Henslow read extracts from Darwin's letters at a meeting of the Cambridge Philosophical Society.[20] Published in a pamphlet two weeks later, they described Darwin's studies in marine zoology, his overland excursions, and his fossil finds from the beginning of the voyage up to his trip across the Cordillera earlier that year. Although the extracts were heavy on description, Henslow also included some of Darwin's interpretations of the phenomena he had studied, such as his criticisms of Lamarck on the natural history of zoophytes and his conjectured history of the Andes. Henslow told Darwin's father when he passed along copies of the pamphlet of letters that he believed Charles would "take [his] position among the first Naturalist[s] of the day."[21] Forty-eight hours after Henslow's presentation in Cambridge, Darwin's other mentor at the university, Adam Sedgwick, discussed the content of these letters at length before the Geological Society of London.[22] Impressed by the industrious labor they described, Sedgwick wrote to Darwin's former schoolmaster to exult, "He is doing admirably in S. America, & has already sent home a Collection above all praise.—It was the best thing in the world for him that he went out on the Voyage of Discovery—There was some risk of his turning out an idle man: but his character will now be fixed, & if God spare his life, he will have a great name among the Naturalists of Europe."[23]

Thanks to the efforts made by Henslow and Sedgwick, anticipation of Darwin's return began to spread well beyond the circle of people who knew him personally. The weekly *Athenaeum*, which had commented favorably on the expedition's prospects at the beginning of the *Beagle* voyage, reported on Darwin's work in South America. Lyell's former mentor, Oxford geologist William Buckland, wrote to thank Henslow for sending him copies of "Mr Darwins very interesting Notices" and expressed the hope that they were mere "prelude to more detailed & methodical communications."[24] At home, Darwin's sisters avidly cataloged these signs of Charles's "fame & glory" and delighted in his future prospects.[25] Meanwhile, Eras-

mus Darwin discovered that his society acquaintances in London were becoming interested in his younger brother's travels.[26]

At the Admiralty, Francis Beaufort had begun to think of him rather
grandiosely as "Dr Darwin," the *Beagle*'s "geologist and philosopher
general," and was encouraging Beechey, the surveyor who had studied the
coral islands of the Low Archipelago in the late 1820s, to take "such a per-
sonage" along when he set off to continue FitzRoy's survey on the coast of
South America.[27] This letter indicates that the high esteem in which Beaufort had come to hold Darwin by the fourth year of the voyage, presumably
as a result of FitzRoy's favorable reports, had begun to change the very
conditions under which naturalists might travel on British survey vessels.
While there certainly were precedents for Darwin's participation in the
Beagle voyage, dating back at least to the participation of Joseph Banks
and Daniel Solander as self-funded naturalists on James Cook's first voyage to the Pacific (1768–71), it has, I think, become easy for those who are
familiar with that tradition to underestimate how much Darwin helped to
invent the role into which he was retrospectively cast.[28] As a young Joseph
Hooker retorted (years before becoming Darwin's closest friend), when
told that in order to be appointed as a ship's naturalist he would have to be
such a person as the renowned Darwin, "What was Mr. D. before he went
out ? he, I daresay, knew his subject better than I now do, but did the world
know him ? the voyage with FitzRoy was the making of him (as I had hoped
this exped[ition] would me)."[29]

The 1835 "Coral Islands" essay was one of the documents Darwin revisited in the closing stages of the voyage as he readied his notes for publication. He had his servant, Syms Covington, recopy the manuscript into finer
handwriting, and FitzRoy served as a preliminary audience. The captain
annotated the fair copy with a question about the effect of "earthquake
waves" and a note about channels that interrupt encircling reefs.[30] That
Darwin planned to use this manuscript as the basis for a future publication is evident from the several notes he added alongside FitzRoy's. Some
illustrate how he intended to improve the clearness and fluency of his argument, such as his reminders to "give the reason first" and "amplify the expression."[31] Another note, "Here perhaps introduce the sentence of Polypi
making a monument," reveals that he must have been pleased with the sentence he wrote in his diary after visiting South Keeling, which said, "Under this view, we must look at a Lagoon Is[land] as a monument raised by
myriads of tiny architects, to mark the spot where a former land lies buried
in the depths of the ocean."[32] He now expanded this thought on the sheet

facing the final page of the manuscript, jotting, "Polypi [are] historians . . . not only of time, but of . . . movem[ent]. a point on which evidence [is] so deficient."[33] As he explained to Caroline, "The subject of Coral formation has for the last half year, been a point of particular interest to me. I hope to be able to put some of the facts in a more simple & connected point of view, than that in which they have hitherto been considered." He declared that the "idea of a lagoon Island, 30 miles in diameter being based on a submarine crater of equal dimensions, has always appeared to me a monstrous hypothesis."[34] Now he was in a position to offer Lyell and Beaufort a superior alternative.

Yet at this late stage of the voyage Darwin also still saw himself as a future expert on the zoology of corals, as is proved by several pages of notes from that time discussing the physiological relations between the separate polyps of a colony.[35] Geology was Darwin's first love, but remedying apparent mistakes by Cuvier and Lamouroux was evidently an attractive proposition as well. Of the multiple sciences toward which his study of corals might lead, it remained an open question which would prove most welcoming.

Lyell as an Author

No man of science awaited Darwin's return more eagerly than Charles Lyell, who in his first annual "anniversary" address as president of the Geological Society said, "Few communications have exerted more interest in the Society than the letters on South America addressed by Mr. Charles Darwin to Professor Henslow."[36] Lyell's personal views on the widespread elevation of land had been disputed by George Greenough, a previous president of the society, and one point of particular contention was whether the coast of Chile had been uplifted by a much discussed earthquake of 1822.[37] Darwin's letters to Henslow implicitly supported Lyell by maintaining that uplift had indeed occurred in 1822. Lyell had also learned, from Darwin's South American acquaintance Robert Alison, of FitzRoy's documenting further elevation of that coastline after the earthquake of 20 February 1835.[38] What was more, Darwin's letters contained explicit praise for Lyell's *Principles* and revealed that, despite having received the second and third volumes only during the voyage, he had already adopted Lyell's new terminology for describing the age of tertiary deposits.[39] Lyell took great relish in restating Darwin's findings at considerable length in his anniversary address and reminding the society that "Mr. Darwin supposes that an upheaval to the amount of 1300 feet has been owing to a succession of small elevations, like those experienced in modern times in Chili."[40]

In the years ahead, Lyell's mentorship played a crucial role in shaping Darwin's attitudes toward the practice of science. But even before Darwin returned, the words of the presidential address presaged what Lyell would have at stake in their relationship. Though Lyell was a major figure in British geology (indeed, he remained the sitting president of the Geological Society during Darwin's first several months back in the country), he was a polarizing individual whose penchant for ambitious theorizing earned him criticism from more empirically minded men who viewed Lyell as engaging in irresponsible speculation. One such colleague was the cartoon-sketching Henry De la Beche, who caricatured Lyell as a gentleman in formal clothing offering tinted spectacles of "theory" to an astonished companion who was outfitted with the tools for studying geology in the field[41] (see fig. 22, p. 116).

De la Beche had prefaced his 1830 atlas, *Sections and Views Illustrative of Geological Phaenomena*, with a sharp assessment of fellow geologists' compulsion to theorize. "The complacent manner in which geologists have produced their theories has been extremely amusing," he noted, "for often, with knowledge (and that frequently inaccurate) not extending beyond a given province, they have described the formation of a world with all the detail and air of eye-witnesses." De la Beche acknowledged that the "collision of various theories" could benefit geology by spurring disputants to further investigations and sharper ideas, but he feared that overhasty theorizing would distort subsequent fieldwork. For so long, he insisted (in language reminiscent of his Lyell caricature), had geologists been viewing the earth's crust through a "theoretical medium" that "descriptions of countries so take the colour of the medium, that it becomes no easy matter to separate what is imaginary from what is real."[42]

Lyell's correspondence with his publishing house, John Murray, sheds considerable light on his attitudes toward scientific ideas and their prospective audiences. Founded in the eighteenth century, the venerable press was run during Lyell's career by John Murray II and his son John Murray III. In the 1820s Lyell announced his geological ambitions to a well-heeled audience by writing essays for the Murrays' influential Tory-leaning periodical the *Quarterly Review*, and the house would go on to publish all three of Lyell's major books, in their multiple volumes and many revised editions.[43] From his earliest days corresponding with the Murrays, Lyell revealed himself to be obsessed with priority, to have a rather complex view of the reading "public," and—underlying both of these things—to have an ardent desire to maximize the sale of his books.

Lyell desired priority not simply in being first to possess a new fact or theory, but in beating potential competitors to print. He believed his articles for the *Quarterly Review* would "prepare [him] for the work" of writing a marketable geological treatise, for which he was convinced there was strong demand. "I have ascertained," he exulted, "that the only 4. Or 5. Eng[lis]h Geol[ogis]ts who are sufficiently up to the present state of the subject to anticipate me, are prevented from doing so, by the necessity they are under of keeping their original information & ideas for their <u>lectures</u>."[44] Two years later, in 1829, he told Murray, "Mr [William] Broderip is one of the very few who have the two necessary qualifications the knowledge, & the power of communicating it in a good style of writing, but whether he has a third viz <u>leizure</u> [*sic*], <u>is the question</u>."[45] By then Lyell was expressing urgency to get his book finished both because his material on the geology of Sicily "may be anticipated in Paris" and because "I do not think it likely that there will ever be a better opening than just at this moment for an elementary book on Geol[og]y."[46]

As these remarks imply, Lyell ardently desired to write books that would sell. His letters to Murray indicate that he had rather sophisticated conceptions of the composition and desires of the reading "public" and of the best ways to market books. For example, he was attentive not only to the sales price of his books but also to Murray's advertising and the size of print runs. To maximize sales of each edition of each book, he gave considerable thought to distinguishing the topic and intended readership of the *Principles* from those of his *Elements of Geology* (first edition 1838; discussed further below), and he helped Murray synthesize decisions about print numbers with plans for revising so that he would not release new editions until existing stock had nearly sold out.

Master and Student

What initially attracted Lyell to Darwin was an urgent desire to incorporate his observations from critical sites in South America into the proprietary synthesis Lyell was trying to build. The early indications of the phenomena Darwin had experienced and the way he understood them suggested to Lyell that Darwin could be an ally in the lively and sometimes contentious community of geologists. In November, after Darwin's work had been communicated to the Geological Society, Lyell told Sedgwick, "How I long for the return of Darwin! I hope you do not mean to monopolise him at Cambridge."[47] In the meantime he appealed to the hydrographer, Beaufort, for more advance news of Darwin's findings, knowing that Beaufort was dispatching Beechey to continue the South American survey where FitzRoy had left off. In December 1835 Lyell wrote, "If Capt. Beechey can see & communicate with Mr Darwin who was with Captain Fitzroy in the Beagle beg him to learn from Mr D[arwin] all the latest intelligence of the Geology of Patagonia & Chili which he can."[48] By that month, however, the *Beagle* was already beyond Tahiti and Darwin was at work on his first essay on coral reef formation.

Like most of his contemporaries, Darwin acknowledged that in natural history there was a substantial gulf between the act of collecting and the interpretive work of a philosophical naturalist.[49] Thus when the *Beagle* returned to Britain he acted very quickly to connect himself and his collections with specialists who possessed the proficiency and the stature to illuminate their significance to the expert scientific community.

Darwin disembarked at Falmouth on 2 October 1836 and traveled overland to visit his family in Shrewsbury while the ship and crew continued to the mouth of the Thames and up the river to Woolwich, where the cargo would be unloaded and the crew paid off. From his father's house he wrote to ask Henslow's "advice on many points," but chiefly for help in identifying zoologists, botanists, and mineralogists who would be willing to describe and classify the contents of the collections he had sent home early as well as the specimens that remained aboard the *Beagle*.[50] From Shrewsbury Darwin proceeded to Cambridge, where he saw Henslow in person, and thence to London, where he immediately began "calling on various naturalist people."[51] He told his sister, "I do not think mortal man ever talked more than I have done during the last three days," but he found that his plans had "only become more perplexed instead of any clearer."[52]

Despite the interest showed in him and his collections, he found very few people willing to volunteer for the laborious task of ordering the mass of material he had gathered. He began to realize with dismay that "the collectors so much outnumber the real naturalists, that the latter have no time to spare." Darwin's old zoological mentor Robert Grant showed a short-lived interest in examining Darwin's corallines.[53] Historians Adrian Desmond and James Moore have speculated that Darwin rejected Grant for this job, because he did not want to be associated with Grant's radical politics and indelicate behavior.[54] The comparative anatomist Richard Owen, whom Darwin met at a tea party hosted by Lyell, was willing to dissect

Figure 22. "Theory." A caricature of Charles Lyell by Henry De la Beche. The field geologist on the right listens in startled distress to the geologist on the left (representing Lyell, and dressed in garb more appropriate for the city), who is holding a book titled *Theory of the Earth* and offering a set of tinted spectacles whose lenses would intervene between the responsible field geologist and his true objects of study. "Take a view, my dear sir, through these glasses, and you will see that the whole face of nature is as blue as indigo." De la Beche notebook, British Geological Survey.

some of his vertebrate specimens. But most of the "great men" were overwhelmed with their own work, and the museums were already bulging with animal specimens that had yet to be cataloged. Disappointed, he reported to Henslow that he was "out of patience with the Zoologists," who cared so little for specimens and who seemed, when he visited the Zoological Society, to spend their time "snarling at each other, in a manner anything but like that of gentlemen."[55]

This experience made Darwin identify even more closely with the geologists. He got a "most cordial reception" from William Lonsdale, secretary of the Geological Society, and an especially gratifying response from the society's president, Lyell. As Darwin told Henslow, "If I was not [already] much more inclined for geology, than the other branches of Natural History, I am sure Mr Lyell's & Lonsdale['s] kindness ought to fix me."[56] Darwin was delighted to find that Lyell had decided, "in the <u>most</u> goodnatured manner, & almost without being asked," to make Darwin his protégé. The author of the *Principles* was offering a truly enticing wealth of intellectual and social opportunity while counseling Darwin against accepting too many official responsibilities when there was so much important work to be done.[57]

There is a striking parallel between the way Darwin distributed specimens to zoologists like Richard Owen and John Gould and the way he began to let Lyell examine his geological theories. Both kinds of exchanges offered the potential for mutual benefit, and in both cases certain kinds of intellectual credit accrued to the metropolitan expert rather than the traveler. While Owen got Darwin's spectacular fossils from South America, Lyell for his part wanted Darwin to enter the debate over the elevation of that continent. What makes the Lyell-Darwin relationship distinctive is the way Lyell helped Darwin obtain a kind of expert status in geology that he never pursued in comparative anatomy, ornithology, or botany. Darwin's dream of becoming the geological authority on South America was perfectly aligned with Lyell's need for an ally in the ongoing dispute, which was partly about the recurring earthquakes in Chile that Darwin had become so familiar with during the voyage, and partly about Lyell's larger claims concerning the gradual vertical movement of continents. It would be immensely valuable for Darwin to broadcast what he had seen, especially because he was disposed to interpret his observations in Lyell's favor. Lyell could see that if Darwin was deemed a credible geologist he would be a strong advocate for the contention that whole continents could be created and upraised by geological causes no more potent than those operating in the present. Whereas Lyell, Greenough, and others had based their disagreements about the interpretation of Chilean earthquakes on secondhand reports, Darwin could argue the case using evidence he had gathered himself.

Less than two months after his return, Darwin was voted into the Geological Society, where his affinity for Lyell quickly became common knowledge. As William Whewell wrote at the beginning of December to John Herschel (who remained at the Cape of Good Hope, where the *Beagle* had called in June), "Darwin, who was with Capt. Fitzroy, and who visited you,

is come home. He has made great natural history collections, and is become an extreme Lyellist in geology."[58] The evidence on which Whewell
based this claim was probably private knowledge—from Lyell and perhaps
via Henslow—of the active role Lyell was taking in Darwin's life.

With his new patron urging him on, Darwin announced his plan to "set
to work: tooth and nail at the Geology."[59] He had not yet dispensed to Lyell
his views on coral reefs, but he was working on the topic in private and
seeking comments on his theory from a less daunting audience. He convinced his brother, whose house in London he shared for eight weeks after
the voyage, to translate for him the latest work by Christian Ehrenberg
about the formation of reefs in the Red Sea, and he allowed his family to
read the section of his *Beagle* diary in which he had summarized his views.
His cousin Hensleigh Wedgwood responded, "I liked your account of Keeling island, but your theory of the lagoon islands seemed to us not quite
clearly enough explained."[60] By late December he had finished writing a
paper on the elevation of South America, which he sent to Lyell so that he
could review it before it was to be read aloud at the upcoming Geological
Society meeting.[61] Lyell responded with delight, saying, "The idea of the
Pampas going up, at the rate of an inch in a century, while the Western
Coast and Andes rise many feet and unequally, has long been a dream of
mine. What a splendid field you have to write upon!"[62] He invited Darwin
to visit him at home in London before the society meeting so they could
discuss the paper in person, Darwin having taken lodgings in Cambridge
earlier that month. Lyell wanted him to elaborate on several passages and
to alter "a word or two" of the paper before the public presentation. Lyell's
personal notebook for the period contains a list of questions and suggestions under the heading "Darwins Paper."[63] From this initial engagement
the two began what would be an enduring practice of offering comments
(and in Lyell's case advice) on one another's unpublished writings. They
met on the second day of the new year, and it was almost certainly during
this discussion that Darwin first told Lyell about his reef theory.

Why had Darwin neglected to talk to Lyell about coral reefs sooner? It
seems likely that he was hesitant to jeopardize their new rapport by arguing
against Lyell's version of the crater theory, even though (as he had reported
home from Mauritius) he privately thought it a "monstrous hypothesis."
However, with Lyell so enthusiastic about his views on the elevation of
South America, Darwin must finally have revealed that coral reefs seemed
to show that there had been compensatory subsidence in the Pacific. Lyell
was stopped in his tracks, reportedly doubling over to rest his head on the

seat of a chair while he absorbed what the younger man had said.[64] He then sprang up in a "state of wild excitement" and, to the nervous Darwin's immense relief, began to "encourage [him] with vivid interest."[65] Indeed, Lyell put off other activities so he could begin working with Darwin on a new statement of the coral reef theory. He wrote to Charles Babbage on 6 January 1837 saying he had not had time to begin offering comments on a draft of Babbage's Ninth Bridgewater Treatise because "I have been working so hard both with Darwin's paper & since with his new views on Coral reefs."[66] This hard work, it seems, consisted of helping Darwin reshape the argument of the 1835 "Coral Islands" essay to align it more closely with Lyell's arguments about the geological history of continental landmasses. That it superseded Babbage's treatise indicates how significant Lyell took this new opportunity to be.

Lyell recognized that Darwin's account of reef formation was consistent with his own well-known claim that whole continents had risen above the oceans and then gradually fallen below them, in cycles that continued into the present. As he told Darwin, "I could think of nothing for days after your lesson on coral reefs, but of the tops of submerged continents.... Your lines of Elevation & subsidence will deservedly get you as great a name as De Beaumont's parallel Elevations, & yours are true, which is more than can be said of his."[67] Lyell saw the possibility of a glorious bargain: he would sacrifice his opinion that annular reefs were formed atop the craters of submarine volcanoes in order to champion a theory that supported his much grander speculations. For seven years he had battled shorthanded on behalf of his *Principles*, and now Darwin had come direct from the other side of the world to argue beside him for the incremental oscillations of the earth's crust.

Lyell's language indicates that he was casting Darwin in the role of his own younger self, as a believer in what he would later sardonically describe as "my heretical doctrines."[68] Lyell's congratulation to Darwin bore the traces of his own struggle: "It is all true, but do not flatter yourself that you will be believed, till you are growing bald, like me with hard work, & vexation at the incredulity of the world."[69] Little did Lyell know that before spring was out he and Darwin would have enacted a strategy that enhanced the persuasiveness of Darwin's theory while strengthening Lyell's position in the Geological Society.

Darwin was tugged in several directions during the first half of 1837. While Lyell was eager for him to publish on coral reefs, others were fascinated by Richard Owen's new announcements about the gigantic mammalian fossils Darwin had found in South America. Lyell praised Darwin and Owen in his second annual presidential address to the Geological Society, on 17 February 1837, announcing that their "striking results" illustrated a law of morphological relations between the present and extinct mammals of a given locale.[70] The London zoologists were encouraging Darwin to draw together in a single treatise all the descriptions being made of his animal specimens from the voyage.[71] In March he moved from Cambridge to London, where his most pressing task was to finish turning his *Beagle* diary into what he envisioned as "a kind of journal of a naturalist, not following however always the order of time, but rather the order of position."[72] He made efficient use of his time by writing "abstracts" about fossil mammals and coral reefs that would be inserted into the journal and could also be circulated separately.[73]

Further light is shed on the nature, significance, and evolution of Lyell and Darwin's face-to-face interactions in these years by Lyell's personal notebooks, which still remain in the hands of the Lyell family.[74] The contents show that he and Darwin met frequently in London during the first few months of 1837 and continued doing so (when Lyell was in residence) until Darwin and his young family moved to Kent in 1842.[75] Much of the discussion was about the geology of South America, for Lyell was eager to learn Darwin's opinions on a range of phenomena the voyager had witnessed, from glaciation to earthquakes. On 6 April Lyell recorded a set of "Memoranda—Darwin" showing that he was gripped by the traveler's conclusions about the movement of the crust. Lyell was also grappling with Darwin's counterintuitive use of the term horizontal elevation to refer to the equable *vertical* movements that left horizontal beds level. "Generally speaking," Lyell learned, "the elevat^s in the Andes have been horizontal movements & Darwin imagines that the inclination & disturbance of beds in the Andes & other mountains took place when they were submarine & that upheavals 20,000 ft. have taken place by insensible upward movements."[76] Lyell's notes include résumés of their conversations, memoranda of questions he intended to ask Darwin at subsequent meetings, annotations containing the responses to these queries, and comments on Darwin's essays. These records, along with the letters the two exchanged, attest to the intensity of their interactions and to the intellectual benefit each derived.

Lyell took care to dispense strategically the knowledge he gained from conversations with Darwin, making a clear distinction between backstage and front-stage interactions. This is evident from an exchange between

Darwin and his sister Caroline, who wrote with excitement in February, buzzing to know "on what points it is that Lyell 'fully agrees with your [Charles's] views.'" She was certain that "the Coral islands I know was one subject."[77] Darwin replied that Lyell had said less than he might have about Darwin in his anniversary address to the Geological Society because "of course he could only allude to published accounts."[78] While it might seem implausible that Lyell would have considered mentioning unpublished work, he had in fact done precisely that in the previous year's anniversary address, coincidentally when discussing coral formations. "I may mention a discovery made by Mr. Lonsdale during the last summer," he said on 19 February 1836, "and which he has permitted me to announce, that our common white chalk . . . is full of minute corals, foraminifera, and valves of a small entomostracous animal resembling the Cytherina of Lamarck."[79] Lyell could therefore have counseled Darwin to let him use the 1837 address as an auspicious moment to reveal the new theory. Instead it was another three months before, as we shall see, Lyell actively stage-managed the circumstances in which Darwin's theory was announced to the society.

The Primacy of Geology in Darwin's Private, as Well as Public, Activities

It was at this moment of the most intense public activity of his life that Darwin's jottings in the Red Notebook began to include speculations that animal and plant species might be mutable.[80] Yet in this theoretical notebook the entries on the movement of the earth's crust and the underlying constitution of the globe continued to predominate. It was while pondering geological and geographical change that Darwin contemplated the succession of one organic type by another, either across horizontal space, like the two species of South American rhea that occupied distinct ranges, or across time, as with the extinct and living species of guanaco.[81] As these examples suggest, Darwin's thoughts on the creation of new species during the first half of 1837 were stimulated primarily by questions about their geographical distribution. These in turn sprang from his reflections on changes to the earth's geography, such as his queries on how newly elevated islands and continents became inhabited by plants and animals. But his main preoccupation was to determine the underlying cause of these physical changes. The notes show that Darwin's primary reason for trying to determine the laws underlying the succession of species was that doing so might add a new type of evidence to his study of geographical change. If organic forms had a determinate response to changes in the earth's crust, then the history of animals and plants could reveal the history of the globe. The line of thought that we know—in retrospect, of course—would lead Darwin to his species theories was originally subsidiary to his effort to understand geographical and geological change.

The lawlike response of coral reef shapes to geographical changes already offered a promising way to characterize the globe's internal workings. In early 1837, about the time he and Lyell had been discussing the implications of his coral reef theory, Darwin noted that coral reefs outdid Lyell's best-known illustration of the gradual rising and falling of the crust, the Temple of Serapis at Pozzuoli, Italy. Evidence from Pozzuoli indicated very localized oscillations, but Darwin believed that in South America and the Pacific he had seen evidence for vertical movements so widespread that they could have been effected only by a profound "final cause," such as the "circulation of [a] fluid nucleus" inside the globe.[82] Thus he wrote, "The great movements . . . agree with great continents." Darwin considered the idea that these movements were not limited to "mere patches as in Italy" to have been "proved by [the] Coral hypoth[esis]."[83] Meanwhile, Lyell was contemplating the same matters in his own notebooks, writing, "How large a part of Europe is the area now rising in Chili equal to" under the heading "April 8. 1837 Queries Darwin." The very next entry in the notebook is a list of tasks headed by "1. Invit[ation] to Darwin for Friday tomorrow Ev[enin]g or Sat[urda]y." At this and other engagements it appears they discussed the topics of volcanism, elevation, and coral reef formation, including the question of whether corals derived the carbonate of lime in their skeletons from volcanic matter. Six years earlier, in the second volume of the *Principles*, Lyell had sought to use both the history of organisms and the present-day appearance of coral reefs as clues to decoding the movements of the earth's crust. Now, under Lyell's guidance, Darwin was working feverishly to understand these same phenomena for the same reason.

How much Darwin's coral reef theory contributed to Lyell's high estimation of the young man is evident in a letter Lyell wrote to his friend Herschel. "I am very full of Darwin's new theory of Coral Islands," he exclaimed.[84] Darwin had convinced Lyell that "I must give up my volcanic crater theory for ever, though it costs me a pang at first, for it accounted for so much." After reviewing the evidence in favor of his former view, Lyell admitted, "My whole theory is knocked in the head, and the annular shape & central lagoon have nothing to do with volcanos, nor even with a crateriform bottom." Echoing the language of Herschel's *Preliminary Discourse on the Study of Natural Philosophy,* Lyell indicated that Darwin's theory was based on a deeper phenomenon, a *vera causa*.[85] "Perhaps Darwin told you when at the Cape what he considers the true cause?" He explained to Herschel the factors that limited the growth of corals, and with the aid of three diagrams explained the effects of subsidence and elevation on a hypothetical "granite island round which coral is growing." As evidence in favor of Darwin's theory, Lyell pointed out that reefs were known to exist in all intermediate states, and he cited Darwin's "proof" that encircling reefs did not exist in sites of elevation. "So then," Lyell concluded admiringly, "the coral islands, are the last efforts of drowning continents to lift their heads above water. Regions of elevation and subsidence in the ocean may be traced by the state of the coral reefs." Lyell reported that he had "urged [William Whewell, his successor as president of the Geological Society] to make him read it at our next meeting," and he "hope[d] a good abstract of this theory will soon be published. In the meantime," he encouraged Herschel, "tell all sea-captains and other navigators to look to the facts which may test this new doctrine."[86]

Through Lyell's intervention, Darwin did indeed arrange to make his first public comments on coral reefs at the next meeting of the Geological Society. Lyell's letter to Herschel indicates that he was already intimately familiar with Darwin's very latest progress on his coral reef theory. It also provides evidence that Darwin's theory was evolving. The hand-drawn diagrams and accompanying explanations of the initial growth of a shore reef that Lyell transmitted to Herschel reflected a view contrary to Darwin's 1835 essay, one prompted only by the late-voyage study of the Mauritius reefs. Likewise, any "proof" Darwin had of encircling reefs' being excluded from areas of elevation was the result of work done after the voyage. As we shall see, Darwin was shortly to make the results of recently conducted geographical comparisons into the centerpiece of his published theory. The key beneficiary of this development would be the man who had been overseeing his work and who scheduled its publication: Darwin's master, Lyell.

7

Darwin's Audacity, Lyell's Choreography

The 1837 paper in which Darwin made the first public statement of his coral reef theory has received very little attention since it was superseded by his longer 1842 book on coral reefs. It would be easy to dismiss this paper as little more than a provisional summary of ideas he went on to express in their full and authoritative form five years later. On closer examination, however, this proves emphatically not to be so. Darwin's coral reef paper was a work of stunning ambition that deserves to be remembered as one of the most provocative performances of his career. Not only did it present a compelling new answer to a question that had animated geologists, navigators, and a wider intellectual public for more than half a century, it argued that understanding the formation of coral reefs was the key to interpreting the geological history of vast segments of the earth's surface. Unlike Darwin's later book, it also pushed well beyond these claims, for he closed by asserting that his new theory of coral reefs might lead to solutions for two of science's most intractable problems: the internal composition of planet Earth and the origin of species.

Understanding the potential impact of Darwin's coral reef paper on the London geological community requires making the effort to forget the outcome of his career, to imagine a scenario in which he had not yet put forward any big and original ideas.

We must overcome what historian Alain Corbin has called "psychological anachronism," in this case the anachronism of identifying the name Charles Darwin with scientific authority. Although he had begun to publish descriptions and interpretations of far-flung phenomena encountered during the voyage, he was still just the twenty-eight-year-old traveler from whom Buckland had hoped to see "more detailed and methodical communications" than were contained in the letters he sent to Henslow from the *Beagle*. Thus it was to the utter astonishment of Darwin's audience that at 8:30 p.m. on 31 May 1837 he stood before them and contradicted the Geological Society's most ambitious and pugnacious theorist on the notorious puzzle of coral reef formation. Darwin's listeners were even more flabbergasted when that theorist, Charles Lyell, did not even hesitate before capitulating in a manner they had never before seen. It appeared that in the course of a single evening this young Darwin had convinced Lyell he should reconsider a dearly held theory. Yet the light cast by the evidence of Darwin and Lyell's private interactions reveals that this was a most strategic retreat.

Going Public

On the night Darwin revealed his coral reef theory before the Geological Society at Somerset House, it had been just eighteen months since he had gazed on the island of Eimeo from the heights of Tahiti, barely a year since he had pounded the corals of Mauritius with the bell-shaped sounding lead, and less than eight months since he had returned to England. With his friend and former shipmate B. J. Sulivan attending as a guest, Darwin gave a presentation that bore distinct marks of the time and the venue in which he delivered it and, also, of the specific objectives Lyell and Darwin had for this piece of work. As is evident from both the abbreviated version that appeared in the *Proceedings of the Geological Society* and the full text that was published after two years' delay in Darwin's narrative of the *Beagle* voyage, he went to great lengths to praise Lyell and to show how the form of coral reefs reinforced lessons great and small from the *Principles of Geology*.[1] As indicated by the title of the paper, "On Certain Areas of Elevation and Subsidence in the Pacific and Indian Oceans, as Deduced from the Study of Coral Formations," Darwin implied that it was the study of coral reefs that had led him to broader conclusions about the vertical motions of the earth's crust. This runs contrary to the tale he told decades later in his autobiographical recollections (on which see chapter 11), of developing the theory in South America before he had seen any coral reefs, but it also differs noticeably from the argument of his 1835 "Coral Islands" essay. Part of this change can be attributed to Darwin's intervening visits to Keeling and Mauritius, but the strategy of presentation bears strong traces of Darwin's collaboration with Lyell in the previous eight months. More broadly, it illustrates how Darwin reshaped the work of the voyage for a variety of audiences in order to serve specific vocational or intellectual ambitions.

The differences between the essay Darwin wrote on the *Beagle* and the one he prepared for the Geological Society consisted not only of rearranging evidence but also of reorganizing the line of argument. In 1835 he had argued from a geographical premise to an explanation of reef forms. He began that essay by describing various Pacific island chains that shared a geographical orientation even though some contained high land and others consisted merely of coral reefs. He had gone on to illustrate the similar shapes of individual Pacific coral reefs that encircled high land and those that encircled only lagoons. From there he proposed subsidence as the mechanism by which reefs encircling high islands could become circular reefs enclosing empty lagoons. When he wrote the 1835 essay, of course, he had viewed reefs from ship and shore, but he had yet to set foot on a coral island or to take soundings on the outer margins of a reef.

In contrast, he began his 1837 paper for the Geological Society by discussing a topic on which he was unquestionably the society's foremost expert: the growth of reef-building corals.[2] Advertising his personal experience of "carefully examining the impressions on the soundings" at Keeling and Mauritius, he argued that the genera of corals capable of forming a reef could not live below ten or twelve fathoms.[3] Given this fact about the natural history of corals, Darwin explained, the structures of lagoon islands, encircling reefs, and barrier reefs were all extremely puzzling. None of those types of reefs could be expected to have grown up from their present foundations, because their outer margins stood in water too deep for corals to grow. Darwin claimed that these three classes of reefs were structurally identical, so that the only way to distinguish between them was "in the absence or presence of neighbouring land, and the relative position which the reefs bear to it."[4] He contrasted these forms with a fourth type, which he called "fringing reefs." Like encircling reefs, fringing reefs were also rings standing some distance from the shoreline of an island, but the difference was that they "extend only so far from the shore, that there is no difficulty in understanding their growth."[5] That is to say, fringing reefs had foundations limited to the shallow depths in which corals could be expected to flourish.

In his 1835 essay Darwin had assumed that if an island provided the necessary physical conditions, "Corall would immediately commence to grow on the shore." When he subsequently visited Mauritius, he had been deeply puzzled to see that corals did not in fact appear to grow in this manner: "Reef very seldom attached to shore. . . . I do not understand this."[6] If corals never grew up to the shore, even on a "fringing reef," this meant that the only distinction between encircling and fringing reefs was whether the depth of the water outside the reef exceeded the limit of coral growth. In principle, therefore (though Darwin did not say it explicitly), the classification of a given reef was entirely dependent on the accepted value of the depth limit, and any change in this value would require fringing and encircling reefs to be reclassified. No wonder, then, that he began his paper for the Geological Society by establishing his own expertise on the zones of coral growth and by setting the depth limit in his terms.

Having described these four types of reefs and their distinctive relation to any adjacent land, Darwin argued that "no explanation can be satisfactory which does not include the whole series."[7] He seems to have been implying that the striking similarities between types demanded a common explanation. Thus he proposed subsidence as a mechanism that could explain the shape of each of the first three reef types, lagoon, encircling, and barrier. This had the added benefit of placing the four types in a genealogical sequence that explained both their similarities and their differences. He gave his most succinct description of his "theory" of coral reefs as follows: "[The reefs'] configuration has been determined by the kind of subterranean movement."[8]

Darwin went on to argue that changes in the level of the land, such as would transform one type of reef into another, were likely occurrences. As evidence he cited his own earlier paper on the elevation of South America, claiming that if the continent was rising insensibly, it was not improbable that the floor of the Pacific was subsiding in the same manner. He also cited the chapter on coral reefs in Lyell's *Principles*, echoing Lyell's point that subsidence was the most likely reason the Pacific contained so little land even though both volcanoes and corals tended to create it. Yet Darwin went beyond Lyell, indicating that subsidence was "rendered almost necessary" by the "inconsiderable depth at which corals grow."[9] The alternative, that every lagoon island was underlain by a submarine mountain just a few fathoms underwater, was just too implausible. This shows Darwin using a method of reasoning "by exclusion" that he used frequently but notoriously disavowed later in life in reference to one specific puzzle: the formation of the "parallel roads" of Glen Roy, Scotland.[10]

When he read the coral reef paper to the society, Darwin apparently used cross-sectional diagrams of different types of reefs to illustrate the effect of subsidence on a growing reef. "A simple fringing reef," he explained, "would thus necessarily be converted by the upward growth of the coral into one of the encircling order, and finally, by the disappearance through the agency of the same movement of the central land, into a lagoon island."[11] If the reef-fringed shoreline of a continent subsided, the result would be a barrier reef, making it simply an "uncoiling [of] one of those reefs which encircle at a distance so many islands."[12] Although the diagrams were not published in the *Proceedings* or the first edition of the *Journal of Researches*, they evidently constituted a substantial component of his argument at the Geological Society. When he published his book on coral reefs in 1842, this point was indeed illustrated by a pair of cross-sectional diagrams. I have found evidence in a letter Lyell wrote the next day (which happened to be published in the *Times* of London ninety-eight years later by the great-grandson of its recipient) suggesting that Darwin also displayed coral specimens at the society.[13] Recently Brian Rosen and Jill Darrell have in turn used this evidence to argue convincingly that tags written by Darwin and still attached to coral specimens now at the Natural History Museum in London were prepared for this meeting.[14]

At the Geological Society, Darwin sought to convince his audience that there were gradations in nature between his taxonomic classes of encircling reefs, barrier reefs, and lagoon islands that suggested transformation from one to another. In doing so he followed closely the argument he had written in his 1835 essay, claiming that "there exist every intermediate form between a simple well characterized encircling reef, and a lagoon island," as well as links between encircling and barrier reefs.[15] As evidence that subsidence had occurred at the locations where such reefs were found, he mentioned four points. First was the "juxtaposition" of the reef types he associated with subsidence: for example, the ocean beyond the Australian barrier reef contained encircled islands and "true lagoons." Second, his personal examination of South Keeling had revealed superficial evidence of subsidence: trees whose roots were undermined by seawater and a tide-washed storehouse that had stood above the high-water mark when it was built seven years earlier. Third, he described reports of earthquakes from the encircled island of Vanikoro and the lagoon island of South Keeling, which he took to be caused by episodes of subsidence. In the case of South Keeling, he linked local earthquakes to those felt six hundred miles away at the high, reef-fringed island of Sumatra, where there was evidence of elevation. "One is strongly tempted to believe," he argued, "that as one end of the lever [Sumatra] goes up, the other [Keeling] goes down: that as the East Indian archipelago rises, the bottom of the neighbouring sea sinks and carries with it Keeling Island, which would have been submerged long ago in the depths of the ocean, had it not been for the wonderful labours of the reef-building polypi."[16] Fourth, and perhaps as the result of a recent insight (for it is not mentioned in the summary version), Darwin recast the work of the French crater theory advocates Quoy and Gaimard to show how their observations could be taken as evidence in favor of his theory. He marveled that although they had crossed both the Pacific and Indian Oceans, every one of the reefs they described happened to be associated with high land and skirted the shore closely enough that all must (in Darwin's taxonomy) be considered fringing reefs.[17] When Quoy and Gaimard described locations that illustrated "the general structure" of reefs, the Frenchmen had mentioned independently "in different parts of [their] account" that the islands fringed by those reefs had recently been elevated. That Quoy and Gaimard had documented the coincidence of fringing reefs with areas of elevation, independent of Darwin's prediction that they should be found together, carried "the same weight as positive evidence."[18]

With the subsidence theory fully sketched out, Darwin anticipated his critics by addressing another of the issues that had exercised him before and during his visit to South Keeling. "It may be said," he acknowledged, "granting the theory of subsidence, [that] a mere circular disc of coral would be formed, and not a cup-shaped mass." This, in other words, was the question of why a lagoon would remain open after the encircled island had finally subsided entirely beneath the ocean. Again Darwin drew attention to his own field study of the Keeling reef, which had given him firsthand experience examining a growing reef unrivaled by any European except Christian Gottfried Ehrenberg, the German naturalist who had made an extensive study of Red Sea corals. Darwin's explanations for the persistence of lagoons were identical to those he had established in the Indian Ocean in 1836. Fringing and encircling reefs always had at least some channel between them and the shore, and the strongest reef-building corals (those he had called the "bulwark" corals during the voyage) grew only on the outer reef. These factors encouraged the presence of a lagoon, while two others worked to keep it from being filled in. The closer the delicate corals of the lagoon came to filling it in by their own growth, the less favorable the conditions they lived in and so the slower they grew. Meanwhile, the absence of high land inside the reef meant there was no source of inorganic sediment to silt up the lagoon.

Putting the Coral Theory to Work

Confident that he had established the plausibility of his theory, Darwin continued his paper by making an important shift from trying to explain the form of reefs to using these forms to explain other phenomena. Whereas he initially sought to make a convincing case that certain reef shapes were the product of subsidence, in the second half of the paper he began to take subsidence as a given. To this moment, Darwin had barely mentioned the geographical distribution of islands, which had been his point of departure in the 1835 essay. But now, having used a discussion of coral growth to establish subsidence as the most likely cause underlying the development of encircling, barrier, and lagoon-encircling reefs, he introduced his geographical evidence as an independent "test" of "the truth of the theory." By implying that geographical evidence was unnecessary in the first place as proof for his explanation of reefs, he could suggest that it was legitimate to use reefs to explain geographical patterns. His theory predicted that reef types characteristic of subsidence would be found together, in areas where there were no islands showing evidence of elevation such as dry-land beds of recent shells and corals or the growth of "mere skirting reefs" close to shore.

He declared, "I think it can be shown that [the predicted distribution] is the case in a very remarkable degree; and that certain laws may be inferred . . . of far more importance than the mere explanation of the origin of the circular or other kinds of reefs."[19] Now Darwin began to pursue what he called "the main object of the paper," which was to use the form of coral reefs as an index to the movements of the Pacific and Indian Ocean floors, which he argued were, like the elevations of South America, taking place "over wide areas with a very uniform force."

Thus Darwin began cataloging areas of elevation and subsidence in the ocean, a task he argued was now possible with the use of his independently formed theory of coral reef formation. To illustrate how such an enterprise would work, Darwin systematically described the distribution of reefs from east to west, starting at the west coast of the Americas. In his presentation at the Geological Society he displayed a map with their locations marked, although no such illustration accompanies the full version published in the *Journal of Researches*.[20] The written text instead guided readers on a hypothetical journey, "commencing on the shores of South America," "passing over the space of ocean," "continuing with our examination," and so on, all made possible by the detailed charts he had encountered as they were being made and used during the *Beagle* voyage.[21] Properly viewed, such charts allowed viewers who had never left Britain to experience the same perspective he had enjoyed when he stood high on Tahiti and imagined the island of Eimeo to be sinking: "Now if we look in a chart," he urged his readers, "at the prolongation of the reef towards the northern end of New Caledonia, and then [mentally] complete the work of subsidence . . ."[22] He speculated about the kind of landforms that underlay various reefs, judging that the lagoon islands of the Low Archipelago had each been "moulded round the flanks of so many distinct islands," while the whole group of the Maldives seemed to sit atop a single multipeaked island that "formerly occupied that part of the ocean."[23]

Darwin claimed that plotting reefs by type across the Pacific and Indian Oceans revealed groupings that could be divided into seven linear bands, four of them areas of subsidence and three of elevation. The bands of movement were roughly parallel, running southeast to northwest, with the regions of subsidence separated by the tracts that showed signs of elevation. Through the course of describing the limits of each of these geographic zones, he repeatedly demonstrated that "the three classes [of reefs] supposed to be produced by the same movement [subsidence] are found . . . in juxtaposition." He likewise illustrated that such reefs were rarely found closely juxtaposed with any signs of elevation. In areas of the world that contained no lagoon islands despite the known presence of reef-building corals, he was able to demonstrate the likelihood that elevation had recently occurred. He urged, "Excepting on the theory of the form of reefs being determined by the kind of movement to which they have been subjected[,] it is a most anomalous circumstance . . . that the lagoon structure being universal and considered as characteristic in certain parts of the ocean, should be entirely absent in others of equal extent."[24]

The two key features of these "linear spaces of great extent" were that the bands of elevation and subsidence alternated and that they were "undergoing movements of an astonishing uniformity."[25] This could have been construed as a victory for Lyell, for the term uniformitarianism had by this time been attached to his geological approach in a review by Whewell.[26]

What Darwin meant to indicate by "uniformity" in this context was that up-or-down movements of a given magnitude had occurred over a vast area.[27] And while such "uniform" episodes of subsidence might have been geographically widespread, their vertical magnitude had not been great enough or rapid enough to draw the living part of the reef beneath the shallow zone of coral growth. Darwin therefore employed his coral reef theory to rule out the possibility that paroxysms had caused sudden, large episodes of subsidence. Downward movement of the crust sufficient to draw entire islands or continents to unfathomable ocean depths had been produced by the accumulation of small changes each incapable of drowning reefs, indeed of changes no greater in magnitude than those produced by the recently witnessed earthquake in Chile.

To those in the inner circle of the Geological Society, the similarity must have been obvious between this claim and Lyell's statements earlier in the decade on hypothetical changes of level. When Lyell had written, "let a series of two hundred earthquakes strike [a] shoal, each raising the ground ten feet; the result will be a mountain two thousand feet high," he had also posited that elevation and subsidence were compensatory movements that occurred simultaneously on different parts of the globe.[28] Now Darwin too proposed a causal relation between elevation and subsidence. Such a view was supported by two kinds of evidence generated by the coral reef theory. First was the general pattern of alternation between upward and downward movements shown by plotting reef types on the globe. Second were specific cases like the relation between Keeling Island and Sumatra, where simultaneous earthquakes occasioned opposite movements in these islands several hundred miles apart. These points, combined with the apparent unlikelihood that there was any alternative explanation for reef shapes, led Darwin to insinuate that he had revealed a "general law" underlying the formation and distribution of coral reefs:

> When we consider the absence both of widely-encircling reefs and lagoon islands in the several archipelagoes and wide areas, where there are proofs of elevations; and on the other hand the converse case of the absence of such a proof where reefs of those classes do occur; together with the juxtaposition of the different kinds produced by movements of the same order, and the symmetry of the whole, I think it will be difficult (even independently of the explanation it offers of the peculiar configuration of each class) to deny a great probability to this theory.[29]

That Darwin limited his use of geographical evidence in the first part of the talk suggests he was aware that this argument might be viewed as a tautology.

In this paper Darwin advertised a preference for "general" explanations, suggesting that the more phenomena a theory could explain, the more likely it was to be correct (about any of them). The type of proof Darwin was aiming at bears some similarity to the "consilience of inductions" described by William Whewell, who was presiding over the meeting. The polymath mathematician-cum-moral philosopher from Trinity College Cambridge was at this point already at work on his *Philosophy of the Inductive Sciences* (1840), in which he asserted that "if we take one class of facts only, knowing the law which they follow, we may construct an hypothesis . . . which may represent them . . . [and] when the hypothesis, of itself and without adjustment for the purpose, gives us the rule and reason of a class of facts not contemplated in its construction, we have a criterion of its reality, which has never yet [failed]."[30] If this was the type of argument Darwin and his contemporaries would have approved, it helps explain why Darwin worked hard in 1837 to portray geographical zones of elevation and subsidence as a class of facts he had not contemplated in constructing his explanation of reef forms despite the essential role they had played in the private origin of his theory during the voyage.

Darwin stated these ideas in an almost aggressively Lyellian language, mixing direct references with allusions that those in attendance would easily recognize. For example, Lyell's desire to determine the historical circumstances in which geological formations had been laid down was not intrinsic to the British tradition of stratigraphy, which for most of its practitioners had been an exercise in describing and classifying formations rather than explaining their origin. Lyell, though, had opened the first volume of the *Principles* with a lengthy analogy between the work of a geologist and that of a historian.[31] At the Geological Society Darwin portrayed reefs as legible historical records. He modified his earlier personification of corals as "historians . . . not only of time, but of . . . movem[ent]" in order to echo Lyell's construction. Thus he claimed that the "importance [of the coral reef theory], if true, is evident: because we get at one glance an insight into the system by which the surface of the land has been broken up, in a manner somewhat similar, but certainly far less perfect, to what a geologist would have done who had lived his ten thousand years, and kept a record of the passing changes." Darwin's reading of this historical record had produced a "law almost established, that linear areas of great extent undergo movements of an astonishing uniformity, and that the bands of elevation and subsidence alternate."

This conclusion pressed him to venture into speculation about the globe itself, musing that the alternating bands were caused by "a fluid most gradually propelled onwards, from beneath one part of the solid crust to another."[32] The Cambridge mathematician William Hopkins had lately been striving to model the mechanical effects of such a fluid on the overlying crust, while Lyell's allies Herschel and Babbage were known to be speculating on the effects the topography of the crust had on the distribution of temperature within this molten layer.[33]

Young Darwin closed with a series of breathtaking claims and speculations of the sort that might have seen him branded a crank if not for some sign that his judgment could be trusted. He began the conclusion by enumerating the geological and zoological lessons that could be "deduced" from studying the growth of corals. He demonstrated that every active volcano in the Pacific and Indian Oceans lay in one of his areas of elevation, and he proposed another "law," that volcanism and elevation were linked consequences of the same "propulsion of fluid matter" to particular locations beneath the crust. This meant that stratigraphers—that is to say, almost everyone in the audience—could draw from Darwin's coral reef theory "a means of forming some judgment of the prevailing movement [elevation], during the formation of even the oldest series where volcanic rocks occur interstratified with sedimentary deposits." To this new conclusion formed from his study of the comparative distribution of reefs and volcanoes, Darwin added an insight that he originally conceived on the west coast of South America, pointing out that "we may feel sure, where a great thickness of coral limestone occurs, that the reefs on which the zoophytes flourished, must have been sinking." Darwin also seemed to imply the possibility of correlating the organic remains in a thick limestone deposit with those of other well-characterized formations and thereby judging "what were the prevailing movements at different epochs."[34] Three months earlier, in his presidential address at the society's anniversary meeting, Lyell had publicly reminded his colleagues that "evidence of a sinking down of land, whether sudden or gradual, is usually more difficult to obtain than the signs of upheaval."[35] Now, in a paper Lyell had helped him fashion, Darwin reiterated that "any thing which throws light on the movements of the ground is well worthy of consideration; and the history of coral reefs may . . . elucidate such changes in the older formations." He once again asserted the significance of his findings for his colleagues' work, arguing that the general laws he had adduced from studying coral reefs would make it more feasible to "speculate with . . . safety on the circumstances under which the complicated European formations . . . were accumulated."[36]

Species

As if interpreting the record of crustal movements and deducing the constitution of the earth's nucleus were not reasons enough to care about coral reef formation, Darwin closed by conjecturing that this topic might also provide the key to unraveling the history of life. First his theory suggested an explanation for the "uniformity" of flora between remote islands of the vast "Indio-Polynesian" region, a puzzle that had been raised by René Lesson after the voyage of the *Coquille* in the early 1820s. Resurrecting the cherished "monument" metaphor from his *Beagle* diary, Darwin explained that the problem was less intractable "if we believe that lagoon islands, those monuments raised by infinite numbers of minute architects, record the former existence of an archipelago or continent in the central part of Polynesia, whence the germs could be disseminated."[37] Speaking in global terms, Darwin believed such insights would allow his coral reef theory to "illustrat[e] those admirable laws first brought forward by Mr Lyell, of the geographical distribution of plants and animals [being] consequent on geographical changes."[38] In this light it is noteworthy that Darwin's solution to the origin of the Indio-Polynesian flora did not claim the former presence of a land bridge between continents currently divided by the Pacific but rather posited a continent or group of islands with its own flora that had stood in the Pacific's present location. In volume 2 of the *Principles*, Lyell had claimed that individual continents were transient features of the earth's crust and that their emergence and disappearance—shifts in the relative location of land and ocean—produced shifts in climate that ushered in new epochs of life. Darwin argued to his geological colleagues that his new method for determining whether a given island was in an area of elevation or subsidence "will directly bear upon that most mysterious question, whether the series of organized beings peculiar to some isolated points, are the last remnants of a former population, or the first creatures of a new one springing into existence."[39]

This sentence was almost certainly a reference to "the replacement of extinct species by others," which John Herschel had called "that mystery of mysteries."[40] Darwin was privately working out a theory of transmutation at this time, but he had not yet conceived of natural selection, which occurred to him only after he read Thomas Robert Malthus's *Essay on the Principle of Population* in October 1838. In mid-1837 his view was that the origin of new species was caused by geographical change. What Darwin seems to have been alluding to in his Geological Society talk, judging by the details of this first of his evolutionary theories, was that using the shapes of coral reefs as a key to understanding past vertical movements would in turn give insight into the changes that had driven the production of present species. The key quotation for understanding his mind-set is found in his B Notebook, which he opened in July 1837. There he wrote, "Species [are] formed by subsidence . . . elevation & subsidence [are] continually forming species." An example of the way he envisioned this happening was for a piece of land to subside until it had been divided into two islands, each possessing members of an original single species whose two sets of descendants might diverge over time in response to their island's distinct living conditions. If the conditions produced sufficient change, the new types would "keep distinct" even if the islands were later "elevate[d] & join[ed]." Such a process would result in "two species made."[41]

Improbable as it may seem in retrospect, the B Notebook reveals that Darwin was just as eager to use species as a gauge of geographical change as to use geographical knowledge derived from coral reefs as a key to the origin of species.[42] "If my [species] theory [is] true," he noted, "we get (1) a horizontal history of earth . . . (2^d) By character of any <<two>> ancient fauna, we may form some idea of . . . connection of those two countries."[43] Probably writing in early 1838, he explained that "with [the] belief of ~~change~~ transmutation & geographical grouping we are led to endeavour to discover causes of change . . . change of species does not measure time but physical changes."[44]

Thus Darwin's bold paper on reefs was his first public allusion to the possibility of developing an evolutionary theory. Keep in mind just how high he had claimed the stakes were in understanding coral reefs; it will help make sense of the anxiety and paralysis Darwin seems to have felt when he faced the task of elaborating and supporting all the speculations he had offered that evening. What startling ambition this had been for a twenty-eight-year-old who was freshly arrived from a circumnavigation: to propose solving two of the biggest questions in all of science, the structure of the earth and the origin of species . . . and to do so by studying coral reefs.

Darwin closed his address with a sentence that simultaneously reminded listeners of his own field experience and implied that self-gathered empirical evidence was the source of his allegiance to Lyell's geological program: "The traveller who is an eyewitness of some great and overwhelming earthquake, at one moment of time loses all former associations of the land being the type [epitome] of solidity, so will the geologist, if he believe in these oscillations of level (the deeply-seated origin of which is betrayed by their forms and vast dimensions), perhaps be more deeply impressed with the never-ceasing mutability of the crust of this our World."[45] When this sentence appeared in an abstract in the Geological Society's proceedings, it was rendered in an even more recognizably Lyellian idiom, referring to these crustal modifications as "the endless cycle of changes."[46]

From start to finish, then, Darwin's first public account of his coral reef theory was consciously styled to do homage to Lyell's principles of geological work and to stand as an extension of Lyell's specific views on the causes and consequences of geological change. The organization of the paper mirrored that of Lyell's *Principles* in proceeding from a catalog of present-day organic and inorganic processes to speculative geological reasoning. And although Darwin contradicted the crater theory that Lyell had championed, he went out of his way to explain why that theory had formerly made sense. He even suggested that the crater theory was rendered implausible only by his own recent determination of corals' depth limit (even though, by this logic, the narrower depth limit proposed by his predecessors Quoy and Gaimard made the theory even more implausible). He referred to Lyell in connection with the likelihood of Pacific subsidence and in discussing the recent elevation around the Red Sea, along with praising his "admirable laws" of organic change. Of the eighteen sources of information named in the paper, Lyell was the only one who had never seen a living reef.[47] Even more significant than these explicit references was the undisguised and unapologetic insistence on seeing and describing the world in Lyell's terms, from the "symmetry" and "astonishing uniformity" of ongoing crustal motion to the unceasing "cycle of changes" that produced it. Darwin's paper delivered more, therefore, than a theory of coral reef formation and a road map to its implications. It was also a partisan statement of allegiance to Lyell and a declaration that in the territorial game of geology, the Pacific and Indian Oceans must now be tinted in "uniformitarian" colors.

It is impossible to know the full extent to which Lyell helped Darwin compose the 1837 coral reef paper. It is clear, as I described above, that he had previously offered editorial help on what was to be Darwin's first paper for the Geological Society, on the elevation of Chile. Lyell's January letter to Charles Babbage about "working so hard [on] Darwin's [Chile] paper" and trumpeting Darwin's "new views on Coral reefs" suggests that the coral reef paper too was being developed collaboratively.[48] Darwin's kid-glove treatment of the crater theory and his gratuitous references to the *Principles* hardly require explanation beyond the fact that Lyell was his new social and professional patron. But there were more subtle changes too in Darwin's presentation of the theory that made it more sympathetic to Lyell.

Noticing these shifts in Lyell's favor depends on recognizing how Darwin's accumulation of new evidence about the kinds and distribution of reefs had forced him to change exactly what he was arguing. The 1835 essay was limited geographically to the Pacific, whereas Darwin's 1837 scheme extended the analysis of reefs and the classification of vertical movement clear across the Indian Ocean, encompassing, as he proclaimed, "more than a hemisphere."[49] In 1835, moreover, he had held the view that almost all Pacific reefs were of the encircling or lagoon-island varieties, and he implied that the "great extent of the Northern and Southern Pacific" was subsiding essentially as a single unit, in a way that "compensat[ed]" for the "general horizontal uplifting" of the "greater part of S. America." In 1837, having spent much more time examining other travelers' charts and accounts, he offered a more precise description of the Pacific as a zone that contained multiple alternating regions of subsidence and elevation. Describing these areas as "symmetrical," however, saved the argument that elevation and subsidence were *compensatory* crustal motions of the sort he had initially envisioned, and of the sort that underlay much of Lyell's conjectured physical and organic history of the earth.[50]

There was also a noticeable shift in the way Darwin characterized the areas that had sunk. Despite now claiming that subsidence affected discrete bands of the Pacific Ocean rather than the entire basin, he flirted in 1837 with the new idea that entire continents had been submerged. In the 1835 essay he stated that groups of lagoon islands indicated the presence of "a chain of Mountains [that] had there subsided."[51] In the 1836 diary entry written when leaving South Keeling, he imagined an "island" subsiding until only the coral reef "mark[ed] the spot where a former land lies buried."[52] Later in the voyage, or shortly after arriving home, Darwin began to contemplate the possibility that there had been a "continent of which Tahiti was a peak."[53] Such thinking would have appealed best to Lyell, for whom the richest potential payoff of Darwin's coral reef theory would be the demonstration of large-scale revolutions in geological history, of entire cycles in which the earth had been "remodeled" by the replacement of oceans with continents and vice versa.[54] As his letter to Herschel had phrased it, "coral islands, are the last efforts of drowning *continents* to lift their heads above water."[55] In the Geological Society paper, as we have seen, Darwin also began to speak of "continental subsidences" and the "former existence of an archipelago or continent." Drowned continents and drowned individual, or grouped, islands were not mutually exclusive, of course. The highest peaks of a sinking continent would necessarily become separate islands before they disappeared altogether. However, it was only once he started working side by side with Lyell that Darwin stopped defining subsided land as part of an oceanscape of former islands and began to speak of the ocean itself as an impermanent feature of the earth's surface.

An Astonished Response from the Geological Elite

With its compelling new solution to the problem of reef formation and its audacious theoretical conclusions, Darwin's paper struck the society's insiders like a blow from a geological hammer.[56] Discussions at the Geological Society were notoriously lively, so their contents were closely guarded. It was strictly forbidden to publish the debates that followed the reading of a paper. On this foundation of formal privacy lay a universally acknowledged strength of the society, that the meetings included truly substantive discussion. As Roderick Impey Murchison said in his presidential address of 1832, "The ordeal . . . our writings have to pass through in the animating discussion . . . within these walls, may be considered the true safeguard of our scientific reputation."[57]

We would not know about the immediate impact of Lyell's grand gesture of surrender had Herschel not been away charting the skies of the Southern Hemisphere. It is thanks to the combination of letters he was sent before the meeting (from Lyell) and after (from Murchison) that Lyell's scheme comes into view. In his letter Murchison exclaimed, "Last Wednesday's Geological [Society meeting] brought with it a paper from Mr C Darwin . . . which astonished us all." Murchison's words made it clear that Darwin had not previously established himself as an insider, referring to the paper's author as "this Darwin" and "Capt[ain] Fitzroy's friend, a naturalist." Yet based on this latest paper, Murchison declared the modestly accomplished traveler an "immense addition" to the geological community. Why? Because Lyell's capitulation seemed decisive to members of a group that was wholly unaccustomed to seeing him yield on matters of theory.[58] As Murchison told Herschel, "This Darwin is an immense addition to our stores, & so Lyell thinks, for he abjured on the spot all his dear theory on this subject." Murchison was not simply lauding the arrival of this particular new theory, he was acknowledging that Darwin turned out to be someone whose ideas deserved to be taken seriously.

To be clear, there were people whose ideas were *not* taken seriously by men like Murchison and Lyell even though they were universally acknowledged to be legitimate and important participants in the overall enterprise of the sciences. Such individuals might be authorized to testify to matters of fact about the fauna or flora of a place they knew well, to collect specimens, or to make and operate instruments, but they were not warranted to interpret broader classes of facts. Darwin's undergraduate mentor, the Cambridge botany professor John Stevens Henslow, had attributed just such status to Darwin when making him aware of the opportunity to join the *Beagle* voyage. Henslow explained that he had recommended Darwin for the position "not on the supposition of y[our] being a finished Naturalist, but as amply qualified for collecting, observing, and noting any thing worthy to be noted in Natural History."[59] This was an allusion to the perceived division of labor between collectors and describers and the philosophical naturalists who might perform such tasks but could also theorize. Before leaving for the voyage Darwin had no reputation as a philosophical naturalist; when he returned it was still an open question whether he deserved one. The events of this evening helped establish for the members of the Geological Society that Darwin was more than a mere collector of facts.[60]

Although it might seem that Darwin had triumphed at Lyell's expense, the opposite is true. He had in fact triumphed to Lyell's credit. We can begin to understand how this could be by looking further at Murchison's letter to Herschel. He described Darwin's topic as "the formation of coral reefs" and credited him with having "proved, that they were always formed by subsidence of continents & islands, & never upon elevating points of volcanic & other rocks as had been supposed." Here Murchison was drawing a contrast between Darwin's and Lyell's respective explanations of the conditions under which certain kinds of reefs might form. But Murchison reported that Darwin had gone beyond explaining reefs for their own sake; he had "summed up by laying it down, that all the land of the Corallian Seas was sinking [and] all that of S[outh] America rising; the one compensating the other." This move was the key to making the paper a victory for Lyell as well as for Darwin. In the context of British geology in the year 1837, claiming that large sections of the earth's crust were moving gradually in opposite vertical directions was a direct endorsement of

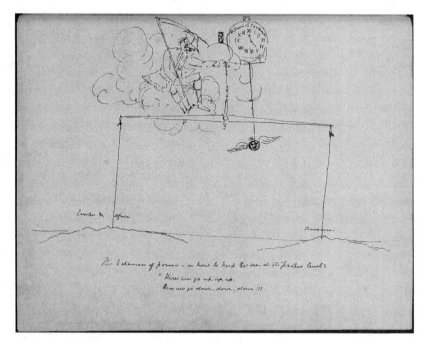

Figure 23. "The balance of power—or how to keep the sea at its proper level." This cartoon by Henry De la Beche caricatures Lyell's profligate use of vast geological time in his theories and his idea of compensatory vertical movement of continents. A figure representing time and wearing tinted glasses (cf. previous figure) holds a balance with Europe and Africa on one side and America on the other. The clock above ticks out "Millions of Centuries." When Darwin argued on his return from the *Beagle* voyage that the floor of the Pacific Ocean had sunk reciprocally with the elevation of South America, his geological audience understood him to be endorsing Lyell's geological system. De la Beche notebook, British Geological Survey.

the conclusions of Lyell's *Principles of Geology*. Such compensatory oscillations were, indeed, such a distinctive feature of Lyell's geological system that they too had been caricatured by De la Beche (see fig. 23).

Murchison seems not to have realized that Lyell was already deeply familiar with Darwin's coral reef theory before that evening, and he betrayed no knowledge of Lyell's active role in reshaping Darwin's paper and choreographing the events at the meeting. Sulivan's recollection of the evening corroborates the impression given by Murchison's account that Lyell's conversion appeared spontaneous, for he wrote how unexpected it was to hear Lyell "[giving] in his adhesion to [Darwin's] views at once."[61] Knowing that Lyell was the foremost advocate of the "monstrous hypothesis" that the *Beagle* shipmates had sought to undermine at South Keeling, Sulivan perhaps expected a more adversarial response from the author of the *Principles*. Thus the young officer's overwhelming memory of the meeting had been, as he told Darwin, witnessing "your views being *so readily* accepted."[62]

The other crucial letter to Herschel makes it clear that Lyell had indeed premeditated the entire affair. Just the previous week he had written to tell the great astronomer how "very full of Darwin's new theory of coral islands" he was and to declare that he had "urged" Whewell "to make [Darwin] read it at our next meeting."[63] The scheduling of the paper was key, because 31 May was the lone remaining gathering of the Geological Society before Lyell was to leave Britain for a summer of field research.[64] And he knew before arriving at the society's rooms in Somerset House that he would recant his own coral reef theory after Darwin had finished reading a paper they had actually prepared together.

Why would Lyell recant so dramatically? Because it allowed him at a single stroke to establish his young ally as a formidable theorist and to enroll the new theory in support of views that were inextricably linked to Lyell's name. It was no coincidence that the paper seemed to support Lyell's broader agenda, or that Darwin's coral reef theory appeared "more Lyellian than Lyell's." On the contrary, this was the intended message of a presentation fashioned intentionally to be mutually beneficial to the aspiring newcomer and the established man. Authorship and credit did not have to be zero-sum games.

The one person who was not surprised to hear Lyell's hearty endorsement was Darwin himself. Three days before the meeting, Darwin wrote in gay spirits to tell Henslow what would happen when he shared his coral reef theory with the Geological Society. "I am going to read a short account of my views of the whole affair, and Lyell I believe intends giving up the crater doctrine.—so that I am just at present full of interest on the subject."[65] Lyell's plan was successful not only at burnishing Darwin's reputation in the eyes of the society's most influential members, but also at raising Darwin's own estimation of his prospects. "I have read some short papers to the geological Soc, & they were favourably received by the great guns," he wrote to his cousin William Darwin Fox a month later, "& this gives me much confidence, & I hope not a very great deal of vanity; though I confess I feel too often like a peacock admiring his tail." Darwin felt a genuine thrill at having been taken seriously by the great guns, telling Fox "I never expected that my geology would ever have been worth the consideration of such men, as Lyell." The steps Darwin knew Lyell had taken to improve his odds of a good reception at the society seemed themselves a mark of the esteem in which Lyell held him. As Darwin explained it to Fox in a modest understatement, Lyell "has been to me, since my return a most *active* friend."[66]

Darwin's Emergence as a Practitioner of Lyellian Geological Speculation

A curious aspect of the history of Darwin's 1837 coral reef paper is that, for reasons I will explain below, it remained unpublished (except as an abstract) for more than two years. Much more rapidly, however, two important figures in the Geological Society used their own works to describe both the content of the paper and the style of theorizing it represented. The first instance was an extensive discussion that John Phillips managed to include in his 1837 *Treatise on Geology*; the second was William Whewell's year-in-review anniversary address to the Geological Society. The text of both these descriptions indicates that Darwin's grand claims had succeeded in drawing attention both to himself and to Lyell.

In terms of output, Phillips was a promoter of geology to rival Lyell. In the year 1837 alone he published the first half of a two-volume *Treatise on Geology* and a separate, similarly titled single-volume work, *A Treatise on Geology*. The latter doubled as the entry on geology in the seventh edition of *Encyclopaedia Britannica*, while the multivolume *Treatise* was part of Dionysius Lardner's "Cabinet Cyclopaedia" series, in which Herschel's *Preliminary Discourse* had appeared seven years earlier. Darwin made his appearance in the Lardner volume in the section titled "Marine Deposits in Progress," which Phillips introduced with a reflection reminiscent of Lyell's reverie on the amphibious being. "We cannot help feeling regret," Phillips began, "at the limited means which man possesses of penetrating the great deep, and watching the phenomena which happen on its quiet bed." The hypothetical opportunities Phillips described bore a striking resemblance to those Darwin received from the hydrographers. "There [on the sea floor] we should behold, it is probable, a number of circumstances connected with the life of marine Mollusca, radiaria [*sic*], crustacea, fishes, which would throw quite a new light on many of the problems of old geology; inform us of the probable depths, distance from the shore, and river mouths, and other conditions, most important for us to know in

Figure 24. These engravings of Lyell (left) and Darwin from the same year, 1849, illustrate their relative ages. Lyell, born in November 1797, was just over eleven years older than Darwin. National Library of Medicine.

constructing trustworthy inferences regarding the formation of the fossiliferous rocks."[67]

Drawing on his personal "notes taken during the reading of Mr. Darwin's paper to the Geol. Society," Phillips devoted several pages to the new theory and its implications. Notably, Phillips echoed Darwin's stated direction of reasoning, writing that "Mr. Darwin has recently been conducted, by a consideration of the structure of coral islands . . . to a remarkable general speculation . . . [revealing] long narrow spaces of ocean, in which the land has undergone and is still suffering gradual depressions."[68] Phillips's text contained two original woodcuts to illustrate the subsidence theory in action. These were, to my knowledge, the first visual depictions ever published of Darwin's new explanation for the origin of barrier reefs and lagoon islands.

Phillips listed a series of "very important points" that emerged as a consequence of Darwin's study of coral reefs. His manner of describing them indicates one practicing geologist's perception of the relation between observing, generalizing, and speculating. He described "Darwin's investigations" as yielding, first, a pair of generalizations: "1. That linear spaces of great extent in the equatorial regions are undergoing movements of an astonishing uniformity, and that the bands of elevation and subsidence alternate. 2. From an extended examination, the points of volcanic eruption all fall on areas of elevation." He continued, "The importance of this [latter] *law* is evident, as affording some means of *speculating*, wherever volcanic rocks occur, on the changes of level even during ancient geological periods."[69] Two further steps of speculation were also possible, both of which again had been stated by Darwin himself: "3. Certain coral formations acting as monuments over subsided land, the geographical distribution of organic beings is elucidated by the discovery of former centres, whence the germs could be disseminated. 4. Some degree of light might thus be thrown on the question, whether certain groups of living beings peculiar to small spots are the remnants of a former large population, or a new one springing into existence."[70] It is striking that Phillips reported Darwin's speculations on the distribution and possible origin of animal and plant populations but did not comment on his similarly bold conclusion that the study of coral reefs might give insight on the internal composition of the globe.

Not long after Phillips's treatise was published, William Whewell used his February 1838 anniversary address to the Geological Society to explain how Darwin's coral reef paper factored into the geological achievements of the previous year.[71] Whewell's remarks confirmed how successful Lyell and Darwin had been in their efforts to advance Darwin's candidacy as a legitimate geological theorist and, moreover, to marry his theories to Lyell's established theoretical framework in a way that would bring credit to them both.

Ever the analyst of the practice of science itself, Whewell liked to distinguish two enterprises within the study of geology. These were "descriptive geology," which was the practice of "[cataloging] the strata and other features of the earth's surface as they now exist," and "geological dynamics," the science of "examining and reducing to law the causes which may have produced such phaenomena."[72] He likened the contemporary state of geology to the science of astronomy in Kepler's time, with true theory beginning to emerge out of a "vast store of facts of observation."[73] The eventual goal of geology, as Whewell saw it, was to develop a fully mathematized science of physical geology, by which terrestrial processes would be reduced to the orderliness of celestial mechanics. "There can be no doubt," he admitted, "that the greater part of us shall be more usefully employed in endeavouring to add to the stores of descriptive geology, than in [the] abstruse and difficult investigations [of geological dynamics]."[74] Among the year's contributions to descriptive geology he cited Darwin's first two papers to the society, on South American elevation and the extinct Mammalia, which led Whewell to remark that "I cannot help considering his voyage round the world as one of the most important events for geology which has occurred for many years."[75] Indeed, he announced that the society had awarded its Wollaston Medal to Owen "for his general services to Fossil Zoology, and especially for his labours employed upon the fossil mammalia collected by Mr. Darwin."[76]

Whewell believed Darwin's most impressive achievement lay not in the descriptive realm, however, but in the realm of causal explanation demanded by geological dynamics. Whewell listed gradual, long-term elevation and subsidence of the crust as principal examples of the "proximate causes" of geological phenomena. He imagined that the future of geological theory lay in the mathematical analysis of "ulterior causes," which might be construed as the "subterraneous machinery" that drove the proximate causes "by which islands and continents appear and vanish in the great drama of the world's physical history."[77] Referring "especially to his views respecting the history of coral isles," Whewell credited Darwin with illuminating the proximate "class of events, its evidence, extent, and consequence . . . with a clearness and force which has, I think I may say, filled all of us with admiration."[78]

Whewell cited Grant, Henslow, and Sedgwick by name for their roles as Darwin's instructors, but he reserved special praise for Lyell despite having been a thoughtful but sharp critic of the system-building of the *Principles*.[79] He implied that it was not Lyell's treatment of the subject matter of coral reefs that had led Darwin forward. Rather, Darwin's advances were a product of his adherence to Lyell's geological method, and the result was that his insights were applicable to geological theory making in the broadest sense. "Guided by the principles which he learned from my distinguished predecessor in this chair," Whewell judged, "Mr. Darwin has presented this subject under an aspect which cannot but have the most powerful influence on the speculations concerning the history of our globe, to which you, gentlemen, may hereafter be led."[80] Although it is difficult to judge from the written word, the connection Whewell drew between the Lyellian approach to geology and the act of "speculation" may have come across as a subtle censure of master and student. Although the president continued by praising "the large and philosophical views" expressed at the close of Darwin's paper, wherein he had gestured toward "the laws of change of climate, of diffusion, duration and extinction of species, and other great problems of our science which this voyage has suggested," Whewell implied that the foundation of these speculations was not yet secure. Instead, these tantalizing comments led the president to "look with impatience to the period when this portion of the results of Captain Fitz Roy's voyage shall be published, as the scientific world in general looks eagerly for the whole record of that important expedition."[81] In other words, he awaited Darwin's geological *book*.

Darwin's coral reef paper and the acclaim it earned him marked the public high point of his first year back in England. By demonstrating in the most audacious manner possible what was at stake in Darwin's understanding of reef formation, it had served as an advertisement for the geological book Darwin intended to write. Whewell's address made it clear that the paper also brought attention to Lyell's geological principles, both by illustrating their pedagogical value in guiding Darwin's reasoning and by reinforcing the evidentiary basis for Lyell's largest speculations. This course of action ensured that Darwin's coral reef work, including his views on the habits and distribution of coral animals, would be identified as a geological undertaking. In the process he had hinted at many of his boldest private speculations. Lyell had catalyzed all the elements of what was, for the moment, a great triumph. He had identified Darwin as a potential supporter, helped to raise his status, and reframed his work as a contribution to Lyellian geology. In the process he had in effect co-opted the *Beagle* voyage as an event whose major scientific impact was for the moment being felt in the science of geology.

Developing and publicizing the coral reef theory this way, however, created expectations that began to trouble Darwin acutely in the coming years. Through the long process of gathering and interpreting data for a more extended publication on coral reefs, Darwin began to recoil. As I argue in the next chapter, he was to renounce the provocative tone and ambitious theorizing that had seemed so attractive and come so easily in 1837, and in the process he repositioned himself against the very model of scientific authorship that Lyell had persuaded him to embrace.

8

Burned by Success

In the wake of his coral reef theory's audacious debut, Darwin began to behave all the more like an elite member of the geological community. He became Lyell's fully trusted deputy, acting as his authorized geological spokesman in various settings, and he began to contribute to the tacit policing of boundaries between nonspecialists and geologists of the kind he had become. He drew more deeply from his *Beagle* work to unleash another bold paper at the Geological Society, and he made a sizzling debut at the Royal Society by showing how the geological perspective he had acquired during the voyage could be used to reinterpret one of the best-known puzzles of British geology.

In these successes Darwin was sowing the seeds of his future anxieties, for they added pressure on him to demonstrate that he could produce the kinds of fact-filled geological works most admired by British geologists. By mid-1839 the foundations of Darwin's self-confidence had begun to crumble. The good fortune of achieving notoriety so soon after the voyage turned into an accursed obligation to be judged according to the bold claims he had made. He learned that he did not relish scholarly conflict the way Lyell did, and in turn he began to realize that Lyell's approaches to authorship and theorizing were not models he could bear to emulate. And he underestimated the time it would take to write a geological book, with his delay adding another layer of pressure. A different burden was imposed by Lyell himself, who urged Darwin to finish writing his book of *Beagle* geology and in the meantime did not hesitate to incorporate Darwin's unpublished material into his own books at a dizzying rate.

A handful of events gave Darwin pause to second-guess his powers of theorizing, but far more important was the larger confluence of circumstances that made him revise his notion of the ideal authorial persona. By a quirk of publishing, his *Journal* of the voyage remained unpublished for two years, so that the audacious 1837 version of the coral reef theory he had included in the text emerged afresh just at the time when he was primed to doubt whether such bold speculations were consistent with the reputation he wished to cultivate.

FitzRoy was the cause of this particular delay, for Darwin's journal was to be published along with FitzRoy's as a volume of the official narrative of the *Beagle* voyage. While awaiting FitzRoy's volume Darwin in fact had to withdraw the coral reef paper from consideration for full publication by the Geological Society in order to avoid preempting the official record of the voyage.[1] And of course it delayed the affirming experience of seeing his narrative of the voyage released for sale. As he told Lyell in September 1838 on learning that Lyell's father had been shown some pages of the unreleased narrative, "I need not say how pleased I am to hear that Mr Lyell likes my Journal—to hear such tidings is a kind of resurrection, for I feel toward my first born child [the unpublished book manuscript], as if it had long since been dead, buried & forgotten."[2]

Meanwhile, despite being extraordinarily productive across a range of projects, Darwin found he was unable to keep pace on his chief task, the geology book. In retrospect this has been considered the most creative and fruitful period in his life, for he was publicly establishing himself at the center of the scientific community and privately developing the theory for which he is best remembered now. But at the time he felt he was falling short of his self-imposed expectations. Rather than considering his hours "collecting notes" on species to have been wisely spent, he noted these occasions in his journal entries as days "frittered away." As psychologist John Bowlby wrote in his retrospective study of Darwin's anxiety, "For him the word 'work' applies only to pushing on doggedly with whichever publication he is currently giving priority to. All the other activities of a studious scholar are discounted."[3] Moreover, Darwin calibrated his own feelings of physical health and well-being according to the amount of work he had accomplished on his geological publications.

Plate 1. A sounding lead (pronounced *led*, like the metal it is made from) and line. This hydrographic tool became crucial to Darwin's work as a naturalist. A "hand lead" like this one was eight to twelve inches tall and weighed six to twelve pounds. Deep-sea leads were even heavier and broader. The base of all sounding leads featured a depression known as a "well," which would be "armed" with soft tallow or wax. By this method a sample from the seafloor (or an impression of the bottom if it was clean) could be recovered. During the *Beagle* voyage Darwin made innovative use of his shipmates' soundings as a method for acquiring geological, zoological, and botanical specimens from the seafloor. The epistemic value of these specimens was enhanced by data about the depth and geographical location where they had been found, which was appended to them as an intrinsic feature of the surveyors' work. Image, Mariners' Museum, Newport News, Virginia.

Plate 2. The broad, gently inclined valley of the Aconcagua River (Chile) where it emerges from the dramatic western slopes of the Andes. Darwin concluded that the mountainous scene before him had been "marine with Islands" (the islands being the peaks of the Andes) before the former seafloor had been uplifted into dry land. Photograph by the author.

Plate 4. Modern-day view of Takapoto atoll, one of the ring-shaped reefs of the "Low or Dangerous Archipelago" (the Tuamotu Archipelago in French Polynesia), showing from foreground to background (1) the open Pacific Ocean, (2) an islet atop the reef, (3) the water of the lagoon, and (4) islets on the far side of the reef. During the *Beagle* voyage Darwin followed convention by referring to such reefs as "low islands" or "lagoon islands." He was later responsible for the widespread adoption of the Maldivian term *atoll* to describe ring-shaped reefs wherever they were found. Photograph by the author.

Plate 3. "On the road [from Mendoza] there is a very extraordinary view . . . white, red, purple & green . . . sedimentary rocks & black lavas; these strata are broken up by hills of Porphyry of every shade of Brown & bright Lilacs. All together they were the first mountains which I had seen which literally resembled a coloured Geological section." Top: Darwin's watercolored cross section of the continent of South America. Middle: Detail of the cross section above. Bottom: View along the present road from Mendoza toward the Uspallata Pass. DAR 44. Reproduced by kind permission of the Syndics of Cambridge University Library. Landscape photograph by the author.

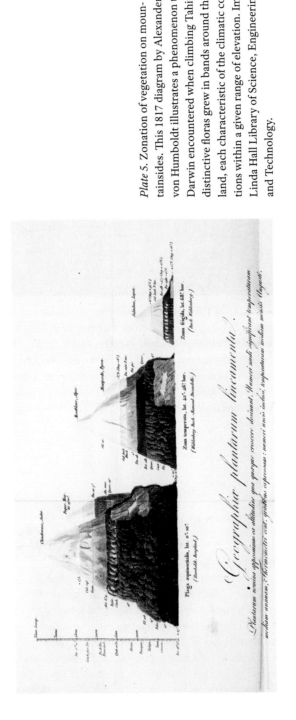

Plate 5. Zonation of vegetation on mountainsides. This 1817 diagram by Alexander von Humboldt illustrates a phenomenon that Darwin encountered when climbing Tahiti: distinctive floras grew in bands around the island, each characteristic of the climatic conditions within a given range of elevation. Image, Linda Hall Library of Science, Engineering, and Technology.

Plate 6. Vegetation at three elevations on Tahiti (from bottom to top, approximately 500 feet, 1,500 feet, and 3,000 feet), illustrating that while climbing Darwin passed through zones like those depicted on the Humboldt diagram. Photographs by the author.

Plate 7. A view of the island of Eimeo (Moorea) as it appears from the neighboring island of Tahiti at an elevation of approximately 2,500 feet. Darwin wrote in his 1835 "Coral Islands" manuscript that seeing the barrier reef encircling Eimeo during his ascent of a Tahitian ridge had sparked his theory of coral reef formation by making him realize that ring-shaped coral reefs could grow up on the circumference of subsiding islands. For viewers of this photo, as for Darwin, the ability to interpret the panorama of distant Eimeo (that it was a mountainous island surrounded, in turn, by turquoise lagoon water, an encircling barrier reef, and darker open ocean water) is made easier because Tahiti and Eimeo share similar geography, meaning that the parallel sequence is visible in greater detail in the foreground and middle distance. Photograph by the author.

Plate 9. Waves meet the living bulwark on the leeward (western) side of South Keeling atoll. Photograph by the author.

Plate 8. Pole leaping at South Keeling atoll in August 2016. Photograph by Ash James.

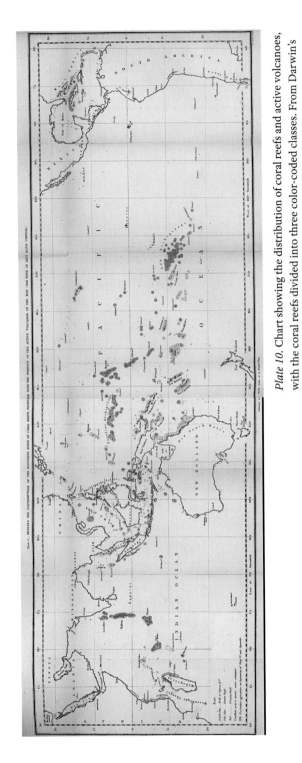

Plate 10. Chart showing the distribution of coral reefs and active volcanoes, with the coral reefs divided into three color-coded classes. From Darwin's *The Structure and Distribution of Coral Reefs* (1842). Image, nasa.gov.

Darwin's New Persona

After the meeting where Darwin presented his coral reef paper, Lyell continued to promote him but also to cultivate him as a semiofficial spokesman for his proprietary method of geological theorizing. While he was away that summer on a field trip to Scandinavia and Germany, for example, Lyell briefed Darwin on how he should respond to criticism that might arise: not criticism of Darwin, but criticism leveled *at Lyell* during the upcoming meeting of the BAAS.[4]

Before leaving the country Lyell also arranged for "the Mr. Darwin who has studied coral reefs so much & who read so splendid a paper on them at the G.S. on Wed[nesday]" to meet missionary John Williams at the Geological Society's Somerset House rooms.[5] Williams, who returned in 1834 from eighteen years in the Pacific, had published his *Narrative of Missionary Enterprises in the South Sea Islands* just weeks before, in April 1837.[6] This work contained an extended geological taxonomy of island types and a genuinely detailed and well-informed argument *against* the "received opinion" that low islands were formed by the growth of corals. Williams pointed out that upraised lagoon islands like Henderson's Island and Mangaia, which consisted of reef-shaped limestone formations that towered up to three hundred feet above the ocean's surface, could not originally have been produced by the polyps that were agreed by "scientific authorities" to inhabit a vertical range no greater than thirty feet. "The inference to be drawn from this," he asserted, "is, that [either] the [coral] insects do exist in greater depths than are now assigned to them, or that these solid masses are not the effect of their labour: the one or the other must be the case." Darwin's recent paper, of course, offered the compelling third possibility that greater thicknesses of coral had accumulated while their foundation was sinking in relation to sea level, and it is easy to imagine him seizing on Williams's apparent paradox as nothing less than an independent confirmation of his own theory that such reefs formed by subsidence.[7] Nor is it difficult to imagine Lyell's amusement at the idea of sending his young bulldog to meet Williams, whose book ridiculed "Lyell['s] reasoning" on the rate of coral growth, which would have required the fantastic sum of "fifty or sixty thousand [years]" to form Mangaia, "and only that portion of [it] which appears above the water!" At Somerset House they would be able to consult Darwin's coral specimens and the sectional reef diagrams that had accompanied his paper, which all remained on display.[8] In the end it proved to be a profitable meeting for Darwin. Although he never engaged with Williams's theoretical reasoning in print, the missionary became one of Darwin's main sources of descriptive data on the geography of the region and of information about the natives' histories of their islands and reefs. Indeed, Darwin would go on to cite either the *Narrative* or personal communication with Williams twenty-two distinct times in his 1842 book.

This episode with Williams illustrates the stratified social relations within British geology in the 1830s. Lyell, as the most recent president of the Geological Society, a former professor of the subject at King's College, London, and the author of an important treatise, along with Sedgwick, Murchison, De la Beche, and two or three others at this time, could presume to interpret geological phenomena that lay outside his own field experience and to arbitrate, along with his mutually recognized elites, controversies over theory or method.[9] As historian Martin Rudwick has pointed out, these were the men, among those Darwin referred to as the "great guns" of science, whose scientific pursuits were conspicuously and almost exclusively geological. Thus individuals like John Herschel and William Whewell, who contributed to geological debate by way of their primary expertise in other sciences, lay outside this exclusive group.[10] Other members of the Geological Society possessed deep but narrow expertise in particular geological formations or geographic regions. Beyond them were the gentlemen who formed the bulk of the society's membership and of the science as a whole, true amateurs of geology who might be relied on for facts by the men of higher scientific status. John Williams lay somewhere on the outer margin of this group, formally registering his interest in contributing to geology by donating a copy of his book to the society's library.[11]

Even though Williams had infinitely more personal experience of coral reefs than Lyell did, however, and though he had read and deployed the same authoritative texts quoted in the *Principles*, Williams was effectively blocked from contributing to theoretical debates. In print, Darwin simply ignored the fact that Williams had offered any analytical ideas at all about coral reefs. Yet Williams's status as a geological *observer* was entirely secure, and Darwin not only relied on him for geographical descriptions but also evidently trusted his judgment on certain questions of lithology.[12] Far from being paradoxical, I see this treatment as a highly conventionalized way of reinforcing relatively new social and literary boundaries. Williams's *Narrative* was part of a genre that also included the *Polynesian Researches* of his missionary colleague William Ellis, which likewise became a source for Darwin's 1842 book on coral reefs.[13] The historian Sujit Sivasundaram has argued that Darwin's use of Williams's observations, among other things, indicates that Williams ought to be remembered as a "man of science."[14] But from the perspective of professed men of science like Darwin and Lyell, Williams held a different status. By tacitly disqualifying Williams and Ellis from the realm of geological theory, Darwin reinforced his own affinities with the geological elite.[15] Williams may have geologized, but he was not acknowledged to be a genuinely philosophical geologist.

This social terrain of British geology shaped the strategy Lyell and Darwin had deployed with the reading of his coral reef paper.[16] Through the course of the voyage, and on the foundation of his training with Sedgwick, Darwin had established himself as a sound observer whose interpretation of the South American strata deserved to be taken seriously. On his return to England, however, Lyell seized the opportunity to chaperone him into the science's elite, making himself, as Darwin had told Fox, "a most active friend."[17] The coral reef paper was Darwin's first attempt to contest matters of general theory, and as we have seen, it was very carefully managed. By making such brash forays into geological theory, Darwin was unmistakably announcing his candidacy for entry into the highest level of philosophical geology. However, being accepted into the geological elite was a process of conscious and conspicuous specialization. This helps explain why Darwin presented his zoological and hydrographic investigations as precursors to his geological arguments rather than as ends in themselves.

As Darwin ascended well above his former station as a traveling naturalist, FitzRoy blamed Lyell for driving a wedge between his protégé and those who had helped him during the voyage. FitzRoy's ire was raised by what he considered scant acknowledgment of the *Beagle*'s officers in prefatory text Darwin had drafted for his publications on the voyage. In an indignant letter he accused Darwin of failing to "listen to the dictates of [your own] heart" and instead acting in adherence to "the partial views and perhaps selfish feelings of persons who neither know, nor feel for, you—or for me—as your Father would feel for either of us." FitzRoy expressed "astonish[ment] at the total omission of any notice of the officers" in the proposed prefaces, reminding Darwin that people "who know anything of the subject" would not forget "how much the Officers furthered your views . . . (especially Sulivan, Usborne, Bynoe and Stokes)." In addition to all the aid I described in part I above, the hydrographers had (as FitzRoy reminded him) given Darwin prime specimens to the detriment of their own natural history collections and in a variety of ways had "held the ladder by which you mounted to a position where your industry, enterprise, and talent could be thoroughly demonstrated, and become useful to our countrymen, and, I may truly say, to the world."[18]

While FitzRoy initially referred only cryptically to having been "derided by a person I had thought your friend," he eventually fingered Lyell as the malefactor responsible for Darwin's betrayal of the officers. "Some time ago," he confessed, "it occurred to me that you had consulted with some person . . . who looked at the subject in a peculiar point of view—and I was informed yesterday, by a conversation with Mr. Lyell—that my conjecture was well founded." It appears that Lyell had, furthermore, berated FitzRoy for delaying the publication of Darwin's *Journal* by failing to complete the other volumes of the *Beagle* narrative. "He does not seem to consider," Fitz-Roy complained, "that the connection of your volume with mine . . . is one of feeling and fidelity—not of expediency." FitzRoy closed the letter with forked-tongued words of conciliation, asserting that "I esteem you far too highly to break off from you willingly," and ominously informing Darwin that he would contact "persons with whom I have conversed on this subject" in order to "remove from their minds any impression which you might wish should not remain."[19] Unsurprisingly, Darwin decided to amend his acknowledgment.[20]

With the *Journal* printed and in limbo, Darwin's focus had shifted to two other publishing projects to emerge from his time aboard the *Beagle*. The zoology of the voyage required him to act as a general editor because the actual descriptions of specimens were furnished by others. Publishing the geology of the *Beagle* voyage, on the other hand, would be the longest and most difficult writing task he had ever undertaken. Already the effort of finishing the journal had been a struggle. In the same letter in which he told Fox how much the favorable reception at the Geological Society had increased his confidence, he confessed to the difficulty of completing the journal and the daunting prospect of shifting to an even tougher book: "I shall always feel respect for every one who has written a book, let it be what it may, for I had no idea of the trouble, which trying to write common English could cost one. . . . [A]s soon as [correcting page proofs of the journal] is done I must put my shoulder to the wheel & commence at the geology."[21] Unlike the journal, which was based on the diary he had kept during the voyage, the geology would aim to synthesize a wide range of his field notes and speculative essays. And as a learned treatise rather than a narrative aimed at a wide audience of scholars and general readers, the geology would be judged by its comprehensiveness and reliability. Five years earlier, during the first year of the *Beagle* voyage, he had written to Fox that the thrill of making and testing geological hypotheses was like "the pleasure of gambling."[22] Now, with a reputation and a family to protect, he began to wilt at the thought of being considered too speculative.

At this stage he envisioned his geological treatise as a single volume despite its broad theoretical ambition, but by the beginning of 1838 he was canvassing Henslow for advice on dividing the material.[23] Shortly thereafter, the publishing house of Smith, Elder and Company advertised a forthcoming work by Darwin titled *Geological Observations on Volcanic Islands and Coral Formations*, indicating that he had abandoned for the moment all his work on South America, and with it his vision of a single, synthetic geological text.[24] Almost immediately, though, he began to find, "rather to [his] grief," that the projects were expanding beyond any schedule he had imagined.[25]

In mid-1838 he began withdrawing from social life because he was embarrassed that his book was not finished. With Lyell restlessly wondering when it would appear, Darwin feared running into him without good news to deliver.[26] "I am very much obliged to you for sending me cards for your parties," he told their mutual friend Charles Babbage, "but I am afraid of accepting them, for I should meet some people there, to whom I have sworn by all the saints in Heaven, I never go out, & should, therefore, be ashamed to meet them."[27] He told his childhood friend Charles Whitley, "Of the future I know nothing[.] I never look further ahead than two or three Chapters—for my life is now measured by volume, chapters, & sheets & has little to do with the sun."[28] Yet, as his correspondence of the following four years reveals, the book consistently remained three or four maddening months from completion.[29]

Darwin had a surplus of reasons for struggling to deliver. Whatever the cause, his health began to deteriorate in 1838, and he lost months at a time to illness. By the time his coral reef book appeared in 1842 he and Emma Wedgwood had married and he had become a father twice over. He was dividing his working hours between innumerable projects. Besides the writing on coral reefs and volcanic islands, he worked up some of his own animal specimens and superintended the publication of the *Zoology*, composed a preface and addenda to his *Journal*, made brief geological field trips, and wrote a paper on the parallel "roads" of Glen Roy, Scotland. Quietly he also began to accumulate enormous masses of data on the origin of species. He told Lyell in September 1838, "I have lately been sadly tempted to be idle, that is as far as pure geology is concerned, by the delightful number of new views, which have been coming in, thickly & steadily, on the classification & affinities & instincts of animals—bearing on the question of species—note book, after note book, has been filled."[30]

It is easy to imagine in retrospect not only that Darwin recognized the immense significance of his nascent work on species but also that he was comfortable redirecting his energies toward the topic we now take to have been the most imaginative and important work of his career. On the contrary, however, Darwin's private and public demeanor at this time was that of a man who considered himself to be misdirecting his energies. His prospective audience was awaiting (and would soon enough be actively demanding) the promised volume on geology. His inability to satisfy this public pressure contributed much more to Darwin's notorious descent into ill health and anxiety than did the distant prospect of publishing his exciting private work on species.

The Obligations of a Student to His Master

It is no coincidence that it was Lyell to whom Darwin made the confession that his species work was distracting from his schedule of geological writing. Lyell remained Darwin's constant taskmaster and intellectual mentor. He had more than one reason for wanting the young man to marshal his evidence and get it into print. He not only was keen to nurture his protégé in the difficult task of producing a book, he also continued to identify Darwin's coral reef work as an important part of his own larger program of research and publication. While Lyell was cultivating Darwin's career in 1836 and 1837 he was also working to produce a textbook, the *Elements of Geology*.[31] Whereas the *Principles* had focused on the present-day physical and organic *processes* that were shaping the earth's crust, the new book was a systematic description of the types of rocks and fossils that made up the geological record. After the *Elements* came out in 1838, he continued to work at revising his *Principles* for what would be the sixth edition of that theoretical treatise (1840). Like any master, he was eager to strengthen his own compositions by incorporating the best of his student's work.

Lyell's personal notebooks reveal that he was demanding more of Darwin in other ways as well. Immediately after the voyage Lyell's queries for Darwin almost always focused on places Darwin had visited or phenomena he had seen during the *Beagle* voyage. However, as their unequal partnership became better established, Lyell increasingly treated Darwin as something like a research student, imposing on (or assigning) him to find answers to scientific questions that Darwin was not intrinsically better equipped to answer than Lyell himself.

The way Lyell appropriated Darwin's work for the *Elements* is particularly fascinating because it illustrates both how readily the coral reef theory could be used to shed light on a wide range of puzzles and how quickly Lyell was able to adopt it. Three months before he and Darwin had ever even spoken about coral reefs, in October 1836, Lyell had already completed his first draft of the *Elements*. Yet by the time it was printed in 1838, the subsidence theory appeared in support of several themes Lyell developed in the text. In the preface he reported that he had consulted Darwin's fully printed, but not yet published, journal of the voyage (which, recall, contained the full text of Darwin's 1837 coral reef paper). Lyell introduced Darwin's coral reef theory when explaining how to interpret the presence at high altitudes of rocks that contain marine fossils, framing it as a corollary to Darwin's views on the elevation of South America.[32] But this only foreshadowed the use to which Lyell put Darwin's expertise on coral reefs.

As a textbook and manual for beginning geologists that was written in terms borrowed from the *Principles*, the *Elements* was meant to inculcate the lessons of Lyell's more overtly theoretical work among members of a different book-buying audience.[33] Over the course of more than a hundred pages in the second half of the *Elements*, Lyell defined and analyzed the Secondary beds of Europe in ascending order of age, from the Chalk to the Old Red Sandstone.[34] The Secondary deposits were a series of sedimentary formations that ranged in composition from the nearly pure carbonate of lime of the Chalk formation to the alternating clay, limestone, and sandstone of the Oolite and Lias groups. The group was staple fodder for early-Victorian geologists, most of whom had heavy appetites for stratigraphy.

The Chalk, a formation that could be found in deposits all over Europe, posed a thorny problem for Lyell.[35] According to the basic principle of Lyell's geological method, the origin of sedimentary rocks should be explained by analogy with present-day processes of deposition. Yet because of the Chalk's wide range and homogeneity, he explained, "geologists have often despaired of finding any analogous deposits of recent date."[36] In a paper read to the Geological Society in November 1837, Darwin had commented that he believed large portions of the Chalk might have been formed from the powdered coral produced by the feeding of reef fish and the boring of mollusks. (Incidentally, it generated another warning against being too speculative. In his referee's report to the Geological Society, William Buckland had advised Darwin to withdraw the passage because it "introduc[ed] a very disputable matter into a paper that is otherwise unexceptional, & which if establishd, would be well deserving to form the Subject of a separate Communication.")[37]

Lyell duly cited Darwin's description of the fine coral mud in the South Keeling lagoon, along with similar samples collected at Bermuda by navy lieutenant Richard Nelson, as a demonstration that deposits almost indistinguishable from the Chalk were still being laid down.[38] To draw the connection more firmly, Lyell pointed to a limestone in Denmark that contained fossils characteristic of the Chalk but that "consists of an aggregate of corals, retaining their forms as distinctly as the dead zoophytes which enter into the structure of reefs now growing in the sea."[39] Regarding the wide geographical extent of the Chalk, he drew on a recent private exchange with Darwin. Previously (in the first edition of the *Principles*), Lyell had claimed that the calcareous formations currently being laid down in the coral areas of the Pacific were "the most extensive of the groups of rocks which can be demonstrated to be now in progress."[40] His preferred explanation for the thousand-mile extent of the Chalk was a direct analogy with the present-day coral seas. But Darwin had warned him, "It will be difficult for you to talk of great areas abounding with corals.— People's ideas of the Pacific are *most false*.— In the thick archipelagoes—in a long days sail, you will often only see one or two islands."[41]

Although Darwin believed corals had built masses that achieved great vertical thickness during periods of subsidence, he knew from experience that the combined surface area of all the Pacific reefs paled into virtual insignificance compared with the vastness of the ocean. Heeding this warning, Lyell quoted almost verbatim (but without attribution) information from Darwin's letter about the geographical range of the great coral archipelagoes, and he acknowledged that "the islands in these spaces may be thinly sown."[42] But he preserved his analogy between modern coral formations and the Chalk by pointing out that there was no evidence that the Chalk had ever existed as a continuous deposit covering the whole vast region within which various Chalk outcrops might be found. There might rather have been only "patches of [Chalk], of various sizes, throughout the area," Lyell argued, just as the analogous deposits here and there being laid down around widely spread coral islands might be identical in character without being continuous in extent.[43]

In the climax of Lyell's disquisition on the Secondary strata, he used analogies with present-day physical and organic processes to reconstruct the history of Europe as it had been recorded in what are highly fossiliferous deposits. Lyell made Darwin's observations of living reefs, and his conclusions about the ongoing subterranean movements implied by reef shapes, integral to this impressive demonstration of reasoning from actual causes. To explain why extensive calcareous beds had been interstratified (appearing in an alternating vertical sequence) with thick deposits of what appeared to be shallow-water mud, Lyell invoked extended periods of slow elevation and subsidence that had borne and extinguished entire continents. In lighthearted anticipation of his critics, Lyell acknowledged that "we are half tempted to speculate on the former existence of the Atlantis of Plato." Turning serious, he asserted that "the story of the submergence of an ancient continent, however fabulous in history, may be true as a geological event."[44] Darwin's coral reef theory emerged to provide compelling evidence for this claim. "If we now endeavour to restore, in imagination, the ancient condition of the European area at the period of the Oolite and Lias," he explained, "we must conceive a sea in which the growth of coral reefs and shelly limestones, after proceeding without interruption for ages, was liable to be stopped suddenly by the deposition of clayey sediment." Alternations of elevation and subsidence, producing and denuding new continents, accounted for these changes in conditions. Thus, "In order to account for [a] great formation, like the Oxford clay . . . covering one of coral limestone, we must suppose a sinking down *like that which is now taking place in some existing regions of coral between Australia and South America* [until] the occurrence of subsidences, on so vast a scale . . . caused the bed of the ocean and the adjoining land throughout the European area, to assume a shape favourable to the deposition of another set of clayey strata."[45]

In making the observation I have italicized, Lyell credited Darwin's unreleased *Journal* in a note. But Darwin's work had not simply provided a modern analogue of prehistoric subsidence on the order of thousands of feet: his explanation of the limits on coral growth had dictated that such subsidence must be slow and uniform. This in turn helped Lyell explain why entire faunas had come into and passed from existence in the period when the Secondary deposits were laid down. "Both the ascending and descending movements may have been extremely slow, like those now going on in the Pacific; and the growth of every stratum of coral . . . may have required centuries for its completion, during which certain organic beings may have disappeared from the earth, and others have been introduced in their place; so that, in each set of strata, from the Upper Oolite to the Lias, some peculiar and characteristic fossils were embedded."[46]

Lyell gave the very first copy of the *Elements* to Darwin, whose reaction seems thick with subtext. "I read with much interest your sketch of the secondary deposits," he told the author. "You have contrived to make it quite 'juicy', as we used to say as children of a good story." There can be no doubt that reading the book was suspenseful indeed for Darwin, because it offered the chance to learn just how extensively Lyell might have cited the *Beagle* work. When Darwin discovered that Lyell had leaned heavily on his observations and theories at key points, it was a source of validation but also unease. After all, although Darwin's *Journal* had been printed by the publisher a year earlier, it remained unreleased because of FitzRoy's delay in completing the other volumes of the *Beagle* narrative. Now a broad readership would encounter Darwin's arguments not in his own words, but in Lyell's. Darwin's response to Lyell was ambiguously worded, suggesting at least a subconscious ambivalence about finding that Lyell had taken such an imperative: "I am in a fit of enthusiasm; & good cause I have to be, when I find, you have made such infinitely more use of my journal than I could have anticipated."[47]

In this statement, as in the continuing correspondence between Darwin and Lyell, I find evidence that Darwin was torn between delight that Lyell found his work so useful and terror on realizing that Lyell was chipping away at the potential novelty of Darwin's own as yet unpublished works. In exchange for the attention Lyell's widely read publications would bring him, Darwin was losing control of how his ideas would be stated and losing the opportunity to present himself as their author. Although there is no doubt that his feelings of affection and indebtedness to Lyell remained strong, Darwin's concern may have been heightened by the past disappointment of having his first publication preempted by his original scientific mentor, Robert Grant. If Lyell perceived Darwin's ambivalence, there is no sign that it discouraged him from continuing to adapt Darwin's work for the enrichment of his own. The pattern of the efficient and officious senior scholar co-opting his student's work continued, but as it continued Darwin began to express misgivings about Lyell's brand of public speculation.

The Beginnings of Darwin's Anxiety about Speculation

Darwin's manuscripts and publications from late 1838 to the middle of 1839 reveal that he was of two minds about theorizing in his geological work. He was immensely confident in the soundness and global applicability of his vision of up-and-down crustal movements, invoking that process in his interpretation of new field sites in Great Britain and entering debates about glaciation in the Northern Hemisphere. But he attempted to discipline his audacity and insulate himself from criticism by studying philosophical writings on scientific method and couching his arguments in the terminology they provided.

Lyell, meanwhile, continued to spur him. No sooner had the *Elements* been published than he began pressing Darwin for information he could incorporate into the latest revision of the *Principles*. In this instance the young man was invited to reason along with Lyell. Calling himself "your adviser," Lyell wrote asking whether the coral reef theory seemed to augur in favor of Élie de Beaumont's theory of mountain building, which stated that chains running in parallel had been elevated in dramatic simultaneous upheavals.[48] Lyell, of course, objected to any doctrine that relied on paroxysmal movements of the crust, so he planned a chapter-long denunciation of the Frenchman's "supposition that nature was formerly parsimonious of time and prodigal of violence." Lyell envisioned two ways Darwin's coral reef theory might help his cause. The first, and less promising, was that it offered a way to cast doubt that parts of the crust undergoing simultaneous upward or downward movement must necessarily be parallel. "If I remember right," Lyell inquired hopefully, "some of your lines are by no means parallel to others, although many are so." The second avenue was more exciting, because it might undermine the Frenchman's reasoning altogether. Élie de Beaumont had argued that the Pyrenees had been elevated rapidly within the short time *after* the end of the Cretaceous, as evinced by the presence of (Cretaceous) Chalk deposits uplifted on their flanks. Lyell had argued that the mountains might have been formed over a much longer period *before* the Cretaceous fauna had died out, with some Chalk formations being uplifted while others continued to form on the seafloor. He believed Darwin's coral reef paper had demonstrated just such a process because it showed that at one geological period (the present one) some reefs were being uplifted into dry land while living reefs in subsiding areas were accumulating strata of indefinite thickness: "Now in your lines of elevation, there will doubtless be coralline limestone carried upwards, belonging to the same period as the present, so far as the species of corals are concerned. Similar reefs are now growing to those which are upraised, or rising." By this "point of view," Lyell contended, "you grand discovery proves . . . in the most striking manner, the weight of my principal objection to the argument of De Beaumont."[49]

Darwin's response to this interrogation reveals that he had become ambivalent about using the coral reef theory in such broad and polemical ways, and perhaps even about his identity as a theorist. "With respect to the question how far my coral reef theory bears on De Beaumont's theory," he quickly replied, "I think it would be prudent to quote me with great caution, until my whole account is published, & then you (& others) can judge how far there is foundation for such generalization."[50] In truth, he hoped Lyell would not have the opportunity to quote him at all before his own book was published, telling him, "I should like my volume to come out before your new edition of Principles appears. Besides the Coral theory,— the volcanic chapters, will, I think, contain some new facts."[51] Darwin was concerned with priority, but he was also increasingly uncertain that his book could support the weight of theorizing his short and ambitious essays had implied. Privately, he was more than happy to agree with Lyell's "generalization." Abjuring his former (but presumably unacknowledged to Lyell) acceptance of the Continental geological theories of Humboldt and Élie de Beaumont, Darwin admitted, "I do not doubt [the] truth [of Lyell's argument]. . . . I do not believe a more utterly false view *could* have been invented than great straight lines, being suddenly thrown up."

Darwin's concern was how far he could afford to stretch his evidence in public, so he sought at the very least to temper Lyell's enthusiasm for quoting Darwin's work in progress. "The extension of any view over such large spaces from comparatively few facts must be received with much caution." This fear—that all his views were based on just a handful of reefs— was the very concern that drove him, as we shall see, to carry out intensive research in the library of the Royal Geographical Society the following winter and spring.

That concern also drove him to begin consulting books that would advise him on acceptable methods of framing a theoretical argument. On 5 October 1838 he noted in his diary that he had begun working on his coral reef writing, but that it "requires much reading." This included topical material about coral reefs, yes, but notably also treatises on scientific methods. The very next entries in his reading notebook (dated 12 October and continuing undated) show that he read several travel narratives, followed immediately by "Whewell's inductive History" and "Herschel's Introd. to Nat. Philosophy" (which, as he noted, he was reading for the second time).[52]

Darwin's new sensitivity to the way he articulated his geological theories may have dated back to March 1838, when he presented a paper that marked the culmination and theoretical climax of the five papers he read to the Geological Society in 1837–38.[53] On this occasion he received his first serious critique, and it stung him badly. As Sandra Herbert and Frank Rhodes, among other historians, have argued, this "Connexion" paper was the closest Darwin came to publicly revealing the full extent of his private speculations on the figure of the earth and the true cause of geological phenomena.[54] In arguing that volcanic eruptions, earthquakes, and the elevation of mountain chains were related secondary effects of the "one motive power" that had uplifted the continent of South America, Darwin was elaborating the hierarchical relation of the geological causes that were hinted at toward the end of his 1837 report on coral reefs. Lyell considered Darwin's 1838 work to be "a paper . . . in support of my heretical doctrines," and he believed it had been an unalloyed success. As Lyell reported to his father-in-law, geologist Leonard Horner, "[Darwin] opened up on [Henry] de la Bêche, [John] Phillips & others . . . his whole battery of the earthquakes and volcanoes of the Andes & argued that . . . all depended on a common cause." In the discussion that followed, Lyell "was much struck with the different tone in which my gradual causes were treated by all, even including de la Bêche[,] from that which they experienced in the same room 4 years ago when Buckland, de la Bêche, Sedgwick, Whewell and some others treated them with as much ridicule as was consistent with politeness in my presence."[55] After years of fighting seemingly alone against the men Murchison called "practical geologists of the highest rank," Lyell was elated to find that Darwin had received only glancing blows from their shared antagonists.[56]

But while Lyell's soft speaking voice belied the fact that he had trained as a barrister and was accustomed to the vigor of Geological Society debates, Darwin was unnerved by this first skirmish.[57] After a Sunday meeting between the two, Lyell was forced to add a surprised postscript to his triumphant letter to Horner: "I found that Darwin, who was with us yesterday evening, had felt very differently in regard to Wed[nesday]'s discussion[,] for[,] not being able to measure the change of tone in the last 4 years[,] he translated de la B's & Co.'s remarks into a vigorous defiance instead of a diminishing fire & an almost beating of retreat." Lyell concluded happily, however, with a Herschelian pun on the righteousness of his geological search for the *vera causa* of geological phenomena, "But I have restored him to an opinion of the growing progress of the *true cause*."[58] Two years later when Darwin sent the paper to John Phillips for review, however, he delicately maintained that "since I wrote it . . . [I] set less value on theoretical reasoning in geology." Nevertheless, he acknowledged, "I even yet think there is some weight in the argument, respecting the necessary slow elevation of mountain chains."[59]

Meanwhile Darwin sought to demonstrate the value of his synthetic geological theory by reinterpreting one of Britain's puzzling physical features, the "parallel roads" of Glen Roy in the Scottish Highlands. These "roads" were horizontal terraces on either side of a valley (Glen Roy), and though earlier visitors had supposed that they must be ancient manmade features, geologists in the past two decades had declared them to be of natural origin. Two Scotsmen, John MacCulloch and Thomas Dick Lauder, proposed in the late 1810s that the "roads" had been cut into the hillsides by standing water and were the beaches of a former Highland lake that had once filled the valley. They supposed the water in the lake to have stood at several distinct levels, each occasion producing one of the terraces.

Darwin's interest in Glen Roy was piqued by the seeming similarity between the parallel roads and a series of terraces at Coquimbo, Chile, that he examined at the height of his geologizing during the *Beagle* voyage. Consistent with his even earlier work on the plateaus of Patagonia, Darwin believed the Coquimbo terraces were former marine beaches that had since been pushed above sea level by the bulging of the earth beneath South America. He traveled to Glen Roy to evaluate whether the parallel roads might also be former sea beaches. If they were, their existence would indicate that Scotland had been elevated from the sea in a manner similar to the process he believed had lifted the continent of South America. In each case, that the terraces remained essentially level would have been evidence of gradual and equable "horizontal" elevation. Darwin dismissed the existing explanation that the roads had been formed by ancient lakes because there was no satisfactory explanation for the temporary damming of Glen Roy, which must have occurred for the valley to fill with water and then be emptied. Instead, he concluded that the roads were indeed the beaches of former arms of the sea, and the terraces at each level represented a former stage in Scotland's emergence from the water.

On 7 February 1839 Darwin read to the Royal Society a paper on Glen Roy that brought his synthetic geology home to Britain by arguing "that the whole country [at Glen Roy] has been slowly elevated, the movements having been interrupted by as many periods of rest as there are shelves." He had to argue hard to advance his marine beach interpretation of the parallel roads. Although this explanation spared him from having to conjure up, as Lauder and MacCulloch had done, the prospect of a past event that could have dammed Glen Roy, it gave him the challenge of explaining why the sea had left no marine fossils on the sides of the glen and why it had not cut similar terraces on other hillsides across Scotland.[60]

He concluded the paper with a section titled "Speculations on the Action of the Elevatory Forces." Before the nation's oldest and most prestigious scientific society he made his argument that the same basic processes that had shaped the geology of South America were also at work in the British Isles. In doing so he was arguing that his authority need not be limited to describing destinations his fellow British geologists had never seen, and asserting the validity and utility of his, and Lyell's, theories of gradual crustal movement.

Darwin had certainly not abandoned speculation altogether, then. His willingness to theorize in private or in direct communication with Lyell was never in doubt. He hypothesized boldly in his personal species notebooks, thrilling at this diversion from the arduous process of drafting a book. In oral presentations and the resulting periodical publications too, he could "choose to enter on speculative grounds," as he acknowledged doing at the end of his Glen Roy paper.[61] However, by indicating that doing so was a "choice," and by stating those public arguments in more explicitly philosophical terms, Darwin was seeking to demonstrate that his judgment had matured along with his powers of reasoning.

Darwin seems to have wrestled with difficulties that were tied to one particular type of public presentation—the book. He earnestly hoped, as the first-time author of a specialist treatise, that it would be an enduring and authoritative work. Lyell, for his part, relished the opportunity to increase the sales of his own books by serially revising them. So recently removed from the voyage, Darwin could not envision himself wanting to revisit his books once they were written. "I have so much more pleasure in direct observation," he wrote (in contemplative notes that are now famous for his weighing the advantages and disadvantages of marriage), "that I could not go on as Lyell does, correcting & adding up new information to old train."[62] (Lyell himself had once felt similarly, judging by an 1832 letter to his publisher in which he referred to the first revised edition of *Principles* as the "permanent" edition.[63] By the time of Darwin's ruminations just six years later, however, the *Principles* was already in a fifth edition with a sixth on the way.) Darwin's attendant desire to be comprehensive and exacting added to the monumental scale of his task. There were, as he had told Leonard Jenyns, an enormous number of "facts" to be dealt with. Even as the planned scope of his first book narrowed from a broad geological treatise to a work on the single topic of coral reefs, his notion of what it meant to write such a treatise expanded to justify this monographic focus. Finally, Darwin began to be haunted by the undisciplined speculations of his 1837 coral reef paper. As a consequence, writing his book carried the added burden of redressing the hastiness with which he had begun revealing his geological theories.

PART III

A Different Approach to Authorship

9

The Life of a Tormented Geologist (and Enthusiastic Evolutionist)

Much has been made of the tension and even illness Darwin started to feel after he began his private work on species in 1837.[1] That work was moving apace by 1838, and many historians, noticing its temporal correlation with the angst and physical unease he began to feel at this time, have concluded that there was a causal relation. But Darwin was under no pressure to produce the species theory, which was at this time almost entirely secret. Too many historians, wishing to explain the development of the species theory that later became Darwin's most famous contribution to science, take it for granted that this theory was the underlying cause of all his decisions and emotions. They tend see Darwin's first geological and geographical theories as mere steps on an inexorable route toward species; often their tone implies that the theories Darwin originally focused on were distractions from the self-evidently important development of his ideas on species.

I am convinced that this is almost exactly the opposite of how Darwin himself felt in the late 1830s and early 1840s.[2] The high stakes of meeting expectations for his geological work drove his anxiety, and the topic of species became a rewarding and indeed exhilarating distraction. Lyell played a key role in Darwin's anxiety for two reasons. First, the bold speculations of the coral reef paper that Lyell had helped craft in 1837 now felt to Darwin like rash promises that he was unwilling and indeed unable to keep. Second, Lyell continually added to the pressure Darwin was under by urging him to finish the book and responding to the delays by continuing to incorporate Darwin's ideas into his own books. Meanwhile, Lyell remained ignorant of any details about the remarkable notes Darwin was compiling on species, which in retrospect appear to document one of the most consequential examples of hard intellectual work in the entire history of science.

Darwin's Turn toward Empiricism and the Ideal of Comprehensiveness

The year 1839 began very, very well for Charles Darwin. He was elected a fellow of the Royal Society on 24 January, and five days later he married his cousin Emma Wedgwood. He abandoned his bachelor accommodations on Great Marlborough Street, and the pair settled into a house suitable for a married couple at number 12 Upper Gower Street. Their first child would be born before the year was out. On 7 February he presented the paper on Glen Roy that signaled his ongoing ambition as a geologist even while his visions of a grand synthetic book had splintered.

It was during the following week, the week of his thirtieth birthday on 12 February 1839, that he put himself back to work on the coral reefs with a diligence that lasted into the summer. His chief task was a map of coral reef distribution that would eventually be the most striking feature of his 1842 book. In the mistaken certainty that a final push was all that stood between him and publication, he put off responding to inquiries about his work. When he finally answered a letter from Leonard Jenyns, who had been classifying Darwin's ichthyological specimens from the voyage, he had to confess, "I admire the ingenuity, with which you perceive a fishy smell about my book, my silence, & [I] daresay the very name of me." In a moment of vulnerability he revealed that his struggle was caused by precisely the challenge of integrating fact with speculation. "I am hard at work, preparing the first volume of my geology—it is very pleasant easy work putting together the frame of a geological theory, but it is just as tough a job collecting & comparing the hard unbending facts." Lately he "ha[d] been for the last six weeks employed over one map to illustrate [his] views on coral formations."[3]

In the twenty months of land-bound time that Darwin ultimately spent in active work on his "coral-volume," he was not simply writing up a longer exposition of the same old theory. Much of his effort was devoted to original research that would allow him to produce a comprehensive account of the world's coral formations. He wanted to classify every known reef according to his taxonomy and to provide an explanation for every apparent anomaly in the patterns of reef growth as he understood them. As he eventually attested to the readers of his 1842 book, he had consulted "as far as [he] was able, every original voyage and map" that contained information on the structure and distribution of coral reefs.[4] Such thoroughness, of course, could broaden his warrant to generalize. After joining the Royal Geographical Society in 1838, he availed himself of its library to study the works of navigators like Beechey, James Horsburgh, Jules Dumont d'Urville, Friedrich Lütke [Fyodor Litke], Jacques de Tromelin, and Louis Duperrey.[5] He also took advantage of Beaufort's willingness to let him study charts at the Admiralty.[6] And his coral research relied, as did so many of his other projects, on correspondents who could provide geographical information or specialized knowledge not available in published works, in much the same way Lyell was doing with him.[7] As Darwin explained in soliciting William Henry Smyth, a naval officer and one of the founders of the Geographical Society,

> I am engaged in drawing up an account of the Coral formations of the Pacific
> & Indian seas, and I observe it is said in Krusen[s]tern's memoir, that you
> were in the [ship] Cornwallis, when Smyth's Isl^d in the Northern Pacific
> was discovered.—I am particularly anxious to know, whether the low islets
> & reefs, of which the group is composed, form a ring surrounding a lagoon, like so many other isl^ds. in the Pacific, and the atolls in the Indian
> ocean:—or, if it has not a lagoon, then is one central island of greater height,
> & apparently of different constitution from the other low islets on the reef,
> & surrounded by a channel of deepish water:—in short whether it has any
> peculiar structure.—As I cannot obtain this information from any other
> quarter, if you would spare me a few minutes & send me an answer, I should
> feel extremely obliged and I trust you will excuse my having ventured so far
> to trouble you.[8]

Darwin was wondering, in short, whether this formation should be classified as a lagoon island or an encircling reef. Chief among the informants for his coral reef work was John Malcolmson, a Scottish surgeon (and fellow of the Geological Society) who had worked in India and visited Arabia and Sinai.[9] In a series of protracted letters written during the summer and autumn of 1839, Malcolmson told Darwin about the composition of islands in the Indian Ocean, informed him about elevated coral beds on the shores of the Red Sea, alerted him to experiments on the rate of coral growth performed by a Dr. J. Allan at Madagascar in the early 1830s (which had been reported in an Edinburgh University thesis), and provided citations to dozens of published references on reefs and other subjects.[10] With the libraries of London at his disposal and the testimony of travelers just a letter away, Darwin was able to compare countless islands he had never seen with the structures he had witnessed for himself at Tahiti, South Keeling, and Mauritius.

Darwin's dry-land coral reef researches were fundamental to the theory he intended to publish. During the whole of the *Beagle* voyage he had seen only a handful of coral formations, yet he planned to apply the theory to every reef in the world. Whether he seemed justified in generalizing in this manner would depend on establishing that the type specimens he had seen in person were truly representative. By my calculations, for every day he had spent examining reefs in the field he spent a full month poring over the particulars described in other travelers' narratives or inscribed on the Admiralty charts produced by other voyages. He compiled notes on every reef and tried to classify each as one of the four types mentioned in his 1837 paper: lagoon island, encircling reef, barrier reef, or fringing reef. These he plotted with a color-coded entry onto his working copy of the distribution map.[11] He had brought this laborious task on himself with the ambitious speculations of his earlier paper. Now the long-term fate of his coral reef theory rested on that unglamorous work of "collecting & comparing the hard unbending facts." By pole-vaulting his way about South Keeling, Darwin had propelled himself to the very limit of what any one naturalist could learn about any one reef. On the other hand, mastering all the reefs of the globe would require him to travel beyond the reach of his leaping pole. Sociologist Bruno Latour adapted Archimedes's famous statement about a pole ("Give me a lever long enough and a fulcrum on which to place it, and I shall move the world") as a metaphor for the way knowledge produced inside a laboratory could speak for, and be applied to, all of nature. As I illustrate in figure 25, Darwin had found just such a high-leverage location half a world away from South Keeling, in the imperial map rooms of the nineteenth century's greatest maritime power. It was there in London, and only there, that he accumulated the knowledge that would let him credibly present his reef theory as applicable to the whole globe.[12]

Figure 25. Track of the 1831–36 *Beagle* voyage superimposed on Darwin's 1842 reef distribution map, in which he used a series of color codes (depicted in full color in plate 10) to classify all known coral reefs into three main types. Note how few of those reefs he had come close to seeing in person. The rest, which is to say the vast majority of the world's reefs, he researched in London using charts and descriptions produced by other travelers. Whereas my earlier maps of the *Beagle* voyage (figs. 1 and 2) were intended to show the sequence and pace by which Darwin experienced a set of places during the voyage, this map is intended to show at a glance all the other places Darwin "traveled" with the scholarly resources available to him in London. Illustration by the author.

The Pressure of Public Expectations

By a twist of fate, Darwin's boldest, Lyell-inspired speculations were thrust before the public just as he was being forced privately to reassess his ability and desire to make good on those claims. The situation could hardly have been better calculated to make him squirm. Recall that Darwin's *Journal of Researches*, his narrative account of the *Beagle* voyage, contained the full text of his audacious 1837 coral reef paper, having been printed shortly thereafter but delayed in publication so that it would not preempt FitzRoy's narrative of the voyage.[13] It was finally released in early June 1839 with a preface that had been printed more recently. It explained his (revised) plans for geological publication: "I hope shortly to publish my geological observations; the first Part of which will be on the Volcanic Islands of the Atlantic and Pacific Oceans, and on Coral Formations; and the second Part will treat of South America."[14] When reviews began to appear the following month, knowledgeable commentators in both the Whig-oriented *Edinburgh Review* and the Tory *Quarterly Review* chastened him for speculating too brazenly. Indeed, both declared that the merit of his coral reef theorizing would be known only when he had brought forward his promised treatises on the geology of the voyage. Thus, while both reviews were full of praise for Darwin's work in the field and at his writing desk, each raised the stakes for the book(s) he had not yet written.

The July 1839 issue of the *Edinburgh Review* contained, as we shall see, a rebuke for the "boldness of [Darwin's] theories." By coincidence it also contained an essay on Lyell that criticized the boldness of *his* speculations and advertised the role Darwin's coral reef theory had played in supporting them. Prompted by the publication of Lyell's *Elements*, this essay on the state of geological theorizing was written (under the veil of quasi anonymity, as most such reviews were) by William Fitton, who like Lyell was a former president of the Geological Society and a confidant of Charles Babbage. Fitton was broadly sympathetic to Lyell's goals in writing the *Elements*, and the *Principles* before that. In fact, in the late 1820s, when Lyell was preparing the *Principles*, he evidently anticipated Fitton's publishing a similar book; he wrote to Gideon Mantell, "Of course you will not attempt to tilt with Fitton and me 'on the general principles of geology', which we mean soon (mine will be soon) to give you."[15] Although Fitton had produced no equivalent opus, he offered a sharp opinion on Lyell's efforts. "The book [*Principles*], at first, appeared to us to be the production of an advocate, deeply impressed with the dignity and truth of his cause. The tone was rather that of eloquent pleading than of strict philosophical enquiry." Lyell had, of course, been trained as an advocate in his career as a barrister. Twisting the knife, Fitton wrote, "Speculation is his great delight, and it is too often indulged in."[16]

In discussing the *Elements*, Fitton pointed out that Lyell's more recent speculation owed a debt to Darwin's reef studies. The "doctrine which [Lyell] has specially endeavoured to establish," he explained, was the "reality" of elevation and subsidence. This belief in extensive vertical oscillations of the earth's crust was, in Fitton's view, "the principal subject of difference between [Lyell] and some of the leading geologists of our day." Fitton insinuated that Lyell's publications offered too little firm evidence in support of that view. However, he drew on the material in the *Elements* to cite on Lyell's behalf the work of "Mr Darwin, the able naturalist, who accompanied Captain Fitzroy in the voyage of the Beagle around the world," who had attested to ongoing elevation in South America and likely subsidence in areas marked by ring-shaped coral reefs.[17]

Whereas Fitton reserved his comments on the ills of speculating for Lyell (and thus implicitly laid responsibility for any misuse of Darwin's speculations at Lyell's feet), the author of the accompanying review of Fitz-Roy and Darwin's *Beagle* narratives offered no such courtesy to the young voyager. The review was likely written by Basil Hall, a naval officer who had published several notable books of travels including a journal of his excursions in Chile the decade before Darwin was there.[18] Hall adopted an acerbic tone whenever he discussed Darwin's interpretation of South American geology, and particularly with respect to the Chilean coast that Hall knew at first hand. Indeed, Hall ridiculed Darwin's zest for theorizing and openly accused him of seeing only what he wished to see. "Mr Darwin, whose faculty of generalization is certainly of no ordinary vigour, has here happily seized the circumstances of superficial configuration, which tend to confirm his theory of elevation." Whereas Hall praised his fellow officer FitzRoy for proving through careful survey work that the 1835 Concepción earthquake had raised the land by as much as eight feet, he condemned Darwin's conclusion that such changes contributed to an ongoing elevation of the continent. "There is reason ... to believe," Hall argued, "that the land so raised again subsides nearly to its former level; so that the permanent encroachment of the land upon the sea, is a slower process than might be inferred from a hasty enquiry into the effects of earthquakes."[19] Hall was even more acid about Darwin's coral reef theory. "If our space permitted, we should willingly extract some of Mr Darwin's remarks on the classifica- tion of coral islands," Hall wrote, without explaining what those remarks entailed. Rather than providing an introduction to the theory for those who had not encountered it, Hall's final words on Darwin were offered as a warning to the author himself. "We submit more cheerfully to the necessity of passing [over Darwin's account of coral reefs] for the present, since he promises a volume specially devoted to the subject; in which, besides some details descriptive of the coral animal, *we hope to find the boldness of his theories a little modified;* and his alternate zones of elevated and depressed coral islands, *resting upon a more solid foundation than the supposed undul- ations of a subterranean fluid.*"[20] Hall was no particular friend of Darwin's, but this was an unkind cut by any standard.

While the review by William Broderip in the December *Quarterly Review* assessed Darwin much more favorably than Hall had, and was in places positively effusive about the scope of his achievements during the voyage, Broderip echoed Hall's sentiment that his theories would best be evaluated in future, more scholarly, iterations. After writing approvingly, but tentatively, about the logic that underlay Darwin's coral reef theory, he pointed out the theory's broader geological implications. "If ... this theory, which includes under one head the origin of the several reefs, be admit- ted, very important deductions must follow from it: for it shows that great portions of the surface of the globe have recently (in a geological sense) undergone movements of subsidence (which it must always be extremely difficult to detect by any direct evidence); and what is even more worthy of note, it shows that the movements have been so far gradual, that no one sinking down has carried the reef below the small depth from which the polypi could rear it to the surface again." Broderip did not spell out the final claims Darwin made, that these implications of the coral reef theory would in turn shed light on the internal composition of the earth and on the history of its organic forms. Instead he reserved judgment on these assertions until they could be considered along with the full weight of evidence that a book would contain. "So far so good," he concluded, "but Mr. Darwin alludes to even more extended inferences, which we shall not notice, as the subject will soon be treated of by him at full length."[21] Broderip did, however, quote a related conjecture Darwin made about the region of Chile where he had witnessed active volcanoes and seen the effects of an earthquake, that "a vast lake of melted matter, of an area nearly doubling in extent that of the Black Sea, is spread out beneath a mere crust of solid land." Here again Broderip pointed out that the theorizing in the *Journal* could not stand alone, confiding in those readers who had not heard Darwin read his "Connexion" paper at the Geological Society, "We have reason to believe that Mr. Darwin means to justify what he has said upon this perilous subject in the forthcoming part of the Geological Transactions." Broderip wrapped up the passage on Darwin's account of the volatile geology of Chile with a blithe quip that may have stung the author: "A pleasant locality this for a building speculation!"[22]

As he made halting progress on his geological book, Darwin felt haunted by the ambitiousness of his earlier generalizations and overwhelmed by the forbidding task of justifying them. He was sickened by the thought of even looking back at the words of the 1837 coral reef paper that was now so conspicuously enshrined as the climactic chapter of his *Journal*. The longer he avoided doing so, the more he came to magnify the flagrancy of his earlier speculations. This is evident from Darwin's reaction when, in February 1840, further queries from Lyell necessitated his rereading the paper. Apparently Lyell had despaired of receiving Darwin's book in time to incorporate its arguments on coral reefs into the revised edition of the *Principles*, and though his letter does not survive, he must have asked Darwin whether to cite the 1837 paper. Darwin was forced to admit once and for all that despite having "set [his] heart" on having the book completed before Lyell's new edition, Lyell was justified in his pessimism.[23] Having no choice but to look back and remind himself of what he had written in 1837, Darwin reviewed the version published in the *Journal of Researches* and offered Lyell a list of the "two or three points, which will be different in my volume." Darwin put on a brave face and reported that "I find I am prepared to stand by almost everything.—it is <u>much</u> more cautiously & accurately written, than I thought." Yet his list of revisions indicated quite the contrary. Darwin had in fact by now adopted a much more conservative position. If it was true that this audacious paper turned out to have been more cautious than he *remembered*, he had clearly come to remember it as extraordinarily incautious.

The preview of the forthcoming coral reef volume that Darwin provided to Lyell revealed that he had abandoned many of his most ambitious conjectures and was now almost exclusively oriented toward bolstering the claim that certain kinds of reefs had formed in areas of subsidence. He had been informed that coral reefs in the Red Sea lived deeper than previously believed, but he assured Lyell that "the argument . . . that there must have been subsidence in the <u>large</u> areas, scattered with reefs, stands firm." After working hard at his reef map, he placed greater weight on the distribution of coral islands, explaining that his subsidence theory would hold "even should coral-reefs be hereafter found to live at much greater depths [than] I suppose; for I find the areas are <u>immense</u> in which every island is low, & of coral-formation." In a little-noticed effort to improve the subsidence theory, he had also refined his "classification of reefs." The first change was one of nomenclature. Darwin resolved to call his first class of reefs "atolls" instead of "lagoon islands," adapting an indigenous name for the annular island groups of the Maldives.[24] Second, he combined encircling reefs and barrier reefs into a single class. The significance of this step should not be overlooked, because it indicates that Darwin's taxonomy was no longer based on the shape of reefs but depended on their proposed mode of formation. Without Darwin's theory there was little to justify placing the annular reefs of the Society Islands and the long, straight barrier of Australia in the same group. Finally, in a stoic understatement he told Lyell, "I shall have only <u>very slightly</u> to modify my general conclusions." In fact, he was retreating from what had been the paper's most strongly worded plaudits for Lyell's *Principles*. Thus he admitted that he would be "speaking rather less positively—& using the words alternate areas more frequently than 'parallel bands,'" and he confessed that he would "not be able to throw any light on [the] distribution of organic forms in the Pacific as [he] had hoped."[25]

This letter shows just how much Darwin's writing project had changed in less than three years. He was now backing away from the bold claims that had exemplified his most ambitious vision for the original single-volume synthetic treatise on the geology of the world. By October 1839 he had despaired of publishing even the coral reef and volcanic island material together. Instead he began to "hope in a couple of months to have a very thin [octa]vo volume on Coral Formations published."[26] The decisions that led him ultimately to divide the grand geology book into a trio of more narrowly focused texts had major implications for what scope of theorizing would even be possible in his geological work. The synthetic vision that spurred his most audacious claims had been predicated on the *connections* between coral reefs, volcanic islands, and continental geology. By dividing these topics he was abandoning the arguments that had prompted Lyell and Darwin to make coral reefs the centerpiece of his geological career in the first place.

However, focusing on this easier assignment fueled two months of hard work on coral reefs, which were followed by two months of illness. Of his working routine during those two productive months he wrote, "One of my days is as like another as two peas. . . . I will give you a specimen; which will serve for every day—Get up punctually at seven leaving Emma dreadful sleepy & comfortable, set to work after the first torpid feeling is over, and write about Coral formations till ten."[27] But after the two unproductive months that followed he was still unwilling to show even the coral material to Lyell because, as he apologized, "My M.S. is in such confusion."[28] The pattern continued as the spur from Lyell prompted another bout of work, but then a full thirteen months of poor health kept him from making any further progress.[29] Just what was the relation between Darwin's health and his productivity? This question has provoked historians just as much as it vexed Lyell, Darwin, and those around them. With no answer forthcoming, Lyell pressed on.

Lyell's Appropriation of the Coral Reef Theory

Left unable to predict when Darwin's book might emerge, Lyell again took the prerogative of reporting his student's coral reef theory in one of his own books. This time, however, and with very little sentimentality, he asserted a claim to authorship of the subsidence theory. Then he proceeded to build on it, treating the theory Darwin continued to struggle with as a fait accompli.

Like the *Elements* before it, Lyell's new edition of the *Principles* was deeply indebted to Darwin's private contributions. It was published in the summer of 1840. Belying what was found in the body of the book, Lyell's preface offered a generous acknowledgment to Darwin. It explained that the latter's "new views . . . have induced me to renounce the hypothesis which I formerly advocated, that [circular] reefs were based on submerged volcanic craters."[30] As he had forecast the previous year, Lyell employed Darwin's reef observations to expose Élie de Beaumont's "faulty induction" about the date and speed at which the Chalk had been uplifted by the formation of the Pyrenees. He also countered the Frenchman by arguing that "all the existing continents and submarine abysses" could have been formed by gradual movements comparable to the subsidence known to be occurring in "parts of the Pacific and Indian oceans, in which atolls or circular coral islands abound."[31] Whereas Darwin had asked to be quoted "with great caution" in relation to Élie de Beaumont, Lyell opted not to mention the young man's name at all in the relevant chapter. Despite containing a line of argument that plainly alluded to the "Connexion" paper, referring to the elevation of South America using words drawn directly from Lyell's first congratulatory letter to Darwin, and even discussing the formation of coral reefs, this chapter did not include a single citation to Darwin's work.[32]

Lyell's main chapter on coral reefs did cite Darwin's forthcoming work, only to strike off independently from it. This chapter was double the twenty-page length of its counterpart in the first edition, and the title was changed from "Corals and Coral Reefs" to "Formation of Coral Reefs." This subtle modification reflected the much greater emphasis Lyell now placed on the mutability of reef structures and on the causes of such transformations. Less subtle was the way Lyell declared his own important role in the development of the subsidence theory. After "abandon[ing]" his submarine volcano theory (but not without explaining all the good reasons why "it was formerly embraced"), he described Darwin's "new opinion" on the formation of reefs.[33] He then rehearsed the supporting evidence from Darwin's Geological Society paper. But after giving this supportive rendering of his acolyte's work, Lyell made a bold declaration of his own priority in this area. Ironically, this very statement employed the term atoll in the generic sense that *Darwin* had proposed in their private correspondence that spring:[34]

When the first edition of this work appeared in 1831, several years before Mr. Darwin had investigated the facts on which his theory is founded, I had come to the opinion that the land was subsiding at the bottom of those parts of the

Pacific where atolls are numerous, although I failed to perceive that such a subsidence, if conceded, would equally solve the enigma as to the form both of annular and barrier reefs.[35]

Lyell proceeded to give three full pages of quotations from his own first edition that he claimed would support this contention. To be precise, however, they showed he had previously surmised that the amount of geologically recent subsidence in the Pacific appeared to have exceeded the amount of elevation over the same period.[36] The selected quotations emphasized "alternate elevation and depression of the same mass," and not—as he now implied—a prevailing "downward movement in the bed of the ocean."[37] He evidently could not help reading and remembering his own past words in light of the new theory.[38] Lyell went on to explain the "important generalization" that Darwin had derived by correlating reef forms with submarine movement: that the globe "might be divided into areas of elevation and subsidence, which occur alternately."[39] Given the geological and zoological significance of Darwin's conclusion, as well as the active role Lyell himself had played in shepherding it to prominence, it is not difficult to see why he wanted to claim the credit he believed he was due.[40]

But with Darwin's full treatment of the coral reef theory unpublished, and indeed unavailable for perusal, Lyell decided not to stop there. "Having laid before the reader this brief analysis of Mr. Darwin's theory," he declared, "I shall next endeavour to trace out some of the other natural consequences to which it appears to me to lead."[41] In a footnote that sounds almost scolding in light of his many private appeals for Darwin to finish his book, he added, "I know not how far the conclusions deduced in the remainder of this chapter may agree with those at which Mr. Darwin has arrived, and which he will explain in detail in his forthcoming work on Coral Formations."[42] In the course of nine pages, Lyell went on to analyze the structure of reefs that might be produced on an island undergoing different types of submarine movement. With the aid of woodcut diagrams he illustrated, for example, that intermittent episodes of rapid subsidence would produce a much smaller atoll than uniformly slow subsidence of the same island.[43] For this reason, he explained, it would be impossible to estimate the dimensions of an atoll's base from its circumference at the surface. He concluded that it was also impossible to "calculate . . . what may have been the height of [an] island now changed into an atoll" or to "estimate the thickness of the coral which has accumulated."[44] Yet by considering the amount of subsidence that would be necessary to submerge the highest points of present oceanic islands like the Canaries, he ventured to guess that if the Pacific were laid dry it would reveal mountains capped with calcareous formations as much as ten or eleven thousand feet thick. "Thus," he pointed out in a silent reference to the debate over the origins of the Chalk, "a recent cretaceous formation may now be in progress in many parts of the Pacific and Indian oceans."[45]

Aside from tantalizing notes in Darwin's copy of the book, there is no direct evidence of his reaction to Lyell's incursion into coral reef territory. Inside the back cover he wrote two sets of notes, one column for himself and another to pass along to Lyell. In the notes to send to Lyell he wrote, "Coral chapt. very satisfactory." But within the text he seemed to take exception to the sentence in which Lyell began to describe Darwin's coral reef theory. Where Lyell wrote, "To explain the phenomena above described, Mr. Darwin supposes that the coral-forming polypi begin to build in water of a moderate depth." Darwin underlined "supposes" and placed a large exclamation mark in the margin.[46] Presumably he was surprised to see that Lyell had not credited him with any more justification for his argument than mere supposition. Such an interpretation is supported by Darwin's reaction to his most famous experience of disappointment on reading one of Lyell's books, the 1863 *Antiquity of Man*, which Darwin had hoped (in vain) would provide a full-throated endorsement of his species theory. As Janet Browne has noted, Darwin annotated his copy of that book with "Oh" in the margin beside Lyell's statement that the gap between beast and man remained a "profound mystery."[47]

Darwin had been writing exclusively on coral reefs from 26 March 1840 until sometime in the summer, when he abruptly ceased. He eventually returned to the manuscript over a year later, on 26 July 1841, noting that it was the first time he had done so in thirteen months.[48] Counting back shows that the moment he abandoned the project coincided with Lyell's publication date of June 1840.

Moreover, the pressure on his geological work continued to mount. Scarcely had Darwin's Glen Roy paper appeared in print than the Swiss geologist Louis Agassiz proposed an explanation for the Roads that had not been considered by Lauder, MacCulloch, or Darwin. Agassiz was convinced that the earth had formerly experienced an "epoch of great cold" and that glaciers had once been much more widespread across Europe. In 1840 he toured locations in Britain with many leading geologists, pointing out how many familiar phenomena could be reinterpreted with reference to the former action of glaciers. In the case of Glen Roy, Agassiz provided the missing component of the lake-beach theory of the formation of the parallel roads. Walls of ice extending across the foot of Glen Roy and through nearby Glen Treig could have dammed the valley and formed a glacial lake like those seen in the present-day Alps.

In the hands of Agassiz and others in the succeeding decades, glacial theory prompted geologists to reappraise much more than the terraces at Glen Roy. Darwin was initially resistant to the glacial explanation for the parallel roads, even as he admitted the action of ice sheets elsewhere. On what turned out to be his final geological field trip, a return visit to North Wales in 1842, Darwin wrote that the signs of glacial action in the valley of Cwm Idwal could not have been more obvious "if it had still been filled by a glacier." Nevertheless he battled mightily to preserve his interpretation of Glen Roy until finally conceding in the early 1860s.[49]

Studying Species as a Diversion from the Task at Hand

Darwin's private preoccupation with species takes on a new color in light of the evidence that he was so tormented by his difficulties as a geological author. Several of the most attentive scholars of Darwin's manuscripts have concluded that his effort on species made him ill, in large part because Darwin often reported being ill or "idle" at the very times when he was compiling notes on species. The correlation exists, but I believe its cause was exactly the opposite: rather than its making him ill, studying species was what Darwin did when he suffered the malady of failing to make progress on his book.

To put it another way, the temporal correlation between feeling unwell and writing species notes suggests that his preoccupation with species was considerably *less* susceptible to being interrupted by illness than were the geological projects that were his main responsibility to Lyell and the publishing house of Smith and Elder. Indeed, some of his statements may be interpreted as evidence that the private and open-ended activity of gathering facts on species was a way of escaping the pressure of his promise to finish writing the geological book(s). Recall the letter to Lyell I have already quoted, in which Darwin described his easy-coming work on species as a sad temptation to idleness "as far as pure geology is concerned."[50] The idea of making a theory of transmutation public may have been something to tremble about, but so it seems was the idea of bringing forth a comprehensive and final statement of a theory on something as metaphysically benign as coral reefs.

Darwin's later recollections must always be treated with circumspection (as I illustrate at the end of chapter 11), but they too hint that this was the case by acknowledging a distinction between coral reef work and species work. Referring back specifically to these years, he wrote, "The greater part of my time, when I could do anything [when illness did not prevent him from doing anything] was devoted to my work on coral reefs." In contrast, he continued, he was persistently "collecting facts bearing on the origin of species & could sometimes do this when I could do nothing else from illness."[51]

I find it particularly telling that there was only one context in which Darwin bothered to fill his journal with information about what he had *failed* to do rather than what he had actually done: when he was failing to make progress on his geological publications. As he wrote in early 1840, "Again became unwell & did not commence <<Coral volume>>." Two years later, by contrast, in early March 1841, Darwin equated another bout of activity on species with indolence: "Was idle & unwell—sorted papers on Species theory."[52]

Knowing in retrospect that the species theory would be so consequential has, I believe, led other scholars to write as though Darwin *ought* to have been devoting his time to that project. This has made it difficult to notice that *Darwin himself*, along with his mentors, felt he should have been doing more pressing things. I have found a formerly overlooked and otherwise unremarkable letter in which Lyell seems to allude to the "idleness" that was so highly correlated with Darwin's actively making notes and sorting references on the species question. This message from Lyell to Darwin's former counselor Henslow was written in April 1841, when it had been almost a year since Darwin had made progress on the long-anticipated coral reef book and a month since the date of his entry about sorting papers on the species theory. After jotting a few lines to acknowledge receipt of a pamphlet from Henslow and signing off with a salutation and his signature, Lyell added a cryptic confidential progress report from one concerned mentor to another: "Darwin is only going on in the same state."[53]

Was Darwin procrastinating? Did he suffer from "writer's block"? Given that he so often described and even defined his illnesses in terms of their effect on his productivity, could it be that he considered his inability to work an illness in itself? I imagine such questions will have occurred to many readers, but I want to sound a note of caution about retrospectively diagnosing Darwin as suffering from states of mind that had yet to be named as such.[54] The phenomenon of describing psychological

impediments as "blocks" was a product of twentieth-century psychoanalytic theory, and the specific term writer's block was not coined until more
than a century after Darwin finally published his geological work. Saying
that Darwin felt blocked would be attributing to him not only a particular
malady but an entire set of anachronistic assumptions about the self, about
the default expectation that work should "flow," and about the (external)
locus of responsibility for his difficulty.[55] Darwin could certainly have been
aware of Samuel Taylor Coleridge's self-reported difficulty finishing the
poem "Kubla Khan," an example of a kind of "self-consciousness" in the romantic literary tradition that predated and, according to Zachary Leader,
underpinned the psychoanalytic theory of writer's block.[56] In their biography of Darwin as a "tormented evolutionist," Adrian Desmond and James
Moore wrote a powerful chapter titled "The Dreadful War" in which they
documented the depth of his psychological turmoil between 1838 and
1842. They argue that most of this anguish was connected to the species
theory while alluding in places to Darwin's other tasks and responsibilities,
recognizing that these efforts also contributed to the strain he was under.
I agree much more than I disagree with their assessment of Darwin's condition, which is to say I believe he was "tormented" at this time. It seems
to me, however, that much of the evidence they present about Darwin's
troubled state of mind describes a man struggling to finish a book for which
his desired audience had a complex and urgent set of expectations, rather
than a man tormented by the pressure of a project nobody was yet asking
him to complete.

10

A Finished Task:
Darwin's Treatise on Coral Reefs

In hindsight Darwin's 1842 book on coral reefs has been portrayed by many writers—scientists and historians alike—as a mere prelude to the more famous and consequential of his theory-driven books, the *Origin*. At the time it was published, however, it felt to Darwin, and appeared to his readers, like a considerable achievement in itself. This was not the grand synthetic geological treatise he had envisioned and even promised five years earlier, but *The Structure and Distribution of Coral Reefs* contained a truly elegant solution to one of natural history's most iconic puzzles.

Darwin was later to describe the *Origin* as "one long argument," and the same description applies to *Coral Reefs*. Whereas the rhetoric and logical structure of the *Origin* have been the subject of intense study over a century and a half, however, the earlier book has rarely been examined from that perspective. The text of Darwin's first specialized scientific book rewards a closer look because, among other things, it illustrates Darwin's first attempt to strike a new balance between fact and theory, description and speculation, moderation and ambition.

The Space between Lyell and Darwin

Not surprisingly, it was a nudge from Lyell that forced Darwin's attention back to coral reefs and led to his finishing the book, but it required some physical distance and perhaps a bit of filial distance as well. Lyell and his wife had made arrangements for a long trip to the United States, where he would give public lectures on geology in Boston, Philadelphia, and New York.[1] He intended one of the lectures in each series to be on fossil and living coral reefs, for which the recent revision of the *Principles* provided plenty of material.[2] To illustrate the alternating bands of elevation and subsidence, he wished to display a map similar to the one Darwin had shown when delivering his 1837 paper to the Geological Society. With this in mind Darwin deposited his color-coded charts with Lyell and left London in late May 1841 to make a recuperative visit to the countryside.[3] In the month before his mid-July departure for the United States, Lyell color-coded a coral reef map based on Darwin's charts so that he might include it among the outsize diagrams and painted landscapes illustrating each of his lectures.[4] Later his exclusive custody of this map came to signify his special position with respect to Darwin's work.

At the moment Lyell was making his copy of the map there remained three groups of problematic reefs that Darwin had not yet classified: those of the Red Sea, the West Indies, and Bermuda. While at his father's house in Shrewsbury, Darwin received a letter from Lyell suggesting he might use a "neutral tint" to demarcate these known but unclassifiable reefs. Apologizing that "I can give you no precise information without my notes (even if then)," Darwin nevertheless proceeded to provide Lyell with detailed analyses from memory of his research on all three locations, elaborating as his recollections became clearer. He discouraged Lyell from prematurely adding a code color for anomalous reefs, explaining that he did not consider them ultimately unclassifiable. "I advise you to leave the Red Sea quite uncoloured," he wrote, "for I have not yet considered all the data I have collected." Nor had he "finally considered [his] portfolio of notes on the West Indies."[5]

Besides illustrating Darwin's practice of collating facts for his coral reef book by region, the letter also revealed his continued ambivalence about the proper venue for theorizing. After offering Lyell a conjectural history of the Red Sea reefs that featured complex local oscillations, denudation, and coral growth that "will I believe make Ehrenbergs, Moresby's & other accounts all harmonize," Darwin admitted he was unwilling to introduce it in his manuscript. "I doubt whether I shall make any allusion to this view [in the coral reef book], as it will appear so hypothetical—though to you & your pupils, as a mere theoretical case, it might have been expected to have somewhere occurred."[6] This confession is noteworthy because it reaffirms that Darwin's aversion was not to private hypothesizing, but to *appearing* "hypothetical" in his book.

Darwin's unwillingness to publicize these views might be seen as a means of putting space between himself and Lyell. The slightly distant reference to "your pupils," moreover, was a marked change from Darwin's former eagerness to describe *himself* as Lyell's subordinate and student.[7] The letter closed on a benedictory note, wishing the Lyells a safe trip to America and offering "my warm thanks for all the friendship you have shown me." The tone was apt, because the Darwins had themselves decided to move away from London, not for a holiday but for a quieter home outside town. The families would never live near one another again. Thus when Lyell departed for his year abroad in the summer of 1841, it marked the end of what had been nearly five years of particularly intensive personal collaboration and mutual inspiration.

Darwin's colored charts were awaiting him when he got back to London from his convalescence in Shrewsbury. Whether he felt unbound by Lyell's departure, stimulated by having to respond to Lyell's questions without digging into the minutiae of his notes, or for some other reason altogether, he immediately resumed work on his coral reef book and stayed with the project until he had finished. Meanwhile the fate of the coral reef chart Lyell took to America offers a most revealing final piece of evidence about the older man's attitude toward Darwin's work during the period before Darwin had finally proved himself by completing the book on coral reefs. Lyell evidently agreed to allow Benjamin Silliman, the professor of geology at Yale, to commission copies of the diagrams and pictures from Lyell's course of lectures. However, the map of reef distribution represented a special case, and Lyell refused to let Silliman adopt it. As Lyell explained, "The map of the coral Islands I could not allow Mr Hall & Emmons to copy [for Silliman] at Boston because it is an unpublished document made by my friend C. Darwin & coming out in his next work."[8] In contrast to nearly all the other diagrams and pictures used to illustrate the lectures, he chose to keep close control over the reef map because it was another person's unpublished work. That Lyell himself was using the map to illustrate his lectures reveals his unique authority relating to Darwin. In preceding years while advocating on Darwin's behalf he had felt it was his prerogative to incorporate his student's unpublished arguments into the *Elements* and the revised *Principles*. Lyell's insistence on his exclusive right to lecture with the unpublished reef map was another reflection of the tacit compact between master and student.

In July 1841 Darwin began what turned out to be the final push on his coral reef material. As Lonsdale reported to Lyell in early December 1841, apprising the American tourist of events on the British geological scene, "The best announcement I can make is the gradual recovery of Mr Darwin. He attends the Council and has been present at our evening meeting. We may therefore look forward to a long series of valuable labours."[9] As in Darwin's private notes, Lonsdale here equated health with productivity on geological work. Darwin managed to devote most of his working hours to coral reefs until January 1842, when he began to send his manuscript to the publisher. On completing his proofreading four months later, he experienced paroxysms of relief, writing to Jenyns, "I have just finished correcting the last Page of Index of my small volume on Coral Reefs, wh[ich] rejoices the inward cores of my heart."[10] Emma and the children had left London for a holiday with her family while he was finishing, so he wrote to tell her the good news that the book he had labored over since before they were married was now complete. "I will give you statistics of time spent on my coral-volume," he told her, "*not* including all the work on board the Beagle—I commenced it 3 years & 7 months ago, & have done scarcely anything besides—I have actually spent 20 months out of this period on it! & nearly all the remainder [on] sickness & visiting!!!"[11]

A Mountain of Facts

Darwin's solution to the problem of appearing excessively speculative is most visible in the overall organization of his 1842 book. *The Structure and Distribution of Coral Reefs* was a dense compilation of geographical facts, zoological and hydrographic information on coral growth, and informants' reports on particular corals and reefs. All this was presented in ostensibly descriptive terms that subtly favored Darwin's theory. To be sure, the book did contain an explicit and vigorous argument on behalf of the subsidence theory, but it had been circumscribed to rule out almost all the wider speculations that had troubled Darwin ever since he included them in his 1837 paper.

In his brief introduction, Darwin laid out three objectives for the book. The first was to describe every kind of reef, with particular emphasis on those in the open ocean. Second, he would explain the origin of their forms. He promised not only to address the widely known enigma of lagoon-island formation, but also to explain the equally puzzling, but little remarked, barrier reefs. Finally, he would examine whether geographical facts supported his "theory of their origin." Roughly the first third of the book was devoted to chapters on each of the three classes of reefs in his newly revised taxonomy: "Atolls or Lagoon-Islands"; "Barrier-Reefs"; and "Fringing or Shore Reefs." The fourth chapter covered the growth of corals and the distribution of coral reefs. In chapter 5 he offered a theory to explain the form of all classes of reefs, and in chapter 6 he revisited the topic of reef distribution, this time "with reference to the theory of their formation." This final chapter was, in effect, a thematic discussion of the color-coded map of the world that was inserted into the book. In a single appendix, which was itself half as long as the rest of the book, Darwin explained the reasoning behind each reef's classification and named the individuals and publications he had consulted in settling each case.[12]

Chapters 1 to 4, then, were a natural history of coral reefs. The first and third chapters began with detailed descriptions of reefs he considered exemplary of their respective classes. In both cases they were the only ones of the sort he had seen himself, the atoll of Keeling and the fringing reef at Mauritius. His information on the outline and composition of these reefs was based largely on his leaping-pole traverses and his examination of the matter imprinted on the armed sounding leads. Darwin's rendering of what he now called the barrier reef (formerly encircling reef) at Tahiti in chapter 2 was comparatively sketchy, reflecting not only the brevity of his examination there, but also the superficiality of that fieldwork. From his mountain vantage point in Tahiti he had acquired a literal overview of the reefs, but he had not had access to soundings there, and his only knowledge of the outer margin came from islanders' descriptions. Thus the uneven accounts in his book mirrored the evolution of his methods of reef study over the course of his visits to Tahiti, South Keeling, and Mauritius. Each of the first three chapters expanded outward from these exemplars to consider the general form of each type of reef and to describe especially noteworthy or problematic examples, such as the dissevered atolls of the Maldives. These discussions revealed Darwin's heavy debt to the charts and descriptions of other travelers. In chapter 1, for instance, Darwin cited fourteen authors or correspondents (including one European resident of an atoll) in his descriptions of thirty-four named reefs and five groups of coral islands.[13]

Darwin's chapter 4 analyzed the distribution and accumulation of coral rock in three dimensions. He began by describing the general distribution of coral reefs across the surface of the globe, concluding that there was no simple rule, whether it related to the variance in water temperature or the quantity of carbonate of lime dissolved in seawater, that could explain why some areas of the tropical oceans contained reefs and others did not.[14] He then turned to the distribution of corals on reefs, arguing against Quoy and Gaimard's claim that stony corals flourished only where the water was calm. "This statement has passed from one geological work to another," Darwin complained, even though "the protection of the whole reef undoubtedly is due to those kinds of coral, which cannot exist in the situations thought by these naturalists to be most favourable to them." The problem, according to Darwin, lay in the tendency to confuse the diversity (as it would now be called) of a location's coral species with the vigor and structural strength of the corals themselves: "If the question had been, under what conditions the greater number of *species* of coral, not regarding their bulk or strength, were developed, I should answer,—probably in the situations described by MM. Quoy and Gaimard."[15]

Expanding on this critique of the most widely cited experts on reef-building corals, Darwin argued the need for a more complex understanding of the relations *between* the organisms whose growth contributed to reef structure. "In the vegetable kingdom every different station has its peculiar group of plants," he explained, in a silent reference to the works of Humboldt and of Augustin Pyramus de Candolle, "and similar relations appear to prevail with corals."[16] He spent several pages describing the living economy of a coral reef and explaining that different "zones" on the Keeling reef were characterized by their own particular kinds of corals and algae. He argued that the way reefs responded to a disturbance—for example, an earthquake that caused them to subside—might be determined by the happenstance of which groups of corals prevailed in the new conditions. "In an old-standing reef, the corals which are so different in kind on different parts of it, are probably all adapted to the stations they occupy, and hold their places, like other organic beings, by a struggle one with another, and with external nature." (Later in the book he introduced a second claim that also appears to have been stimulated by his species-oriented reflections on the struggle for existence. In a long footnote on page 94 he pointed out what he considered a particularly absurd shortcoming of Forster's theory that corals formed annular reefs by instinct: "According to this latter view, the corals on the outer margin of the reef instinctively expose themselves to the surf in order to afford protection to corals living in the lagoon, which belong to other genera, and to other families!")

The counterpoint to the fact that various species and genera were finely adapted to their locations on the reef was that any vertical movement of a reef's foundation would disrupt the whole economy of its animals and plants. If the South Keeling atoll were to subside by just a few feet, for example, "the Nulliporae [that] are now encroaching on the Porites and Millepora" would find "that the latter would, in their turn, encroach upon the Nulliporae."[17] Depending on the characteristic bulk and growth rates of the genera that prevailed, a given reef might regain the surface or it might languish underwater because it was "covered with luxuriant coral[s], [that] have no tendency to grow upwards."[18] Darwin pointed out that just these contingencies might explain why, despite occupying apparently identical physical conditions, some atolls of the Indian Ocean remained several fathoms beneath the surface while others grew right up to sea level.[19] Such differentiae might also explain the wildly varying estimates that different authors had offered for the rate at which corals grew and reef rock accumulated. Here he reviewed the evidence presented by other voyagers—who had learned from a combination of direct observation, comparisons with earlier surveys, and interrogation of long-lived residents—and gathered that while local conditions often precluded it, it was possible for reefs to grow rapidly in comparison with "the average oscillations of level in the earth's crust."[20]

He closed the chapter by reiterating and expanding another point whose significance would be revealed when he explained his theory in the following chapter: that the growth of reef-building corals was constrained by water depth. Reviewing particular details offered by hydrographers Robert Moresby and his lieutenant James Wellstead, Phillip Parker King, John Lort Stokes, and the ubiquitous Beechey along with those of the naturalists Ehrenberg and Joseph Pitty Couthouy, he quibbled with the *general* depth limit of thirty feet given by Quoy and Gaimard. Darwin's use of the reef literature was so thorough and up-to-date that he incorporated information from a voyage that had not even ended, the United States Exploring Expedition (1838–42). Couthouy, the erstwhile conchologist to the expedition who had been sent home by Commander Charles Wilkes, learned of Darwin's coral reef theory from Lyell's lectures in Boston in the fall of 1841. This prompted him to send Darwin a paper containing his observations on the Pacific reefs, which Darwin mentioned in footnotes added after the publishers had already begun printing the book.[21] Relying on the soundings he had carried out himself at increasing depths from the coast of Mauritius, he argued that this systematic inquiry established a trend of reef-building corals disappearing below 20 fathoms (120 feet) or so, and that this trend was more important than the exact depth where it happened. He again pointed out the analogy with terrestrial plants struggling to maintain their stations, which (his readers would know) had been described by Candolle.

> The circumstance of a gradual change . . . from a field of clean coral to a smooth sandy bottom, is far more important in indicating the depth at which the larger kinds of coral flourish, than almost any number of separate observations on the depth, at which certain species have been dredged up. For we can understand the gradation, only as a prolonged struggle against unfavourable conditions. If a person were to find the soil clothed with turf on the banks of a stream of water, but on going to some distance on one side of it, he observed the blades of grass growing thinner and thinner, with intervening patches of sand, until he entered a desert of sand, he would safely conclude, especially if changes of the same kind were noticed in other places, that the presence of the water was absolutely necessary to the formation of a thick bed of turf: so may we conclude, with the same feeling of certainty, that thick beds of coral are formed only at small depths beneath the surface of the sea.[22]

"I have endeavoured to collect every fact," he assured his readers, "which might either invalidate or corroborate this conclusion."[23]

The Theory Emerges

In chapter 5 Darwin offered a theory to explain all the foregoing facts. Having primed his readers with a strategic version of the natural history of coral reefs, he was able to dismiss his predecessors' theories of atoll formation in the space of a page.[24] Having also delivered some of his most important evidence in the descriptive chapters, he was able to introduce his own alternative as an inescapable consequence of established facts about coral growth and reef distribution:

> What cause, then, has given atolls and barrier-reefs their characteristic forms? Let us see whether an important deduction will not follow from the consid-

eration of these two circumstances,—first, the reef-building corals flourishing only at limited depths,—and secondly, the vastness of the areas interspersed with coral-reefs and coral-islets.[25]

What, he asked, could provide the foundations for the atolls that formed the great archipelagoes of the Low Islands, the Gilberts, the Marshalls, the Carolines, and the Laccadives? He began to eliminate alternative explanations: banks of sediment; chains of broad-summited mountains that all reached within 180 feet of the surface; and the leveling-off of individual mountains by waves as each was elevated close to the surface.[26] Showing his willingness to reason by exclusion, he embraced what he saw as the single remaining possibility: "If, then, the foundations of the many atolls were not uplifted into the requisite position, they must of necessity have subsided into it; and this at once solves every difficulty."[27]

As he had done in his private 1835 essay and in his Geological Society paper of 1837, Darwin asked his readers to imagine what would happen to a reef-fringed island under a scenario in which coral growth kept pace with a subsiding foundation. In the 1837 paper, and hence also in his *Journal of Researches*, he had admitted that "without the aid of sections [such as he had displayed at the Geological Society] it is not very easy to follow out the result" of such a mental exercise. In the 1842 book, by contrast, he described the hypothetical scenario with reference to a pair of carefully designed sectional diagrams that were included as woodcut prints on the same page as the text[28] (see fig. 26). The first showed the transition from a fringing reef to an encircling barrier reef, and the second showed the transition from that barrier reef to an atoll.[29] The rhetorical key to these diagrams was that the barrier reef stage depicted was not, in fact, imaginary. Rather, it was a vertical section of the recently surveyed island of Bolabola (Bora Bora) in the Society Islands. Thus the conjectural fringing reef and atoll stages were revealed to be just short steps in either direction from the actual conditions of a typical reef-encircled island.

This point about the structural similarity between reef types was driven home by another graphical technique presented in plate I of the volume, titled "Shewing the resemblance in form between barrier coral-reefs surrounding mountainous islands, and atolls or lagoon-islands." Here he juxtaposed charts (views from overhead) of reef-encircled islands with those of similarly shaped atolls (see fig. 27). In all ten examples shown on this plate the reef was tinted orange while the high land was only hatched in black.

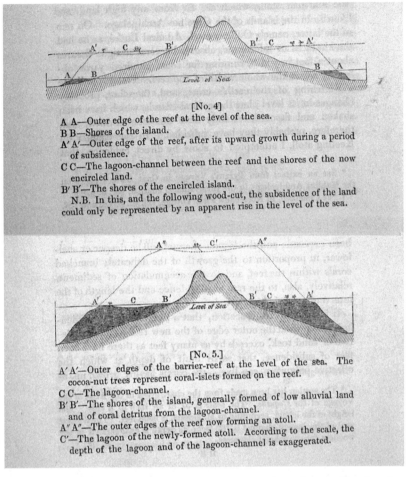

[No. 4]

A A—Outer edge of the reef at the level of the sea.
B B—Shores of the island.
A′ A′—Outer edge of the reef, after its upward growth during a period of subsidence.
C C—The lagoon-channel between the reef and the shores of the now encircled land.
B′ B′—The shores of the encircled island.
    N.B. In this, and the following wood-cut, the subsidence of the land could only be represented by an apparent rise in the level of the sea.

[No. 5.]

A′ A′—Outer edges of the barrier-reef at the level of the sea. The cocoa-nut trees represent coral-islets formed on the reef.
C C—The lagoon-channel.
B′ B′—The shores of the island, generally formed of low alluvial land and of coral detritus from the lagoon-channel.
A″ A″—The outer edges of the reef now forming an atoll.
C′—The lagoon of the newly-formed atoll. According to the scale, the depth of the lagoon and of the lagoon-channel is exaggerated.

Figure 26. Sectional diagrams from Darwin's *Structure and Distribution of Coral Reefs* (1842). The upper diagram shows the proposed transition from fringing reef to barrier reef by the action of coral growth during subsidence. The lower diagram shows the transition from barrier reef to atoll by subsidence. In each diagram the depiction of the barrier reef stage was based on a survey of the real barrier reef and island of Bolabola (Bora Bora), and the other stages were conjectural. Images courtesy of the History of Science Collections, University of Oklahoma Libraries; copyright the Board of Regents of the University of Oklahoma.

The result was to draw the viewer's attention away from the land to the shape of the reefs, which illustrated the striking parallel between the form of barrier reefs and atoll reefs while making the land within the barrier reefs appear ephemeral. This helped him conclude that "the close similarity in form, dimensions, structure, and relative position . . . between fringing

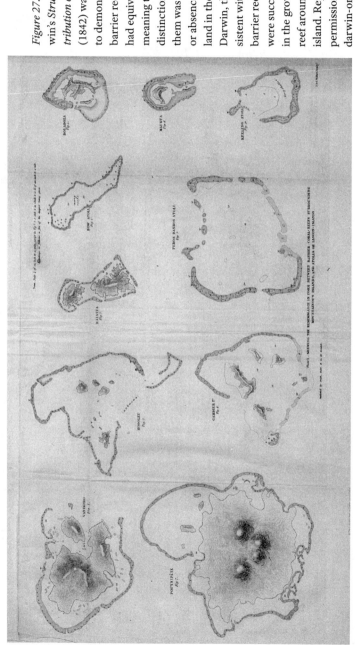

Figure 27. Plate I of Darwin's *Structure and Distribution of Coral Reefs* (1842) was intended to demonstrate that barrier reefs and atolls had equivalent shapes, meaning that the only distinction between them was the presence or absence of high land in the lagoon. To Darwin, this was consistent with his idea that barrier reefs and atolls were successive stages in the growth of a coral reef around a subsiding island. Reproduced with permission from http://darwin-online.org.uk.

reefs and encircling barrier-reefs, and between these latter and atolls, is the necessary result of the transformation, during subsidence, of the one class into the other."[30]

Much of the evidence Darwin adduced to show that subsidence could indeed occur on such a vast scale was new, or newly applied to the coral reef question, since he had made his coral reef theory public for the first time in 1837. Chief among the novelties was a series of observations indicating that the sea had transgressed on present-day atolls (presumably owing to subsidence), some drawn from local traditions reported by Williams and FitzRoy, others showing that the form of islands had been modified between successive visits of European voyagers.[31] He concluded chapter 5 by addressing a series of special cases that might present problems for his theory. He referred to several irregularly shaped islands and degraded reefs whose forms were all explicable, in Darwin's view, by "the action of . . . nicely balanced forces during a progressive subsidence . . . like that [implied] by our theory" but which had simply been "modified by occasional accidents which might have been anticipated as probable."[32] Thus he claimed that virtually every known reef in the world, even those that "differ[ed] from the *type* of the class to which they belong[ed]," could be "included in our theory."[33]

Darwin discussed his third, and most compelling, piece of graphical evidence in chapter 6. It was a large colored foldout map[34] (see plate 10). The base map was copied from the 1835 "Carte hydrographique des parties connues de la terre" (Hydrographic chart of the known parts of the earth), plotted on a Mercator projection by the Frenchman C. L. Gressier. Darwin's heavily annotated working copy of Gressier's map currently hangs in the "old study" at the Down House museum in the Darwin family's home in Kent. Given his close connections with Francis Beaufort at the British Hydrographical Office, it is curious that Darwin chose to use a French hydrographic map. I am not sure why he did so.[35] The version Darwin presented showed the 270 degrees of longitude that encompassed the Indian and Pacific Oceans and the Caribbean, from 30 degrees east of Greenwich to 60 degrees west of Greenwich, with the longitudes explicitly adapted to the English standard. As the map legend indicated, the locations of reefs had been marked and color-coded according to their classification. Because atolls and barrier reefs were indistinguishable "as far as the actual coral-formation is concerned," they shared the color blue, with atolls being marked in a darker tint and barrier reefs denoted by pale blue. Fringing reefs, on the other hand, were colored red. Thus the map was meant to call attention to the division between "two great types of structure": the coral reefs whose foundations lay within the possible depth of coral growth (colored red) versus those whose foundations were believed to lie below that depth limit (colored in one or the other shade of blue). It also depicted the locations of active volcanoes, which were indicated by vermilion spots.

If Darwin's subsidence theory was true, the distribution of the red and blue tints told a larger story. Reefs colored blue had subsided during the time since corals began to grow, whereas the foundations of those colored red had either stayed stationary or been elevated. Analyzing the "grouping" of differently colored reefs, he pointed out that the blues and reds were "not indiscriminately mixed together."[36] Atolls were often seen clustered, as were barrier reefs; these types were also found close together, which "would be the natural result of both having been produced during the subsidence of the areas in which they stand." Only in rare instances were red and blue dots found near one another on the map, and in these areas Darwin believed there was evidence of "oscillations of level," the term he used for *relatively* brief episodes of alternating uplift and subsidence. Darwin treated this apparent orderliness in the distribution of reefs, the systematic segregation between the "two great types of structure" represented by blue/subsiding and red/not subsiding areas, as a prediction made by his theory. He deemed the patterns on the map to be proof that the subsidence explanation was correct. "The grouping of the different kinds of coral-islands and reefs," he was thus prepared to argue, "is corroborative of the truth of the theory."[37]

Despite the well-known centrality of distribution (or biogeography, as it is now called) to Darwin's understanding of the origin of species, this chart of coral reefs was the only distribution map he ever published. By the 1840s, thematic maps of geology and of plant and animal geography were firmly established parts of the naturalist's repertoire. As historian Jane Camerini has demonstrated, there were many precedents for Darwin's use of this technique, though applying it to the distribution of coral reefs was novel. Like other creators of thematic maps, Darwin relied on the existence of satisfactory base maps on which to plot information. Camerini emphasizes in her examination of what she calls Darwin's "visual thinking" (by which she means cartographical or geographical thinking in particular) that, while he *published* only a single distribution map, Darwin often referred to imagined "mental maps" as he formulated, tested, and presented his theories of coral reef formation and the origin of species.[38] He also, of course, marked up maps for his own use, as well as publishing a handful of maps to accompany his *Beagle* works.[39] However, Camerini's concern with evidence of spatial *thinking* in addition to published and unpublished maps themselves has led her to dispute claims by fellow Darwin scholars David Stoddart and Martin Rudwick that Darwin's mode of thought was relatively nonvisual. Whereas Stoddart has argued that Darwin was "the least cartographic of men," Camerini argues that the many verbal references to maps and distribution in Darwin's notes and publications count as visual thinking.[40] She is therefore arguing a slightly different point than Stoddart, who was concerned with the fact that Darwin rarely *made* maps and that when he did they were rarely cartographically innovative.[41]

Whether it is seen as exceptional (Stoddart) or as the most obvious manifestation of his deeply ingrained geographical sensibility (Camerini), it is no surprise that Darwin's sole published distribution map has attracted considerable scholarly attention. What is difficult to determine, however, is the role this map played for Darwin himself. At various times Camerini refers to it as "the visual representation of his theory," as "proof of the theory, rather than a stimulus for it," and as the "consummate test of Darwin's theory."[42] This ambiguity reflects a tension between the role the map plays in the argument of the book itself, where it was indeed offered as proof of the theory, and the role of the concepts portrayed on this finished map, namely the location and relative orientation of coral reefs. Darwin contemplated these matters as early as the 1835 essay, in which he discussed what would be found "in looking at a chart" showing coral islands. My own view is that Darwin acquired what I would call a *geographical perspective* (consonant with Camerini's definition of visual language) during the *Beagle* voyage through his firsthand acquisition of practices, data, and specimens from the surveyors.

Darwin's classification of reef types served an important purpose in supporting the theory of their formation. He claimed on the first page of the book that his reef taxonomy reflected travelers' instinctive groupings and thus implied that these classifications were independent of theory. "Without any distinct intention to classify coral-reefs," he averred, "most voyagers have spoken of them under the following heads: 'lagoon-islands,' or 'atolls,' 'barrier,' or 'encircling reefs' and 'fringing,' or 'shore reefs.'"[43] This was literally true, in the sense that each of these six terms had previously been used to describe one reef or another. Yet the order imposed on these terms by Darwin's theory was already evident: not only did he list these ostensibly natural classes in a sequence suggested by his theory, he also implied that certain terms were straightforwardly synonymous, as with "atoll" and "lagoon island." Yet the voyagers Chamisso and Eschscholtz had been uncertain whether annular reefs of the Indian and Pacific Oceans even constituted the same phenomenon. In his second chapter, Darwin admitted that lumping encircling and barrier reefs into a single class was in fact his own taxonomic innovation, one he justified based on their similarities "in structure, and in position relatively [*sic*] to the land."[44]

There were even more profound respects in which Darwin's definitions of reefs made sense only in the context of his theory. The outcome of his deep comparison had been to reveal many fundamental similarities between the structures of fringing, barrier, and atoll reefs. In fact, Darwin excluded altogether from his consideration reefs that did not share these features. This decision reveals the theory-ladenness of Darwin's reef classification and was most evident in a series of statements explaining that certain reefs "resembled" atolls but were not "true" atolls. For him it was axiomatic that atolls were surrounded by deep water. A true atoll was one whose structure posed the question his theory had been designed to answer: How could shallow-water organisms establish a reef in a location where the ocean floor was too deep for them to live? It was indeed possible, he admitted, that reefs forming in shallow seas or around banks of sediment might "sometimes assume, (and this circumstance ought not to be overlooked,) the *appearance of atolls*" if there happened to be "more vigorous growth of coral on the outside."[45] Such reefs had no place in his classification. Indeed, in the appendix, discussing three annular reefs that lie off the coast of what is now Belize, he wrote, "These reefs have so completely the form of atolls, that *if they had occurred in the Pacific, I should not have hesitated about colouring them blue*. . . . [Yet] I consider it more probable that the three foregoing banks are the worn-down bases of upheaved shoals, fringed with corals, than that they are *true atolls, wholly produced by the growth of coral during subsidence*."[46]

This desire that all members of a class share the same mode of formation was evident in his dismissal of the theories of Quoy and Gaimard and Chamisso. Regarding the Frenchmen's crater theory, he explained, "I am far from denying that a reef like a perfectly characterized atoll might not [*sic*] be formed [atop a submarine crater]; some such, perhaps, now exist; but I cannot believe in the possibility of the *greater number* having thus originated."[47] On Chamisso's "earlier and better theory" that reefs assume a ring shape because the sturdiest corals flourish on the rims of coral masses, he acknowledged that "I believe some such exist in the West Indies[, b]ut a difficulty of the same kind with that affecting the crater theory, renders . . . this view inapplicable to the *greater number* of atolls."[48]

In his description of the colored map, he took this reasoning a step further and argued that the only "true atolls" were those his theory applied to: "[Even] if I had means of ascertaining the fact, I should not colour a reef merely coating the edges of a submarine crater, or of a level submerged bank [in the color reserved for an atoll]; for such superficial formations differ essentially, even when [they do] not [differ] in external appearance, from reefs whose foundations as well as superficies have been wholly formed by the growth of coral."[49] In other words, if it could be proved that a ring-shaped reef originated in the way described by the rival theory of atoll formation, then Darwin would consider that reef by definition *not* to be an atoll.

After arguing that his theory was the natural outcome of well-ordered thinking about coral reefs, he turned in conclusion to a discussion of the theory's broader implications. Recall that the 1837 paper had promised that this book, when it arrived, would shed light on the origin of species and the internal structure of the earth. Rather than using the book as an opportunity to expand on the speculative conclusions he had formerly offered, he narrowed the scope of his closing reflections considerably. He addressed two main themes: first was the apparent link between volcanic action and elevation, which he had proposed in the 1837 coral reef paper and elaborated in the 1838 "Connexion" paper, and which was supported by the absence of active volcanoes in the blue-colored areas of the map. The second theme was the relation between the areas of subsidence and of elevation. He pointed out that the portions of the earth's surface that had been raised and lowered were "immense," and he explained that the long, narrow shapes of the blue areas on the map might themselves represent only small proportions of the true geographical extent of subsiding areas. What he considered "perhaps, the most interesting conclusion in this volume" was that "the whole vast amount of subsidence, necessary to have produced the many atolls widely scattered over imense spaces [had been produced by] movements [that] must either have been uniform and exceedingly slow, or have been effected by small steps, separated from each other by long intervals of time."[50]

Darwin stopped short of examining the ultimate causes of elevation and subsidence. He was willing to diagnose movement of the crust, which by Whewell's definition was a proximate cause, but not to explain it. He declined to speculate on the internal matter of the globe, and he spoke of the relation between elevation and subsidence in decidedly measured tones, as when he pointed out that "[a] view of the map will show that, generally, there is a tendency to alternation in the parallel areas undergoing opposite kinds of movement; *as if* the sinking of one area balanced the rising of another."[51] (The 1837 paper, by comparison, had said, "One is strongly tempted to believe, that as one end of the lever goes up, the other goes down.") He summed up chapter 6 by stating that "the subterranean changes which have caused some large areas to rise, and others to subside, have acted in a very similar manner," which indicated that elevation and subsidence were only proximate causes of another phenomenon he would not elaborate on.[52]

In the three-paragraph "Recapitulation" of his argument, Darwin underscored the way his coral reef theory, and the map he had developed along with it, could be used as tools for the imagination. "When the two great types of structure, namely barrier-reefs and atolls on the one hand, and fringing-reefs on the other, were laid down in colours on our map, a magnificent and harmonious picture of the movements, which the crust of the earth has within a late period undergone, is presented to us." Just as he had previously imagined the island of Eimeo in motion, when he looked at his colored reef map he saw this "picture of movements" as though it were animated.

> We there see vast areas rising, with volcanic matter every now and then bursting forth through the vents or fissures with which they are traversed. We see other wide spaces slowly sinking without any volcanic outbursts; and we may feel sure, that this sinking must have been immense in amount as well as in area, thus to have buried over the broad face of the ocean every one of those mountains, above which atolls now stand like monuments, marking the place of their former existence.[53]

Here he signaled the climax of his argument by once again deploying the "monument" image that had pleased him so much ever since he first entered it in his *Beagle* diary six years earlier.

The last sentence of the book declared that he had shed light on such phenomena as were illuminated by his coral reef theory by undertaking a narrowly motivated attempt to explain the characteristic shapes of coral reefs. It was a claim that would completely contradict his later autobiographical account of the origin of the coral reef theory (on which see chapter 11 below). In the 1842 book he wrote:

Reflecting how powerful an agent with respect to denudation, and consequently to the nature and thickness of the deposits in accumulation, the sea
must ever be, when acting for prolonged periods on the land, during either
its slow emergence or subsidence; reflecting, also, on the final effects of these
movements in the interchange of land and ocean-water, on the climate of the
earth, and on the distribution of organic beings, I may be permitted to hope,
that the conclusions derived from the study of coral-formations, *originally
attempted merely to explain their peculiar forms*, may be thought worthy of the
attention of geologists.[54]

This strategic claim, that he had "originally" undertaken the explanation of
reef forms without any more general geological or zoological problems in
mind, was the final thrust of his booklong effort to portray his views on elevation and subsidence as conclusions independent of the theory's original
development. The point was not to give a true accounting of the genesis of
his theory. Rather, it helped him avoid the appearance of a circular argument and so to meet the conditions under which philosophically inclined
thinkers such as Herschel and Whewell would accept that his compelling
conclusions about "more general geological problems" could also stand as
independent evidence of the validity of the coral reef theory.

Note also that this closing sentence made one last allusion to the Lyellian cycle of changes that affected the earth's crust, climate, and organized
beings, and it stated explicitly that Darwin's "study of coral formations" was
intended for the benefit of geologists. This was indeed true of the effort he
had put into writing the book: it was intended to solidify his place in Lyell's
geological world. In doing so it completely effaced Darwin's original ambition (which I described in part I of the present book) to pursue the study
of corals as a zoologist.

To summarize, we can see that Darwin had structured the book carefully to make the theory as compelling as possible. This did not simply
mean arguing explicitly for the theory. It meant framing large sections of
the book as a theory-neutral natural history of coral reefs, much longer
and more detailed than any work previously written on the topic. The descriptive natural history was a Trojan horse, however, for Darwin's reef
terminology instantiated his theory even when it was not being deployed
to make an explicit argument. (A similar point was made in 1789 by a Scottish critic of Antoine Lavoisier's system of naming chemical compounds:
"We cannot speak the language of the new Nomenclature, without thinking as its authors do.")[55] Given that Darwin's book was indeed the most

comprehensive work ever produced on the growth and structure of reefs, it had value well beyond the theory it contained. However, anyone who accepted its framework and its terminology would be primed to find the theory plausible.[56]

The Immediate Reaction to *Coral Reefs*

With the book finally published, Darwin started trying to place it in the hands of his desired readers. Judging by his efforts, he envisioned geologists as the book's primary audience while also demonstrating his debts to, and common interests with, practicing hydrographers. He asked Smith and Elder to send copies for review to the *Athenaeum*, the *New Edinburgh Philosophical Journal*, and the *Philosophical Magazine*, and he told his publisher he was sending some of his presentation copies to "foreigners, who I thought by noticing the work, would aid its sale."[57] The other recipients were libraries and "people who had materially aided" Darwin in producing the book. He drafted at least two lists of intended recipients (some of whom may not ultimately have been sent a copy), illustrating the audience he perceived for the work.[58] Among the eminent foreigners he contemplated were his hero Humboldt; the navigator and cartographer Krusenstern; Lyell's two greatest European rivals, the geologists Léonce Élie de Beaumont and Leopold von Buch; and Christian Ehrenberg, the German naturalist and microscopist who had studied the Red Sea reefs. Absent from the lists were Quoy and Gaimard, the French naturalists who had proposed the crater theory, and Chamisso, who had died in 1838. The institutions to receive the work were London's geographical and geological societies, the Geological Society of Paris, and the "public libraries" of Britain. The individuals Darwin planned to reward for their assistance included the hydrographers FitzRoy, Beaufort, Moresby, and Beechey, along with Sedgwick and the "Dr. Allan" who had studied coral growth at Madagascar. Not coincidentally, Beaufort arranged for three copies of the new book to be delivered to the Admiralty Hydrographic Office.[59]

The reader Darwin most wanted to impress and convince was Lyell. This was as true of his coral reef book as it had been of the material he brought home from the *Beagle* voyage, and as we shall see it proved emphatically so when Darwin was putting the finishing touches on the *Origin* in late 1859. At the moment *Coral Reefs* was published, however, Lyell was still in the United States. For this reason, presumably, Darwin left him off the initial list of recipients, adding his name only to the second list.

Charles and Mary Lyell returned to London at the end of August 1842, three months after the book appeared. Lyell's immediate reaction is lost, but it must have been enthusiastic. "I have just received such a letter from Lyell on my Coral Volume," Darwin reported to Lyell's family-in-law the Horners, "that I have scarcely yet ceased stalking about like a peacock."[60]

Darwin's book received two published reviews before the year was out. One, by Colonel Julian Jackson, the sitting secretary of the Royal Geographical Society, was a highly laudatory piece that recommended the work to "the geographer, the navigator, and the *savant*."[61] Significantly, given Darwin's keen desire to avoid being seen as overly speculative, Jackson framed his review around the question of when theories might legitimately be established to describe collections of evidence. Jackson was ideologically skeptical of the value of undisciplined travel, and his stance within the internal political debates of the Geographical Society was as a strong advocate for systematizing geographical knowledge. Indeed, it was just such undertakings that he felt it was the society's prime duty to support.[62] Thus he was more than just sympathetic to Darwin's use of charts as a way of understanding the order that underlay the distribution of reefs.

Jackson opened his review by pointing out that while it was conventional "to deprecate the precipitancy with which theories and systems are raised upon the insufficient foundation of a few isolated facts," the opposite problem was equally troubling: that "an immense number of valuable observations on the most interesting and important subjects remain dispersed, and therefore almost useless, long after there is more than enough from which to deduce some satisfactory conclusion."[63] Jackson believed Darwin's greatest contribution was his organizing previously disconnected facts. He praised Darwin's mastery of other travelers' accounts and argued that this work of systematizing was an adequate warrant for the theoretical conclusions at the end of the book. "From the manner in which he has grouped the facts, and then reasoned upon them, the mind remains satisfied that he has detected the law, or rather the process, of nature in [coral reef] formation." In describing the structure of the book, Jackson commented that in the first four chapters "Mr. Darwin has confined himself generally to the arrangement and detail of facts." That a reader might consider these chapters to be mere description, given before he laid out the theory itself, was of course Darwin's goal.

Less easily persuaded, or less predisposed to Darwin's manner of theorizing, was Charles Maclaren, a Scots geologist and editor of *Encylopaedia Britannica*. In a review published in the *Scotsman*, a periodical he cofounded, and abridged in the *Edinburgh New Philosophical Journal*, Maclaren drew attention to the fact that the distinction between barrier and fringing reefs discussed in the second and third chapters, far from being pure description, "has reference chiefly to theoretical considerations."[64] Nevertheless, Maclaren accepted Darwin's premises about the ideal characteristics of theories themselves, concurring that it would be preferable in principle to have a single theory that could explain the origin of all types of coral reefs.[65] In explicating Darwin's argument for his readers, Maclaren was also willing to assent that even if the subsidence of reef foundations could not be directly proved, incorporating this cause into the theory was yet acceptable, and "involves no inconsistency," because "geology . . . renders it certain that some portions of the earth's surface have sunk to a lower level."[66]

Both Maclaren and Jackson described Darwin's color-speckled reef distribution map as a collation of evidence from which fresh theoretical conclusions could be drawn rather than as a mere illustration or manifestation of a preexisting theory. This too was a vindication of Darwin's careful segregation of his two main types of evidence in the manner first employed in the 1837 paper, with facts relating to the growing conditions of corals used as support for the subsidence theory and those related to the distribution of entire reefs reserved for interpretation in light of the theory. On Jackson's viewing of the map, "the direction of the spaces, coloured red on the map, and which represent the areas raised [*sic*], is such relatively to the spaces coloured blue and indicating the depressed areas, that their co-relation of effect seems evident on simple inspection, though their synchronism of action cannot in all cases be fully established."[67] Meanwhile, Maclaren himself speculated for several lines on the nature of the landforms that underlay the coral archipelagoes, and on the probable effects of a continued uplifting of the reef-fringed areas of the East Indies, which might one day "unite that vast chain of islands to one another, and to the continent of Asia . . . converting the Chinese sea into a vast inland lake."[68]

Despite believing that Darwin's theory "explains the phenomena under consideration better than any other which has been proposed," Maclaren closed his review by pointing out what he took to be three of its deficiencies. He first mentioned the imperfect segregation on the map between signs of elevation or stasis and signs of subsidence.[69] The point of this criticism was to show that it would sometimes require very selective boundary drawing to make these contrary indicators belong to separate geographical areas. His next objection was based on the rarity of sites that showed stages of transition between living fringing reefs and coral rock upraised on dry land. If so many areas of the coral seas had been elevated, Maclaren protested, there should be more cases where recognizable fringing reefs were found uplifted and in various states of degradation. His final criticism was the one he considered "the most serious objection to the theory." If the subsidence-built foundations of atolls were formed of coral rock two or three thousand feet thick, as Darwin's diagrams suggested, then there should somewhere be upraised masses of coral rock of similar thickness. In the cycle of geological changes the book was premised on, some such reefs ought to be upraised into terra firma. Yet nowhere in the world had "a bed or formation of coral, even 500 feet thick, been discovered, so far as we know."[70]

Darwin's responses to Maclaren's final objection revealed that for all his efforts to avoid overextending his theory, he was still sensitive about appearing too speculative. In a letter to Lyell, who evidently had "alluded to [Maclaren's] criticisms," Darwin admitted it was improbable that coral growth on a given reef could continue uninterrupted over the time required for several thousand feet of subsidence. However, he considered this admission "no ways fatal to the theory," because subsidence on such a massive scale was necessary only if one wanted to account for the disappearance of entire continents beneath the Pacific. "In the areas, where the large groups of atolls stand, & where likewise a few scattered atolls stand between such groups, I always imagined that there must have been great tracks of land, and that on such large tracks there must have been mountains of immense altitudes," he explained. "But now it appears to me, that one is only justified in supposing that groups of islands stood there."[71] In a published reply to Maclaren's review, Darwin conceded that "in my volume, I rather vaguely concluded that the atolls, which are studded in so marvellous a manner over wide spaces of ocean, marked the spots where the mountains of a *great continent* lay buried, instead of *merely separate tracts of land or mountainous islands*; and I was thus led to speak somewhat more strongly than warranted, of the probable vertical amount of subsidence in the areas in question."[72]

If Darwin had been indirect on this point in his book, however, it was hardly an oversight. The nature of "continental" subsidence and the question of what lay beneath the coral archipelagoes had been issues of central theoretical importance in the provocative version of the theory presented in 1837. Of course, at that time it had been Lyell who explicitly interpreted Darwin's coral reef theory as a sign that "the coral islands, are the last efforts of drowning *continents* to lift their heads above water."[73] It was just such extensions of the theory, and the question whether they had a place in learned treatises as well as in short papers, that had so paralyzed Darwin when he wrote the book. After five anxious years of writing and rewriting, the "bold" conclusions of his 1837 paper had become the very "vague" conclusions that he now abjured.

Although Darwin's anxious response to the criticism generated by his papers of 1837 and 1838 had discouraged him from pressing the implications of the theory as far as he had originally imagined doing, he remained devoutly attached to his answer to the limited question of what determined the shape of coral reefs. Darwin's personal faith in this theory, and in the type of theorizing it represented, held strong. As he said in closing his response to Maclaren, "The case, undoubtedly, is very perplexing; but I have the confidence to think, that the theory explains so well many facts, that I shall hold fast by it, in the face of two or three puzzles, even as good ones as your third objection."[74] Far from being empty boosterism, this statement portended the confidence Darwin would place in his coral reef work in the years to come. More significantly, it reveals what might be described as Darwin's pragmatic—but more appropriately perhaps as his Herschelian— attitude toward theories themselves. One might hold on to them in the face of objections because they explained so many facts, and because they remained useful tools for further thinking.

Maclaren's criticisms were among many reasons why, when Darwin decided to issue a second edition of his *Journal of Researches*, he made several noteworthy revisions to the chapter on coral reefs. Like so many of Darwin's opportunities, this one had come thanks to Lyell, who reported that his own publisher, John Murray, was eager to produce a new edition of Darwin's *Journal*. Having had the first edition tethered to FitzRoy's publisher and schedule, Darwin was eager to work with the house that had published Lyell's *Principles*.[75] When the *Journal* reappeared in 1845 as part of Murray's affordable Colonial and Home Library series, it had been condensed in some places, revised and expanded in others, and embellished with more than a dozen woodcut illustrations. Darwin had added an effusive dedication to Lyell "as an acknowledgment that the chief part of whatever scientific merit this journal and the other works of the author may possess, has been derived from studying the well-known and admirable Principles of Geology."[76] As with so many of Darwin's public references to Lyell, this one implied that Lyell was more important as an author than as a face-to-face teacher. With this edition too Darwin could replace his 1839 promise of forthcoming books on geology with an announcement of his geological credentials. "I have myself published," he wrote in the preface, "separate volumes on the 'Structure and Distribution of Coral Reefs;' on the 'Volcanic Islands visited during the Voyage of the Beagle;' and a third volume will soon appear on the 'Geology of South America.' The sixth volume of the 'Geological Transactions' contains two papers of mine on the Erratic Boulders and Volcanic Phenomena of South America. I intend hereafter to describe, in a set of papers, some of the marine invertebrate animals collected during the voyage."[77]

Darwin's now published coral reef book played a major role in his revised *Journal* chapter on coral reefs. Five of the new woodcuts had been taken directly from it: the three cross-sectional diagrams showing reefs above and below water and landscape views of Whitsunday Island and Bolabola. And while Murray did not reproduce the foldout distribution map, Darwin nevertheless devoted three new pages to discussing it. He followed his 1842 book in using the new tripartite classification of reef types and in calling the first of those classes "atolls" instead of "lagoon islands." In contrast to that book, however, he avoided mentioning the possibility that entire continents had been drowned in the region where atolls could be found. Whereas in the first edition he had written of the potential "former existence of an archipelago or continent in the central part of Polynesia," he now, having weathered Maclaren's review, wrote of "the probability of the former existence of large archipelagoes of lofty islands, where now only rings of coral-rock scarcely break the open expanse of the sea." The stated implications of those former lands' existence had also changed since he wrote the first version of that sentence in 1837. His original, remarkable declaration was that "if the [subsidence] theory should hereafter be so far established . . . it will directly bear upon that most mysterious question,—whether the series of organized beings peculiar to some isolated points, are the last remnants of a former population, or the first creatures of a new one springing into existence." Now, after half a decade spent publicly absolving himself of such bold declarations of intent to speculate (and privately speculating aggressively on this very topic) he avoided all implications of transmutation or generation of species. Removing both the allusion (to the mystery of mysteries) and the phrase ("springing into existence") yielded the subtler aim of "throwing some light on the distribution of the inhabitants of the other high islands, now left standing so immensely remote from each other in the midst of the great oceans." By 1845 that topic had become one of the chief preoccupations of Darwin's closest interlocutors, which helped to ensure that the coral reef theory would be a recurring subject of discussion.

A Theory in Use and in Memory

For Darwin, the oceans and shorelines of the world remained colored in reds and blues. The coral reef map was not only, as Jane Camerini has noted, "a representation of the base map . . . that was in his mind during and after the exciting and profound developments in the so-called transmutation notebooks."[78] It was also the resource that he literally unfolded and consulted as his primary reference on matters of tropical geography, geology, and natural history. As his correspondence and notes reveal, he made particularly frequent use of the coral reef distribution map in contemplating the dispersal of animals and plants across the oceans. This was the topic of an involved correspondence in the mid-1840s between Darwin and the young botanist Joseph Dalton Hooker. Darwin advised Hooker, "If you will look at the map in my Coral-volume, you will see that probably much more land existed within geologically recent times than now exists," and he annotated the letters he received with similar reminders to himself, such as "Islands like Mountains—Isl^ds of Pacific most puzzling. . . . Look at my Coral Is^d Map & see whether most peculiar on Blue or Red."[79] This pattern of behavior was repeated for decades by Hooker and Lyell as well as Darwin, each treating the coral reef volume as a source of enduring intellectual stimulation as well as a compendium of data on tropical landforms.[80]

Lyell's scientific journals show that during the years to come he strove to analyze the effects of subsidence that had occurred on the scale Darwin's book implied, even while impishly delighting that Darwin had shown how a "tremendous catastrophe"—the disappearance of vast tracts of land beneath the sea—could be "brought about by what Sedgwick called 'Lyell's niggling operations.'"[81] After the *Origin of Species* was published in 1859, Lyell used Darwin's coral reef theory as a tool for reasoning on the relative speeds of geological and organic change, assuming that species on a subsiding island or continent might become adapted to their new conditions, and writing, "As to atolls . . . if they subsided very slowly, the absence of volant & amphibious forms of reptile & mammifer in such a region proves rate of transmutation slow, even as compared to revolutions in physical geography."[82] For Lyell, the coral reef volume self-evidently belonged to the uniformitarian canon.

It turned out that Darwin himself, as the years passed, could not quite keep track of where he had fallen on the question whether groups of atolls were underlain by mere archipelagoes or entire continents. Although he had followed Lyell by making statements that suggested the subsidence of whole continents, he often went on to claim the opposite, This was in part an ad hoc manifestation of Darwin's opposition in the mid-1850s to claims made by Edward Forbes and others that the former existence of land bridges could explain the distribution of organisms, which Darwin viewed as unphilosophical speculation. He exclaimed self-righteously to Hooker, "I never made a continent for my Coral Reefs."[83] Later he told Lyell, "With respect to ~~permanence~~ <<long endurance>> of our existing continents, I formed my opinion chiefly from facts of geographical distribution, to which I allude in Origin—& partly from views given under Coral Reefs."[84] Hooker, for his part, could "not find a reference to the permanence of continents in [Darwin's] "Coral Reefs",—a book by the way that shook my confidence in that theory more than all others put together, & the effect of which it has required years of thought to eliminate or rather to overlay."[85] No doubt Darwin's protestations were partly a case of revisionist history, of remembering what he wanted to remember. But they also illustrate one consequence of recasting and reassessing his theory during and after the *Beagle* voyage: he ended up producing statements over the years that were in tension or even downright contradictory.

Despite the fondness he expressed for the coral reef book throughout his life, he returned periodically to considering whether he had been overzealous in his speculations. While each recollection was aimed at a different recipient or audience and meant to serve a different purpose, it is nevertheless instructive to survey how Darwin managed this issue. In some instances he defended his rigor in guarding against error. In 1843, when Ernst Dieffenbach was preparing a German translation of the *Journal of Researches*, Darwin sent him a copy of the recently published *Coral Reefs* to review. Thinking of the coral reef chapter in the *Journal*, which had been taken verbatim from his 1837 Geological Society paper and whose bold speculations had caused him so much anxiety, he added, "Perhaps you will be good enough to insert a note to the effect that the chapter in my Journal is only a *rough* sketch of the facts & details given in full in this work."[86] He explained to the retired army officer Charles Hamilton Smith in 1845 that he had taken a highly skeptical approach to the sources he used in compiling his reef distribution map because "every one knows how greedily a theorist pounces on a fact [that is] highly favourable to his views."[87] (This, incidentally, is the earliest instance I have seen of Darwin's employing the term theorist, which was not in wide use in the early or middle nineteenth century.)[88] When Henry De la Beche, whose sketches had lampooned Lyellian theorizing two decades before, asked during his 1848 presidential address to the Geological Society whether atolls might in some instances form atop banks of sediment, Darwin responded with a letter restating the arguments made in the book and advertising the judicious approach he had taken there. "I remark," he reminded De la Beche, that "'the evidence from a single atoll or a single encircling barrier-reef, must be received with some caution, for the former may possibly be based on a submerged crater or bank, and the latter on a submerged margin of sediment or of worn-down rock.'—Whether you consider my remarks satisfactory or not, I trust that you will find that I have not proceeded without consideration of the sources of error: I assure you, I did not spare time or labour in examining thousands of charts & all voyages.—But forgive the length of this letter; a man is as tender of his theories as of his child<<ren>>."[89]

When, on the other hand, Darwin was responding to new reef studies that seemed to support his theory, he was then all too willing to acknowledge that he had speculated, and done so perceptively. Writing to Joseph Beete Jukes, whose examination of the Australian barrier reef had led him to declare Darwin's "the true theory of coral reefs," he affected this confessional tone. "I have always felt that my coral-reef book was too bold & speculative & therefore you will not easily imagine how gratified I am when anyone, who has had opportunities of observation, does not give his verdict against it."[90] After James Dwight Dana of the United States Exploring Expedition reported on the findings of that most comprehensive trek through the Pacific islands, Darwin gloated to Lyell, "I am astonished at my own accuracy!! if I were to rewrite now my coral book, there is hardly a sentence I sh^d. have to alter. . . . Dana talks of agreeing with my theory in most points; I can find out not one in which he differs.—Considering how infinitely more he saw of Coral Reefs than I did, this is wonderfully satisfactory to me; though really I think it some little reflection on him, that he did find other & new points to observe."[91] Then three days later, when he had read further in Dana's book, Darwin himself began to adopt a judgmental tone. Writing again to Lyell, he griped that "Dana is dreadfully hypothetical in many parts. . . . He strikes me as a very clever fellow; I wish he was not quite so grand a generaliser."[92]

Darwin's apparently inconsistent message on theorizing reflects the tension between his desire to adhere publicly to his colleagues' strictures against overzealous speculation and his desire personally and philosophically to identify laws and construct theories. Whenever he looked ahead, anticipating the prospect of publishing a theory, the social norms of the British scientific community were foremost in his mind. Once he had gone forward with an act of generalizing—that is, when he had published a theory—he seems to have prioritized defense of the theory over defending the act of theorizing. That was presumably because being correct—or more properly speaking, producing a *useful* theory—was the strongest possible justification for daring to speculate.[93] Nevertheless, when Darwin looked ahead to the prospect of publishing *another* grand theory—his species theory—he knew that his positive and negative experiences offered the most valuable lessons for how—and how not—to proceed.

PART IV

Writing the *Origin* with His "Fingers Burned"

11

Atoning for the Sin of Speculation

The personal experience that loomed largest in Darwin's mind when he considered publishing the theory of evolution by natural selection was his earlier struggle to become a geological author. In this final chapter I argue that Darwin's well-known authorial decisions before publishing *On the Origin of Species* were attempts to avoid repeating, and ideally to compensate for, the missteps he believed he had made in his first forays into theoretical authorship. And, indeed, if writing on coral reefs and global geology had been the turbulent experience I have claimed it was, we would hardly expect Darwin to proceed as though his earlier approach had been an unalloyed success when he made plans for his next theory.

Darwin's reluctant approach to publishing the species theory is generally understood to have been due to some combination of innate scholarly caution and fear of social repercussions from advocating the controversial doctrine of transmutation. Darwin did of course recognize that promulgating a theory of transmutation would be controversial, especially if he included the origins of mankind in his evolutionary account, but I have argued that a considerable portion of Darwin's well-known anxiety—in the several years around 1840—was in fact brought on by his geological publishing projects rather than by his private conclusions about species. In those years the species project was an escape from the effort to complete his geological manuscript and meet the expectations surrounding it. No doubt Darwin's long-term plan to publish a theory of transmutation had the effect of privately raising the stakes for his first scientific treatise, making him doubly sensitive to accusations (aimed at his already published work) that he was prone to premature speculation. But in this chapter I demonstrate both that Darwin persisted in feeling he had erred in revealing his coral reef theory and other geological speculations too early and that this conviction had real consequences for his approach to publishing on species. At key moments when he might have chosen a more aggressive publishing strategy, he took measures to avoid being seen as overly or prematurely speculative. We are now in a position to recognize that this desire was neither an inherent quality of Darwin's character nor a fear stimulated by the specific topic of transmutation. Rather, it was a consequence of the devastation this criticism had wrought on his morale when it was leveled at his initial geological publishing strategy, and it persisted long after his eventual geological publication strategy and subsequent work on barnacles had helped everyone else to forget that Darwin had ever been quite so zealous.

Recent years have seen debate among historians of science about whether Darwin "delayed" publishing his species theory.[1] If we mean this in the extremely restricted sense that he neglected to send for publication a statement of the theory he had written for that purpose and did not intend to continue working on, then the evidence indicates that Darwin did not delay. On the other hand, that is an artificial criterion leaving open a range of ways Darwin did postpone (or fail to elevate to top priority) writing up a publishable expression of the theory. Thus the more compelling question in my view is why, in practice, from day to day and month to month, Darwin felt it was important to treat the species theory as a long-term project and to work toward publishing it in the most comprehensive and ambitious form he could imagine. This is not to say he ever made a single, inviolable choice to work steadily and dispassionately on writing a "big species book." Rather, through the accumulation of many small decisions, he revealed that he was less comfortable with the option of publishing than with the option of continuing to work.

The combined effect of these decisions was that he accepted the risk of losing his priority in order to avoid the consequences of publishing prematurely. But there were specific moments when he was required, either by circumstances or by an explicit intervention from a colleague, to address directly the question of when and in what form he would publish the species theory. At such times he repeatedly chose the long course, making it clear that he had not been born cautious. He had adopted caution as a publishing strategy in response to earlier disappointments. These moments of reflection and decision, which reveal so much about his attitude toward the production of scientific knowledge in the wake of his earliest publishing attempts, are the subject of this chapter.

Balancing Speculation with Facts

On 5 July 1844 Darwin committed to paper a "most solemn & last request." Dots of ink left by the nib of his pen show that he paused after writing each word of the salutation:

> My. Dear. Emma.

"I have just finished my sketch of my species theory," he explained, referring not to the "pencil sketch" he had written two years earlier but to a recently completed 230-page essay.

> If, as I believe that my theory is true & if it be accepted even by one competent judge, it will be a considerable step in science. I therefore write this, in case of my sudden death, as my most solemn & last request, which I am sure you will consider the same as if legally entered in my will, that you will devote 400£ to its publication & further will yourself, or through Hensleigh [Wedgwood], take trouble in promoting it.[2]

These opening sentences alone offer several revelations. They reveal Darwin's utter confidence that his theory was true and that he thus saw as his primary task not confirming the theory but ensuring that it would be accepted when it did appear. This meant the published theory required "promoting." And so the letter went on, with detail after detail providing insight into Darwin's conception of his species theory, his attitude toward publishing theories in general, and his notion of how scientific knowledge was produced.

This letter might be taken as an expression of the severity of Darwin's ill health and his feelings of physical and emotional vulnerability. But Darwin was motivated less by a vague sense of impending death than by the specific conviction that he possessed a valuable piece of intellectual property. "Considerable step[s] in science" were and are made through the *sharing* of ideas and facts. But Darwin was not ready to share his theory with the world. The letter to Emma created space for Darwin to continue to work toward producing the optimal expression of the theory while making sure that his accidental death would not result in the total loss of an idea to which he had devoted so much time and energy.

This letter has been widely described as an instruction for Emma to publish his work posthumously.[3] In fact, though, it made precisely the opposite point. This was a letter in which Darwin instructed his survivors on the necessary way to *continue* his work . . . *before* publishing.

Charles wanted Emma to identify "an Editor" who could take charge of the manuscript. Do not be misled by his choice of the word editor into thinking he wanted someone to tidy his prose and submit the paper for publication. What he sought, in fact, was someone who could take his place and complete the process of authoring the theory, from composition to publishing to promoting. "I wish that my sketch be given to some competent person," Darwin wrote,

> with this sum to induce him to take trouble in its improvement. &
> enlargement.—I give to him all my Books on Natural History, which are
> either scored or have references at end to the pages, begging him carefully
> to look over & consider such passages, as actually bearing or by possibility
> bearing on this subject.—I wish you to make a list of all such books, as some
> temptation to an Editor.

This is nothing less than a syllabus of the work Darwin himself felt obliged to finish, complete with an explanation of the techniques he was using to gather and catalog facts. And he continued with further details of the working method he intended to bequeath, along with his theory, to the editor who would take up the task.

> I also request that you hand over [to an editor] all those scraps roughly di-
> vided in eight or ten brown paper Portfolios:—The scraps with copied quota-
> tions from various works are those which may aid my Editor.—I also request
> that you (or some amanuensis) will aid in deciphering any of the scraps which
> the Editor may think possibly of use.

Darwin was acutely aware of the magnitude of the job he had set himself, and he doubted anyone else could be persuaded to assume it without considerable incentive.

As the looking over the references & scraps will be a long labour, & as the correcting & enlarging & altering my sketch will also take considerable time, I leave this sum of 400£ as some remuneration & any profits from the work. . . . Should one other hundred Pounds, make the difference of procuring a good Editor, I request earnestly that you will raise 500£.

These instructions reveal how deeply obsessed he was with the *style* in which his theory would become public. He had no intention of publishing anything less than a comprehensive statement of the theory, a systematic treatise in which he had considered and accounted for every class of facts that might bear on—or be illuminated by—a theory of transmutation. The best example of such a book, the model Darwin likely had in mind even at this time, was Lyell's three-volume *Principles of Geology*. However, the *Principles* set out to defend a general approach to geological work rather than a specific single theory. Arguably the most notable natural history book that had yet been published *in defense of a single theory* was Darwin's own coral reef book, which, as we have seen, was the consequence of his abandoning the kind of broad ambition that still defined the species project.

If Darwin was so confident that the species theory was correct and that the work ahead of him would be "a long labour," why didn't he begin with the more manageable task of writing a short statement of the theory? It might seem obvious that he should have made his theory public to establish his priority over other potential discoverers of the principle of natural selection. If he had cared only about priority—that is, about ensuring that he would be the first to publish an idea that might in due time be amplified and supported with a greater mass of evidence—he could have put forward a preliminary sketch like the one he had written on coral reefs in 1837. In fact, he wanted his executor to do just the opposite.

So profound was his desire to publish the species theory first *as a book* that Darwin would not countenance having the "editor" of his 1844 sketch make a priority of establishing priority. Only as an absolute last resort, if Emma could find no editor at all, would Darwin accept the idea of publishing what he had already written. "If there sh[ould] be any difficulty in getting an editor who would go thoroughily into the subject & think of the bearing of the passages marked in the Books & copied out on scraps of Paper, then let my sketch be published as it is." In such an event, he insisted, it must carry a disclaimer to insulate him from any accusation that the essay was unphilosophical. Emma was to state "that it was done several

years ago & from memory, without consulting any works & with no intention of publication in its present form."

Examining the qualities Darwin's ideal editor would possess can reveal
what he valued most about himself as the prospective author of his transmutation theory. The editor would have to be "a capable man," but also
"a geologist, as well as a Naturalist." Darwin's first choice was Lyell, who
would not only be the best candidate but would, Darwin believed, "find
the work pleasant and . . . learn some facts new to him." The next best candidates would be the two most recent curators of the Geological Society's
collections, Edward Forbes and William Lonsdale. Whereas Lonsdale was
(as Darwin noted) in ill health, his successor Forbes had recently settled in
London and, in addition to the curatorship of the Geological Society museum, had taken the position of professor of botany at King's College. Like
Darwin he had become enthusiastic about marine zoology as a student at
Edinburgh and had pursued natural history aboard a surveying vessel. Not
coincidentally, he also shared Darwin's interest in marine invertebrates and
their distribution across not only space but also depth. His 1843 report to
the BAAS on the Mollusca and Radiata of the Aegean Sea carried a telling
subtitle: "on their distribution, *considered as bearing on geology*." In pursuing geology via the study of the depth range of living marine organisms,
Forbes had taken up the same enterprise that I have argued was so crucial
to Darwin's own intellectual development.[4]

Lower on Darwin's list of potential editors were his closest botanical
friends: his mentor and his protégé, Henslow and Hooker. But Lyell's special status as grandfather to Darwin's work was emphasized again in closing: if none of the named candidates was willing to accept Emma's commission, it was to Lyell she should turn for advice on identifying an editor
who was, Darwin repeated, "a geologist & naturalist."[5]

The most interesting feature of this letter is the insight it offers into Darwin's attitude toward the relation between "facts" and what he called "my
theory." To reiterate, his opening lines revealed confidence that the theory
was true. He expressed no concern that the editor of his manuscript would
question whether the theory was worth publishing. He had taken pleasure, after all, in imagining that Lyell would "learn some facts new to him"
while completing Darwin's work. Likewise, Darwin revealed no insecurity
about whether the long labors he envisioned for himself or his editor might
weaken or contradict the theory. Rather, some facts might simply prove
irrelevant. "Many of the scraps in the Portfolios," he explained, "contains

[*sic*] mere rude suggestions & early views now useless, *& many of the facts will probably turn out as having no bearing on my theory*."[6]

This phrase is reminiscent of the lesson Darwin had learned at a similar stage in the effort to produce his geological book. In 1839, recall, he had struggled mightily with his initiation to such labor. As he had written at the time to Leonard Jenyns, "it is very pleasant easy work putting together the frame of a geological theory, but it is just as tough a job collecting & comparing the hard unbending facts." What he had meant by *facts* in 1839 were the data about individual coral reefs that he was working to plot on his distribution map. He learned most of these facts, which ended up being enumerated in the appendix to *Coral Reefs*, by consulting charts in London. As he said to Jenyns, he had just spent six weeks gathering these data. Similarly, what he meant by facts in reference to the species theory in 1844 were the fragments of knowledge he had collected from his voracious, theory-motivated reading of journals, books, and correspondence.[7] Now the facts resided in the pen-scratched margins of his books and in the portfolios in which he had begun to collect individual notes. On both occasions Darwin described writing as a process in which *facts followed theory*. This is not to say his theories were pure deductions, prior to any facts. Rather, making a useful theory ready for publication required vetting it against all facts that were possibly relevant. This was the challenging ideal he became committed to after his early, speculative papers had been met with criticism. He had come to realize that what provided a warrant for *public* speculation was the accumulation of an unassailable mass of relevant facts.[8]

Darwin's instructions thus also provide a glimpse into the way the apparent origins of a theory might change in the process of writing. If preparing a theory for the press involved marshaling facts in support of an argument, then the resulting publication by definition did not simply restate the story of his original thinking. If his written theories did not have a place for explaining "early views now useless," there was no reason we, as readers, should assume that such publications contained reliable accounts of his original method of hypothesizing.[9]

Beyond writing and editing lay the necessity—as he told Emma—of "promot[ing]" the book. As early as 1837 Darwin had learned the importance of self-presentation and publishing strategy. Lyell's premeditated recanting of the crater theory had in fact helped Darwin secure a favorable response to the initial product of his zeal for theorizing: his coral reef paper. It was only by 1839, as I have explained, that circumstances conspired to convince him that a paper alone was an insufficient vessel to carry all necessary demonstrations of rigor, and that a theory-oriented book—though challenging to write—was essential for the long-term credibility of both the theory and its author. Darwin had absorbed these lessons, and they are evident in his plan for the species theory. He offered Emma little detailed advice on how his theory should be promoted, saying only that "the Editor is bound to get the sketch published either at a Publishers or at his own risk." But given that Lyell taught Darwin so much about how to advocate for one's position, and had taken such a proprietary interest in promoting Darwin's earlier work as a subsidiary part of his own geological program, he could have been expected to act as Darwin's champion again if he accepted the role of editor. Darwin closed the letter to Emma with a postscript: "PS Lyell, especially with the aid of Hooker (& of any good zoological aid) would be best of all." In other words, Darwin did not necessarily expect Lyell to master the botanical and zoological facts relevant to the species theory, but he had deep faith that Lyell would be an effective advocate.

: : :

Two letters Darwin wrote in the following weeks offer further insight into his attitude toward facts in relation to theories. The first was to Lyell's father-in-law, Leonard Horner, who had written to praise the second book to emerge from Darwin's decision to subdivide the *Beagle* geology, the recently published *Volcanic Islands*.[10] Darwin latched on to a comment Horner had made on Darwin's discussion of the old rivalry between Lyell and Leopold von Buch, responding, "With respect to Craters of Elevation, I had no sooner printed off the few pages on that subject, than I wished the whole erased. . . . I wish I had left it all out; *I trust that there is in other parts of the volume more facts & less theory*."[11] Here again we see Darwin assessing his work's fitness to be published in terms of an appropriate ratio of evidence to speculation.

This criterion for publication was even more explicit in another letter of that same year, to the Swiss geologist Adolph von Morlot. Darwin adopted a manner that was unusually patient by the standards of his dispatches to little-known correspondents but that began to establish a pattern he would repeat whenever he found himself in a position to warn others against being speculative. On each occasion he confessed that he had committed that transgression himself. This time he was advising Morlot how to rewrite a paper on glaciers to make it adhere more closely to the norms that governed geological publishing, both tacitly and "according to the rules of the [Geological] Society."

"Plainly & briefly describe every fact, which you observed," Darwin urged, "& after you have so described your facts, *aid your reader* in drawing his conclusions." Rhetorical strategy was no afterthought. Darwin may have become more cautious since 1837, but advertising caution was a tool of persuasion rather than a theorist's genuine act of apostasy. He made it plain to Morlot that he did not mean to indict the theoretical cast of mind itself, but rather intended to offer a more effective *publishing strategy*. "Pray observe I do not pretend to say your theories are not right, but a substratum of facts ought surely to be first given. . . . I well know that this comes very badly from me, who have dealt so largely in *the sin of speculation*, which I endeavor, though with little success, to check." Was the geological "sin" theorizing? No. The sin was failing to make a conspicuous show of performing the appropriate ablutions beforehand. Evidence could be tainted by premature contact with theory: "Your letters, in which facts are so mingled with speculation, are not fit to be published."[12]

A few years later Darwin told another correspondent, C. H. L. Woodd, "How neatly you draw your diagrams; I wish you would turn your attention to real sections of the earth's crust, & then speculate to your hearts content on them; I can have no doubt that speculative men, with a curb on make the best observers." Once again, Darwin advertised that he was speaking from experience. "All young geologists have a great turn for speculation; I have burned my fingers pretty sharply in that way, & am now perhaps become over cautious; & feel inclined to cavil at speculation when the *direct & immediate* effect of a cause in question cannot be shown."[13] Darwin had not been born cautious. Caution had been drilled into him by the criticism he received in the crucial years between publishing his own speculative papers—on coral reefs, volcanoes, and Glen Roy—and producing the geological books that by their scale had added so greatly to the stock of geological *facts* as well as *theory*.

Darwin's admission that his fingers had been "burned pretty sharply" by his past speculations has, of course, received considerable attention. It has largely been taken for granted that he was referring to his paper on Glen Roy, which he described many years later as a "great blunder." At the time he was writing to Woodd, however, he had not yet ceased to defend his explanation for the parallel roads of Glen Roy.[14] The occasions when his fingers had been burned, as I have shown, were not necessarily ones in which he later felt he had been *wrong*, but ones in which he felt he had opened himself to criticism by adopting the poor publishing strategy of speculating in short-form papers that did not carry the maximum possible weight of evidence. The coral reef theory was the key example from Darwin's early career. And when, in 1856, Lyell pushed him to publish a short-form version of his species theory, it was to the precedent of his coral reef paper that Darwin's mind immediately turned.

Rejecting Lyell's Suggestion to Publish a "Sketch"

The possibility that Darwin would publish a short statement of his species theory in order to establish priority is no mere imaginary scenario. He was in fact forced to make a conscious decision about whether to reveal the theory that way. Darwin had to wrestle with this dilemma because Charles Lyell insisted upon it.

It was May 1856. Darwin had by then published not only his three volumes of *Beagle* geology, the multivolume *Zoology*, and two editions of the *Journal of Researches*, but also a two-volume monograph on living and fossil Cirripedia (barnacles). After nearly twenty years spent working in Britain on projects about, or expanding from, his labors during the *Beagle* voyage, the species theory had finally become his primary occupation. (He wrote in his journal on 9 September 1854, "Finished packing up all my Cirripedes. . . . Began sorting notes for species theory.") He had, in turn, begun to discuss the theory in greater detail with colleagues. Lyell learned the extent of Darwin's views during a visit to Down House in April 1856. According to a note he wrote in his journal on the sixteenth, he was "With Darwin." The topic of discussion was momentous: "the formation of species by natural selection."[15] At the end of the same month Charles and Emma Darwin hosted Hooker, Thomas Wollaston, Thomas Henry Huxley, and their wives for a lively weekend of conversation. The men wrestled with the topic of transmutation, and from what Lyell could gather "they (all four of them) ran a tilt against species farther than I believe they are deliberately prepared to go." It was fully a quarter of a century since Lyell had comprehensively rejected Jean-Baptiste Lamarck's theory of transformism in the pages of the *Principles*. He was no closer to accepting it. "I cannot easily see how they can go so far, and not embrace the whole Lamarckian doctrine," he told his brother-in-law, Charles Bunbury.[16]

Nevertheless, Lyell believed in the value of theorizing, and he wanted to have Darwin's theory as a tool to think with. The question of species origins remained as salient to a range of puzzles in geology, paleontology, zoology, and botany as it had been thirty years earlier when Herschel described it, in the widely circulated letter to Lyell, as "that mystery of mysteries." During this very spring of 1856, for example, Lyell and Darwin were deep in debate with Hooker, Forbes, and others about whether former land bridges could explain the global distribution of flora and fauna. Lyell had recently received an essay from the Swiss paleontologist Oswald Heer positing, based on its flora, that the island of Madeira had at separate times in its history been joined to North America and to Europe. "He makes out a good case in favour of the old union," Lyell told Darwin, "provided one believes in specific centres. According to any other hypothesis I cannot as yet very well see how to bring the geograph[ical] facts to bear one way or the other."[17]

But Lyell knew that Darwin's work on the origin of species would offer an alternative to the old idea of "specific centres" of creation. He wrote to Darwin, "The multiple creation of Agassiz will one day rank with spontaneous generation [as a discredited scientific idea] but Madeira seems to me to favour the single specific birth-place theory & I long to see your application of any modification of the Lamarckian species-making system." Lyell was frustrated that Darwin had withheld his theory when there were puzzles it might solve. "I wish you would publish some small fragment of your data . . . & so out with the theory & let it take date—& be cited—& understood."[18]

This was a significant moment. By continuing to work on the project as he was then envisioning it, Darwin would be making an active choice to disregard Lyell's advice. In his immediate reply to Lyell's letter he acknowledged that the recent consultation with Hooker, Huxley, and Wollaston had produced "much to me most interesting conversation." He crowed, "It is really striking (but almost laughable to me) to notice the change in Hookers and Huxley's opinions on species during the last few years."[19] This might have reminded Lyell of his own confidence in the "heretical doctrines" of uniformitarian geology at the time when he was cultivating Darwin as a disciple and finding that his inexperienced protégé, "for not being able to measure the *change* of tone" in geological debate, had mistaken others' critical remarks as "vigorous defiance instead of a diminishing fire & an almost beating of retreat."[20] Darwin may not yet have published on transmutation, but he had made reassuring progress in persuading others within his coterie to take the prospect seriously.

But Darwin was exercised by Lyell's suggestion to publish immediately.

Having spent years disposing of obstacles and building up his resolve to tackle the "long labour" envisioned in his 1844 letter to Emma, he now found Lyell urging him to second-guess this publishing strategy. He hesitated, wondering whether he should choose alacrity over comprehensiveness, whether the desire to establish priority should supersede his notion of propriety.

Darwin may have been confident in the usefulness, persuasiveness, and indeed the truth of his theory, but it was difficult to hear his mentor once again chiding him for his slow progress on publishing. The tone of Darwin's reply shifted from confidence to diffidence as he addressed Lyell's advice. "With respect to your suggestion of a sketch of my view; I hardly know what to think, but will reflect on it; but it goes against my prejudices." He was being asked to do with his species theory what he had done with coral reefs: to go public with a brief statement of a big idea. Everything in the aftermath of that previous decision, which had itself been urged on him by Lyell, had caused him to second-guess the value of such an approach. A short paper would be all the more audacious for its brevity. Publishing it would invite immediate demands for more work rather than inspiring admiration and assent for work already done.

Darwin balked. "To give a *fair* sketch would be absolutely impossible, for every proposition requires such an array of facts."[21] What remained would be, by definition, an *unfair* sketch, an incomplete one. "If I were to do anything it could only refer to the main agency of change, selection,—& perhaps point out a very few of the leading features which countenance such a view, & some few of the main difficulties. But I do not know what to think."

Darwin could acknowledge at least one merit of Lyell's path. Imagine if the approach he had planned actually cost him his precedence. "I rather hate the idea of writing for priority," he averred, "yet I certainly sh[ould] be vexed if any one were to publish my doctrines before me."[22] That Lyell's mentorship was well intended, Darwin was sure. "I thank you heartily for your sympathy." Signing off, he recalled the role Lyell had played in creating his scientific career:

Farewell
My dear old Patron.
Yours
C. Darwin

How poignant that after twenty years of friendship, with dozens of books and editions published between the two of them, Lyell and Darwin had fallen back into their old roles of master and student. Once again Lyell was giving publishing advice, urging speed and audacity. Once again Darwin was quailing at the thought of going public too soon or in the wrong fashion. Below the signature he scribbled another note to Lyell. "If I did publish a short sketch, where on earth should I publish it?"

There were many reasons Darwin might have felt uncertain where one could publish a theory-oriented paper. As historian Sandra Herbert has pointed out, the primary zoological and botanical journals in Britain rarely published articles on matters of general theory, that is, theories that applied across multiple classes of organisms as Darwin's would.[23] The absence of an English-language venue in which theoretical natural history was respectable shows how relatively little precedent there was for the type of article (or indeed book) Darwin contemplated writing.

Darwin knew, of course, that he would eventually go public in some fashion. He spilled out his uncertainties in a pair of letters to Joseph Hooker, saying, "I very much want advice & truthful consolation if you can give it. I had a good talk with Lyell about my species work, & he urges me strongly to publish something."[24] But which kind of publication should it be? Darwin was dead set against publishing in one of the journals of a scientific society such as the Geological or the Linnean because he did not want to make the organization an accomplice to speculations "for which they might be abused." Nor did he want to expose the editor of an independent periodical to injury. Of course, sparing editors or a learned society from being associated with the publication of his theory also spared Darwin the obligation of making his theory conform to the standards of any periodical or organization. The alternative to publishing in a journal or reading a paper to one of the societies was to publish independently something like a pamphlet or a small book, "a very thin & little volume," containing "a sketch of my views & difficulties."

Darwin allowed himself to envision what a preliminary sketch would look like. To keep it short and write it quickly (he imagined "giving up a couple of months" to the task), he would not oblige himself to include formal references to others' work. The short piece would, however, refer explicitly to the yet-unwritten full-length treatise to which Darwin would remain committed. He thought, however, that such a précis would be "dreadfully unphilosophical." To Hooker he groaned, "Eheu eheu, I believe I sh[ould] sneer at anyone else doing this, & my only comfort is, that I truly never dreamed of it, till Lyell suggested it, & seems deliberately to think it adviseable. I am in a peck of troubles."[25]

Hooker replied in favor of publishing a pamphlet or small volume as opposed to a paper in a society's proceedings; evidently Lyell had advocated the opposite. Darwin told Hooker, "I am extremely glad you think well of a separate 'Preliminary Essay' i.e. if anything whatever is published; for Lyell seemed rather to doubt on this head; but I cannot bear the idea of *begging* some Editor & Council to publish & then perhaps to have to *apologise* humbly for having led them into a scrape."[26] Nevertheless, he said his mind was far from being made up whether to publish a preliminary essay at all, reiterating that it would be "unphilosophical" to promulgate something incomplete and that such an essay might "do mischief by spreading error." On the other hand, he acknowledged, the eventual task of writing a fully comprehensive book on his species theory would benefit from his familiarity with whatever reactions his sketch might elicit from friends and critics.

The most important aspect of this letter to Hooker is Darwin's revelation that he considered his old coral reef theory and his present transmutation theory to pose equivalent publishing challenges. Setting the two experiences directly in parallel, in terms of both his strategy and Lyell's role in shaping it, he recounted to Hooker a story from before the two had become acquainted. Hooker evidently having reasoned against publishing a short essay (in a letter that is no longer extant), Darwin responded, "What you say . . . that the Essay might supersede & take away all novelty & value from my future larger Book, is very true; & that would grieve me beyond everything. On the other hand (again *from Lyell's urgent advice*) I published a preliminary sketch of [my] Coral Theory & this did neither good nor harm."[27] It may have seemed true that publishing a preliminary sketch had not taken away value from Darwin's eventual coral reef book of 1842, but the pressure created by his audacious 1837 paper had in fact contributed substantially to Darwin's inability to produce the bold, synthetic book of global geology he had originally intended to write. Reading out the preliminary sketch of his coral reef theory to the Geological Society began a cascade of events resulting in the philosophical caution that now paralyzed him. Only in considerable hindsight could he have concluded that the decision had done "no harm."

: : :

Later in his life Darwin had the opportunity to counsel one of his own children on whether to publish a "preliminary sketch." In 1873 his twenty-eight-year-old son George was eager to have his father's advice on an essay he had written about prayer and the moral sense.[28] Darwin's concluding sentiment, "my advice is to pause, pause, pause," has been held up as an example of his characteristic caution. But as with his earlier letters to Morlot and Woodd, this letter strikes me as containing rather more autobiographical reflection than other commentators have acknowledged.

Darwin's fatherly advice to postpone publishing rested on two considerations. The first concerned the particular topic of George's essay. Advocating controversial religious views carried the risk of "giving pain to others, & injuring your own power & usefulness." Darwin believed that John Stuart Mill and Charles Lyell had been successful in undermining old orthodoxies precisely because they had avoided making provocative statements about religious doctrine, for example (in Lyell's case), against the literal truth of a biblical flood. Darwin's second argument against publishing was based on what he described as a general "doctrine of mine," that as a young author it was crucial to associate one's name only with excellent and novel work, "so that the public may have faith in [an author], & read what he writes."[29]

It is this second argument that seems to reflect Darwin's own experience. Not (as might be assumed) in publishing *On the Origin of Species* in middle age, but in his decisions as a young author. Here as in his letters to Morlot and Woodd, he seems to be speaking not in hypothetical terms but rather about the hard-learned lessons from a particular mistake in his past. "I am rather alarmed at you getting into the habit of desiring an early harvest," he told George, "frittering away your time on many such subjects or by writing short essays (& therefore temporary) on important subjects." Darwin was not referring specifically to work that might be an affront to religious views, but rather to anything published precociously. This advice to the son invoked, in both its phrasing and its substance, the father's earlier struggles. His choice of words recalls the entry he had made in his private journal thirty-five years earlier: "Frittered these foregoing days away in working on Transmutation theories." And the lessons were the same: that divided attentions made it more difficult to complete comprehensive works such as the geological treatise he had once envisioned; that short essays were just "temporary" contributions to knowledge; and that it was only in big books containing an appropriate weight of facts that one could earn readers' faith and thereby author *enduring* changes to the conventional wisdom. "I wish you were tied to some study on which you could not hope to publish anything for some years," he advised. "Remember that an enemy might ask who is this man, & what is his age & what have been his special studies. . . . This sneer might easily be avoided . . . my advice is to pause, pause, pause."

I suspect that Darwin's triple repetition of the word pause was meant to echo the phrase "action, action, action," that one might imagine a father's using at various points to exhort his children. It was a classical reference likely as familiar to the well-educated George as to Charles. More to the point, Darwin had used the phrase in giving the very same advice three years earlier to a naturalist who was almost exactly the same age then as George was now. "Forgive me for suggesting one caution," he had written in response to an exuberant letter from the young German naturalist Anton Dohrn. "As Demosthenes said, 'action, action, action' was the soul of eloquence, so is caution almost the soul of science." The lesson again was to avoid publishing prematurely. "Pray bear in mind that if a naturalist is once considered, though unjustly, as not quite trust worthy, it takes long years before he can recover his reputation for accuracy."[30] Darwin considered (but evidently rejected) offering Dohrn a specific negative object lesson in the person of Karl Friedrich von Gärtner, who is named in the notes Darwin wrote on Dohrn's letter: "Gärtner made a great blunder when he began his experiments in Hyb[ridization], & even to this day his error is sometimes remembered & his admirable works forgotten."[31] Such were the "sneers" Darwin wanted George to avoid.

And, indeed, when Darwin was George's age he had published words that provoked Basil Hall to sneer that "[his] faculty of generalization is certainly of no ordinary vigour." Make no mistake: Darwin had not always been one to pause. The lessons he dispensed later in life about the benefits of caution and comprehensiveness had been learned the hard way.

Lyell Choreographs Another Debut

In the end it was not Lyell's cajoling in 1856 that finally prompted Darwin to publish a statement of his species theory, it was Lyell's cajoling in 1858. Of course, one thing had changed in the interval. From the other side of the world another naturalist, Alfred Russel Wallace, had written asking Darwin to read the draft of an essay titled "On the Tendency of Varieties to Depart Indefinitely from the Original Type." This essay contained a theory Wallace had developed during his extended natural history sojourn in the Malay Archipelago, and its arrival set Darwin reeling. As is well known, Lyell sprang into action along with Hooker to preserve Darwin's morale and make a public record of his priority. However, in light of the account

I have provided, this episode looks less like the panicked reaction to an extraordinary crisis than the careful redeployment of techniques Lyell had used more than twenty years earlier to cultivate a favorable response for Darwin's coral reef theory.

But to be clear, that Lyell helped choreograph the appearance of Darwin's species theory was not just a function of the surprising turn of events Wallace's paper precipitated. On the contrary, the pair had intended to coordinate their plans since that summer in 1856 when Lyell first learned the details of Darwin's long labor on species. Working on the expectation that the theory would first become public in the form of a book, they planned how to cultivate sympathetic readers. As in 1837, the scheme involved a statement by Darwin and a premeditated response by Lyell, except instead of occurring at a Geological Society meeting this would be performed in a coordinated series of publications.

This original plan is revealed by a passage in one of Lyell's private notebooks. About the end of June 1856 Lyell used the notebook to draft a letter in apparent response to an offer from Darwin to dedicate the species book to him. Lyell's recommendation was rather elaborate. Darwin could write a dedication that would include an "anecdote" revealing that Lyell had confessed that his *Principles* "ought in consistency to have gone for [supported] transmutation [but] that I have uniformly taken the other side" even though "in no book has the gradual dying out & coming in of spec[ie]s been more insisted upon [than in the *Principles*], nor the necessity of allowing for our ignorance & not assuming breaks in the chain because of no sequence & of admitting lost links owing to small area observed or observable." In other words, Lyell would be portrayed simultaneously as a disbeliever in the *phenomenon* of transmutation and as the author of the *scientific approach* that would be embodied in Darwin's species work. Placing words on the tip of Darwin's pen, Lyell urged him to write "that finally you hope your book will convert me wholly or in part."

This would set the stage for a performance startlingly reminiscent of the one the two had enacted at the 1837 meeting of the Geological Society where Darwin read his paper on coral reef formation. Again Darwin would be instructed to champion Lyell's methods while differing only on interpretation. Would Lyell again "abjure on the spot" in a premeditated show of surprised acquiescence to the conclusion Darwin had arrived at by applying Lyell's principles of scientific work? Lyell intended readers to find out by buying one of *his* books. The next step of the plan recorded in his notes said, "I c[oul]d reply [to Darwin's preface] in a new Ed[ition] of

[the] Manual or P[rinciples] of G[eology]—wh[ich] w[oul]d act in setting the case well before the public."[32]

Darwin responded with enthusiasm. "I am <u>delighted</u> that I may say (with absolute truth) that my essay is published at your suggestion." Would it in fact be a mere essay he published, though, or something more? He concluded that his preface "will not need so much apology as I first thought; for I have resolved to make [the species publication] nearly as complete as my present materials allow." So it would be a book after all.[33] And he wanted to avoid appearing to be either a grandstander or a syco-phant in priming readers for Lyell's conversion. "I cannot put in all which you suggest," he demurred, "for it would appear too conceited. . . . I had <u>already</u>, after a few remarks on the Principles, ventured the words [about Lyell]—'and with a degree of almost <u>prophetic</u> caution which must excite the admiration &c&c.'"[34]

To Hooker, on the other hand, he could not resist an unabashed boast. "From Lyell's letters he is coming round at a Railway pace on the mutabil-ity of species, & [he] authorizes me to put some sentences on this head in my preface."[35]

Darwin was feeling rather more somber two years later when Wallace's letter arrived. It was summertime again, but this June was an anxious one at Down House for a very different reason. The Darwins' fifteen-year-old daughter Etty had caught diphtheria, and the baby, one-and-a-half-year-old Charles Waring Darwin, had come down with scarlet fever. Charles and Emma still grieved the loss of another daughter, Annie, who had died in 1851 at age ten; once again they feared the worst. This unhappy domestic scene was further jolted by the arrival of the day's mail on June 18. After reading what had arrived, Darwin began writing to Lyell. "Your words have come true with a vengeance."[36]

It was just as Lyell had warned two summers ago. Wallace's letter en-closed a paper he had written on organisms diverging from their original type. "Forestalled," Darwin sighed. "You said this when I explained to you here very briefly my views of 'Natural Selection.' . . . [I]f Wallace had [seen] my M.S. sketch written out in 1842 he could not have made a better short abstract."[37]

Darwin's plea to Lyell is well known, but it deserves another glance. Here again we have a look through Darwin's eyes at the social life of scien-tific ideas. The letter shows that he saw priority of publication and the im-pact of a published work as distinct: not mutually exclusive but not necessarily identical either. Many have quoted his reflection on the first topic, but fewer have quoted it alongside the next sentence in Darwin's letter. "So all my originality, whatever it may amount to, will be smashed. Though my Book," he continued, "if it will ever have any value, will not be deteriorated; as all the labour consists in the application of the theory."[38]

This attitude is entirely consistent with the one Darwin expressed to Fox when he was first humbled in the 1830s, to Emma and to Morlot in the 1840s, and now to Lyell in the 1850s. Framing theories was comparatively "easy." Applying them, embedding them in fact upon fact to vouch for one's diligence, this was "long labour." Wallace's paper struck a disappointing blow not because Darwin preferred priority to persuasion, but because he had dared to dream of achieving both at the same time. Now it appeared that his "originality" was lost, and what remained available—if he ever could finish his project—was the harder thing to come by: "value."

Was he sure this is how he felt? Perhaps priority could be salvaged as well. Darwin wrote to Lyell in anguish ("at present quite upset," he said) and uncertainty, this time to plead for advice. The question was whether Darwin could transport himself back to 1856, when Lyell had urged him to publish, and make his decision again. Then he had steeled himself in the conviction that it was "unphilosophical" to make haste for the sake of priority, to publish a sketch when one could instead write a grand book. Now he wondered whether the private knowledge that his priority was at risk—private, after all, because Wallace's paper had come directly to Darwin—disqualified him from revisiting that decision. Now Lyell's advice would provide Darwin "as great a service, as man ever did." Now Darwin was quite upset. Perhaps not in 1856, but *now.* . . . Darwin used two strokes of his pen to underline his distress, and one can almost hear his indignant voice: "I sh[oul]d be <u>extremely</u> glad <u>now</u> to publish a sketch of my general views in about a dozen pages or so . . ."[39]

"But I cannot persuade myself," he continued, "that I can do so honourably." It would mean exploiting the fact that he knew about Wallace's work, making—after all—the decision that a mere sketch was enough to start with, precisely because it would be a way to beat Wallace into print.

> Wallace says nothing about publication, & I enclose his letter.—But as I had not intended to publish any sketch, can I do so honourably because Wallace has sent me an outline of his doctrine?—I would far rather burn my whole book than that he or any man sh[oul]d think that I had behaved in a paltry

spirit. Do you not think his having sent me this sketch ties my hands? I do not in least believe that that [*sic*] he originated his views from anything which I wrote to him.

Darwin could see the solution staring him in the face. He simply had to follow Lyell's advice, *all of it*: to let himself publish a sketch and let Lyell take the blame for his doing so. Would Lyell be so kind as to offer his advice one more time? If Lyell would affirm ("for I have entire confidence in your judgment & honor") that Darwin *could* "honorably publish," then that is what Darwin would do. He could explain that receiving Wallace's letter had induced him to publish a "sketch," and in doing so, he told Lyell, "I sh[oul]d be very glad to be permitted to say to follow your advice long ago given." He signed off by asking Lyell to solicit advice from Hooker before giving his verdict, "for then I shall have the opinion of my two best & kindest friends."[40]

Darwin's mind would not rest, though, at the thought of Wallace's reaction. He sent Lyell a postscript in which he imagined an omniscient Wallace saying, "You did not intend publishing an abstract of your views till you received my communication. [I]s it fair to take advantage of my having freely, though unasked, communicated to you my ideas, & thus prevent me forestalling you?"[41]

Events at home were even more upsetting. Etty was rallying against her illness, but baby Charles's condition had deteriorated badly. Two days later he succumbed to scarlet fever. Emma had been stoic while her son was clinging to life, but now she "let her feelings break forth."[42] The grieving father was at a loss, for even at this moment his closest friends were trying to spur him to action. Word of the crisis over Wallace's letter had passed from Lyell to Hooker, whose offer to help arrived the morning after baby Charles's death and loosed its own outpouring of emotion from Darwin. "I daresay all is too late," he lamented. "I hardly care about it." But even so, he sent along his 1844 essay—which Hooker had read and annotated— "solely that you may see by your own handwriting that you did read it."[43] Along with the text of a letter Darwin had written the previous year to the Harvard botanist Asa Gray, which he also sent along, the 1844 essay could stand as proof that Darwin's ideas had emerged independently, and long before Wallace's.

Lyell was springing to action again, this time with Hooker's aid. What he had in mind was to take the matter out of Darwin's hands altogether. The grieving father would surrender Wallace's essay and hand over evidence of his own sustained work so that Lyell could do what he did so well: place work before a scientific audience in the most advantageous circumstances possible.

On 30 June, the very day after Hooker took possession of Darwin's fourteen-year-old essay and less than two weeks after Wallace's letter had arrived at Darwin's door, Lyell and Hooker conveyed a joint letter to the botanist J. J. Bennett, secretary of the Linnean Society. Attached were three papers for Bennett to read before the society; they were enumerated, notably, "in order of their dates." The first paper comprised a set of extracts Lyell and Hooker had chosen from Darwin's 1844 essay; the second—also by Darwin—was an abstract of the letter to Asa Gray, which "repeats [Darwin's] views, and which shows that these remained unaltered from 1839 to 1857." Third came Wallace's full essay, which they described as having been "written . . . for the perusal of his friend and correspondent Mr. Darwin." Darwin, they explained, had followed Wallace's "expressed wish" that he forward the essay "to Sir Charles Lyell, if Mr. Darwin thought it sufficiently novel and interesting."[44]

Lyell and Hooker explained their course of action through a story calculated to validate Darwin's priority *and* to emphasize his commitment to sober, responsible theorizing. Both Darwin and Wallace, they wrote, had "independently and unknown to one another conceived the same very ingenious theory to account for the appearance and perpetuation of varieties and of specific forms [species] on our planet." But while the two authors could each be considered "original thinkers in this important line of inquiry," Darwin had conceived the ingenious theory first. Indeed, Lyell and Hooker had "for many years past . . . repeatedly urged" Darwin to publish his views. But instead, they continued admiringly, the first thing Darwin had seen as fit for publication on this topic was *Wallace's* essay. Lyell and Hooker agreed with Darwin that Wallace's essay was an important work that deserved to be published, but they wrote that they had been willing to usher it into print *only* if Darwin "did not withhold from the public . . . the memoir which he had himself written on the same subject . . . the contents of which we had both of us been privy to for many years." Darwin, they explained, had been "strongly inclined [to withhold] in favour of Mr. Wallace." They implied that he had in fact never agreed to publication himself and had simply given them permission to act as arbitrators of the problem by giving them his memoir along with Wallace's.

By portraying Darwin as a passive bystander to the decision on whether to publish his own work, Lyell and Hooker painted him as a disinterested investigator. This implied that the conclusions drawn in his work were all the more trustworthy. In the end they claimed to have decided to present his ideas to the Linnean Society along with Wallace's, "not solely considering the relative claims to priority" but rather "[in] the interests of science generally." For it was in the interest of the whole "scientific world," rather than that of any single author, that they wished heartily to promulgate Darwin's views: "views founded on a wide deduction from facts, and matured by years of reflection."[45]

Darwin himself could hardly have chosen better phrasing to define the spirit in which public speculation could be justified. If his theory was going to be placed before the public in a form that fell so far short of his (and what he perceived to be the whole scientific world's) ideal genre, then at least it would come with a disclaimer by two of the nation's most eminent naturalists that a grand book's worth of "facts" was forthcoming in what they called "Mr. Darwin's complete work." And his theory was indeed placed before the public, or at least a version of it, the very next day, when the secretary read aloud all three papers, prefaced by Lyell and Hooker's letter, at a meeting of the Linnean Society. That was 1 July 1858. By August 20 the letter and the papers by Darwin and Wallace were published in the society's *Proceedings*. After all this time, the species theory had gone from privacy to print in the space of two months.[46]

The publication of Darwin's theory of natural selection initially caused much more of a stir in his own life than in the rest of the scientific world. The Linnean Society meeting itself could be described as an anticlimax, which of course meant it had been a success. Provoking debate was not the goal of the day. Lyell and Hooker had acted to ensure Darwin's priority and prod him into accelerating his schedule for publishing. As Lyell acknowledged the following year (when Darwin had drafted the manuscript for the *Origin*), "I have just finished your volume & right glad I am that I did my best with Hooker to persuade you to publish it without waiting for a time which probably could never have arrived tho' you lived till the age of 100, when you had prepared all your facts on which you ground so many grand generalizations."[47] If Wallace's letter had panicked Darwin enough to let Lyell and Hooker make his theory public, it was knowing they had done so that gave him a sense of urgency to present the theory on something closer to his own terms.

But it is worth pausing to wonder whether Lyell, if not Darwin, felt a sense of déjà vu about all this. There are striking similarities between the way Lyell choreographed the release of Darwin's coral reef theory to a specialist audience and the way he and Hooker helped Darwin manage the crisis occasioned by Wallace's paper. In each instance an eminent naturalist was presented in private with a compelling but potentially inconvenient theory from a younger man who was known as a traveling collector rather than as an author of theories. When Darwin revealed his coral reef theory to Lyell in 1837 it contradicted the explanation Lyell had published. When Wallace's paper arrived on Darwin's doorstep two decades later, it seemed to steal the thunder from a theory Darwin had not yet published. Yet in each case Lyell was able to see a larger opportunity where others might have seen only an obstacle. In each case he claimed (or claimed on Darwin's behalf in 1858) to have been convinced and chastened by the newly arrived work while imposing nearly complete control over the circumstances in which the younger man's theory would be made public. And in both cases the senior figure (Lyell in 1837 and Darwin in 1858) succeeded in bolstering his own reputation by absorbing the newcomer's findings into a larger project that would carry the better-established man's name.

Did this mean Lyell was exploiting Darwin in 1837 or helping Darwin to exploit Wallace in 1858? If the answer is yes it is a very qualified yes. For as I argued earlier, these arrangements were *mutually beneficial* to both master and student. What Darwin had gained in 1837—access to a privileged pulpit and the endorsement of a member of the elite—Wallace gained in 1858. And Wallace needed these things even more than Darwin had, for whereas Darwin had money, a distinguished name, and a number of other possible patrons including Sedgwick and Henslow, Wallace was destitute, at a distance from London, and at a loss for the sort of patronage that could open doors to his work and minds to his ideas. Indeed, Wallace had sent his essay to Darwin *precisely* to request help in acquiring the kind of support Lyell had formerly given to Darwin. And when news reached him of what had happened to his paper, Wallace's reaction was as effusive as young Darwin's had been two decades earlier. Darwin, recall, had reported to his second cousin and friend Fox, "I have read some short papers to the geological Soc, & they were favourably received by the great guns, & this gives me much confidence, & I hope not a very great deal of vanity; though I confess I feel too often like a peacock admiring his tail.—I never expected that my geology would ever have been worth the consideration of such men, as Lyell."[48] Wallace reported the news to his mother by writing, "I have received letters from Mr. Darwin and Dr. Hooker, two of the most eminent naturalists in England, which has highly gratified me. I sent Mr. Darwin an essay on a subject on which he is now writing a great work. He showed it to

Dr. Hooker and Sir C. Lyell, who thought so highly of it that they immediately read it before the Linnean Society. This assures me the acquaintance and assistance of these eminent men on my return home." And he wrote to a friend, George Silk, urging him "to borrow [a copy of the Linnean Society *Proceedings*] for August last, and in the last article you will find some of my latest lucubrations, and also some complimentary remarks thereon by Sir Charles Lyell and Dr. Hooker, which (as I know neither of them) I am a little proud of."[49]

It would be easy to conclude that this situation was created entirely by the arrival of Wallace's letter in June 1858, but as I noted at the beginning of this section, Lyell and Darwin had *already*, for two years, been planning to coordinate their efforts to maximize the likelihood that Darwin's theory would receive a fair (that is to say, favorable) hearing. Bringing Wallace into the transaction simply added another layer of authority to Darwin's claims, for he was now bolstered not only from above by Lyell's imprimatur but also from below by Wallace's independent findings. Here we have another illustration of the way scientific credit was not a zero-sum game. Leaving aside the question whether Wallace's 1858 paper would have received any notice had he simply sent it for publication in the same fashion as his allied, and largely overlooked, 1854 "Sarawak" paper, his sending it to Darwin gave Darwin the opportunity to perceive his own (broader and better developed) views in the younger man's work and to strategically enroll Wallace as an ally. There are many reasons Darwin's relationship with Wallace did not immediately take the same shape as that between Darwin and Lyell, including Darwin's more retiring social profile and their lack (owing to Wallace's ongoing absence from England) of the face-to-face interaction that had proved beneficial to both Lyell and Darwin after the *Beagle* voyage. But they nevertheless developed a parallel sense of obligation, with Wallace dedicating his 1869 book on the Malay Archipelago to Darwin and naming his 1889 book *Darwinism* and Darwin, late in his life, ensuring that Wallace's financial straits were eased by the award of a government pension.

: : :

In a fitting parallel to the events at the Linnean Society, Lyell also stepped in on Darwin's behalf at their old shared venue, the Geological Society of London. In the spring of 1859 the society awarded Darwin its Wollaston Medal, the same honor it had bestowed on Richard Owen two decades earlier for his work on Darwin's *Beagle* fossils. The council of the society cited Darwin's "numerous contributions to Geological Science," including his "observations . . . on the structure and distribution of Coral-reefs." With Darwin's species theory now public as a consequence of the Linnean Society meeting and with his book on this important topic expected later in the year, awarding the prize was a timely reminder of Darwin's contributions to geology and of geology's important place in his increasingly important scientific career. Notably, the medal was presented not to Darwin himself but to his proxy, Lyell, who reported Darwin's "gratitude for the high honour conferred . . . [which was] the more prized by him in the seclusion in which he finds it necessary to pursue his studies and researches." To those who remembered Darwin's emergence as an "extreme Lyellist" in the society's rooms twenty-one years earlier, it must have seemed apt for the citation of Darwin's achievements in geology to be literally addressed to the author of the *Principles*: "Sir C[harles] Lyell," began the citation by John Phillips, the society's president, "To no one can the Medal which is destined for Mr. Darwin be committed with so much justice as yourself, who, like him . . . have always looked on the phaenomena of nature with a comprehensive survey, a minute attention, and a just appreciation of the dignity of our science." Turning his attention to the recipient of the award, Phillips continued,

> Mr. Darwin, ever since his great abilities became known by the "Researches" during the voyage of the *Beagle*, has never ceased to labour, even in spite of ill health, in the cause of geology. We owe to him the admirable observations on Coral-growth, which led to the grand speculation of alternate zones of elevation and depression in the Pacific and Indian Oceans. He has given us data for the modern elevation of Chili, for the often repeated elevations of the Andes and the bordering regions; through great tracts of America his masterly hands have sketched and measured the prominent structures of the rocks; in the British Islands he has studied the distribution of boulders, the change of level of land and sea, the parallel roads of Glen Roy, the course of ancient glaciers. . . . Let Mr. Darwin be assured that we hope to welcome him often in better health, and personally to renew the congratulations which must be agreeable to him when received through the hands of one of his earliest friends and most illustrious fellow-labourers.[50]

Lyell thanked Phillips and the society "for the kind and complimentary manner in which you have spoken of me and my labours," explaining that

Darwin's absence was due to ill health. More than a month elapsed before Darwin wrote to Lyell offering "very sincere thanks to you for standing my Proxy for [the] Wollaston Medal." In the same letter he told Lyell that the species book at last had a name: "An abstract of an Essay on the Origin of Species and Varieties Through Natural Selection."[51]

Publishing an "Abstract" After All: *On the Origin of Species*

Darwin had spent the previous twenty years anxious lest he be seen as too speculative, so the necessity—following the Linnean Society meeting—of rushing a book on species into print was a cruel irony indeed. Under these curious circumstances, we might expect him to have attempted to forestall the kinds of criticism that had burned him in the past. Indeed, he did so. A key theme of the letters from Darwin I quoted earlier in this chapter is that the topic of any given theory was less important than the general principle of hiding premature, incomplete, or insufficiently grounded speculation from public view. The topics studied by Darwin's correspondents across the decades ranged from Morlot's glaciers to George Darwin's metaphysics, but Darwin's strictures remained the same: pause, add facts, avoid committing the public "sin of speculation." And in the pages of the *Origin* his most explicit statements of self-fashioning were aimed not at mitigating the controversy of being seen as a transmutationist but at defusing the charge that he was prone to hasty speculation.

Given the immediate impact and enduring influence of the book Darwin published in 1859, it can be difficult to grasp the idea that *On the Origin of Species* more closely resembled the inadequate sketch he had refused to write than the grand, multivolume treatise he had been working on all those years. It was, after all, a document that gave scant treatment to some of Darwin's most valued examples and was written without explicit references to other scholars' work. In the course of his exertion during the second half of 1858 and into 1859, however, even Darwin's "sketch" came to take on the dimensions of a substantial book. Writing to Wallace in April 1859, he described the text he was completing as "a small volume of about 500 pages or so," revealing simultaneously the length of the *Origin* as it took shape and, by using the word small to describe it, suggesting something about the evidently massive scale of his original ambition. Even then he protested to Wallace, "You must remember that I am now publishing only an Abstract & I give no references."[52]

Echoing Lyell more than two decades earlier, Darwin cultivated scientific allegiance from Wallace in sectarian terms. "I forget whether I told you," he wrote, "that Hooker, who is our best British botanist & perhaps best in World, is a <u>full</u> convert, & is now going immediately to publish his confession of Faith. . . . Huxley is changed & believes in mutation of species: whether a <u>convert</u> to us, I do not quite know.—We shall live to see all the <u>younger</u> men converts."[53] One might be tempted to assume that Darwin chose this religious phrasing to emphasize anti-Christian implications of the theory of natural selection. But Darwin had used "sin of speculation" to refer to geological work that did not embody special antipathy for the church, and Lyell had of course used "heretical doctrines" to speak of the causal connection between volcanoes and earthquakes.[54] That is to say, agreements and disagreements over all manner of scientific precepts were expressed in religious metaphors. Likewise, we have seen that Darwin feared being deemed a speculative sinner for a broad range of geological ideas, suggesting that the potentially controversial implications of his species theory with respect to literal religious views were not *necessary* components of such language.

The seriousness and respectability of Darwin's accelerated book project were enhanced by an arrangement Lyell brokered between his protégé and his own longtime publishing house, John Murray. Darwin used Lyell as both a go-between and a counselor for his interactions with Murray, and the content of his messages reinforces the idea that Darwin saw his theory's religious implications as a concern for others but not as *his own* primary concern. "Would you advise me to tell Murray," he asked, "that my book is not more ¨un-orthodox than the subject makes inevitable. That I do not discuss origin of man. . . . Or had I better say <u>nothing</u> to Murray, & assume that he cannot object to this much unorthodoxy, which in fact is not more than any Geological Treatise, which runs slap counter to Genesis[?]"[55] Murray did object to the proposed title, "Abstract of an Essay . . . ," as too prolix for his audience. Darwin tried to protest to Lyell but acquiesced: "I look [at the longer proposed title] as [the] only possible apology for not giving References and facts in full.—but I will defer to him & you."[56]

It is clear that Lyell retained a hugely important role as mediator between Darwin and his various audiences, but it is equally clear that Lyell himself was Darwin's most cherished audience. As a consequence of his personal admiration for Lyell's writing and his firsthand experience of the benefits Lyell's scientific endorsement could bring, Darwin devoted considerable effort to shaping Lyell's response as an individual reader. On 2 September, when he had the manuscript ready to share, he urged Lyell,

"Do not, I beg, be in a hurry in committing yourself, (like so many natural-ists) to go a certain length & no further; for I am deeply convinced, that it is absolutely necessary to go [the] whole vast length, or stick to creation of each separate species. . . . *Remember that your verdict will probably have more influence than my Book in deciding whether such views as I hold, will be admitted or rejected at present.*"[57] The last two words are notable here, for Darwin had complete confidence in the ultimate success of the theory, just as he had in 1844 when he left the letter of instructions for Emma. Indeed, he proceeded to tell Lyell confidently, "In the future I cannot doubt about [my views'] admittance, & our posterity will marvel [at] the current be-lief."[58] But Lyell's support was, as ever, critical to cultivating support within the actual social world of mid-nineteenth-century British science.

Darwin need not have worried; Lyell was already poised to act. That same month, September 1859, the British Association for the Advance-ment of Science met for the twenty-ninth time. That year's meeting was in the port and university town of Aberdeen, on the east coast of Scotland, about sixty miles from the Lyell family home at Kinnordy. Lyell addressed the geological section. "Among the problems of high theoretical interest which the recent progress of Geology and Natural History has brought into notice," he asserted, "no one is more prominent, and at the same time more obscure, than that relating to the origin of species." Lyell advertised the significance of Darwin's forthcoming work while praising the sobriety of his reasoning.

> On this difficult and mysterious subject a work will very shortly appear, by Mr. Charles Darwin, the result of twenty years of observation and experi-ments in Zoology, Botany, and Geology, by which he has been led to the conclusion, that those powers of nature which give rise to races and perma-nent varieties in animals and plants, are the same as those which, in much longer periods, produce species, and, in a still longer series of ages, give rise to differences of generic rank. He appears to me to have succeeded, by his investigations and reasonings, in throwing a flood of light on many classes of phenomena connected with the affinities, geographical distribution, and geological succession of organic beings, for which no other hypothesis has been able, or has even attempted, to account.[59]

The words gave Darwin a shot of courage when they reached him in his study at Down House, and they caused his mind to flash back to Lyell's first public display on his behalf. "You once gave me intense pleasure, or rather delight," he recalled, "by the way you were interested, in a manner I never expected, in my Coral-reef notions." By "interest" in this context Darwin meant commitment—the opposite of disinterested impartiality—rather than simple curiosity. Now Darwin knew that Lyell had once more spoken on his behalf (in a venue whose proceedings would receive wide circulation in print), writing that Darwin had thrown a "flood of light" on the mystery of mysteries. The student declared to his master, "You have again given me similar pleasure by the manner you have noticed my Species work. Nothing could be more satisfactory to me, & I thank you for myself, & even more for the subject-sake, as I know well that sentence will make many fairly consider the subject, instead of ridiculing it."[60] In another letter five days later Darwin added, "I do thank you for your euloge at Aberdeen. . . . You would laugh if you knew how often I have read your paragraph, & it has acted like a little dram."[61]

Just as they had planned back in June 1856, Lyell also gave Darwin the text of a statement to include in the book that would prime readers to await Lyell's judgment. Whereas Lyell's former instruction to Darwin had been to write "that finally you hope your book will convert me wholly or in part," he now gave him permission to venture a guess at the outcome. In its eventual published form the text read, "[Our] most eminent palaeontologists, namely Cuvier, Owen, Agassiz, Barrande, Falconer, E. Forbes, &c., and all our greatest geologists, as Lyell, Murchison, Sedgwick, &c., have unanimously, often vehemently maintained the immutability of species. But I have reason to believe that one great authority, Sir Charles Lyell, from further reflexion entertains grave doubts on the subject."[62]

Darwin continued to cultivate Lyell as his own ideal reader by flattering and cajoling him in correspondence even as Lyell was working as Darwin's agent and advocate. We have already seen evidence of the pattern Darwin used, first declaring Lyell the most important arbiter and then emphasizing his own utter confidence in the eventual acceptance of the theory. The audacious message, if Lyell cared to see it, was that while the present scientific community might judge the theory based on Lyell's reaction to it, posterity would judge Lyell based on his reaction to the theory.

"I regard your verdict as far more important in my own eyes & I believe in [the] eyes of world than [that] of any other dozen men, I am naturally very anxious about it," Darwin wrote in one letter before continuing, "I cannot too strongly express my conviction of the general truth of my doctrines." The pattern repeated in the same letter: "I am foolishly anxious for your verdict. Not that I shall be disappointed if you are not converted; for

I remember the long years it took me to come round."[63] Again, five days later, "I care not what the universal world says; I have always found you right, & certainly on this occasion I am not going to doubt for the first time.—Whether you go far or but a very short way with me & others who believe as I do, I am contented, for my work cannot be in vain."[64] And once more, later that week, "I look at you as my Lord High Chancellor in Natural Science. . . . I shall be deeply anxious to hear what you decide (if you are able to decide) on the balance of the pros & contras given in my volume." And yet, "I feel an entire conviction, that if you are now staggered to any moderate extent, that you will come more & more round, the longer you keep the subject at all before your mind."[65]

Of course, Lyell was no more a passive recipient of the *Origin*'s argument than he had been a bystander to the writing of Darwin's geological essays two decades earlier. And he was no lord chancellor, no impartial judge. As ever, Darwin shaped his arguments in conversation with Lyell. Yes, their opportunities for meeting face-to-face were far fewer than when Darwin was living in London after the *Beagle* voyage, but Darwin sent Lyell the manuscript of the *Origin* chapter by chapter as it was drafted and, as we have seen, introduced changes offered by Lyell or suggested by his comments. As Darwin joked on October 11, after Lyell had at last read the entire draft and Darwin reported having "adopted" various corrections small and large, "I thank you cordially for giving me so much of your valuable time. . . . But you are a pretty Lord Chancellor to tell the barrister on one side how best to win the cause!"

When Darwin's book was released on 24 November 1859, it had indeed lost the word abstract from its title, but it retained other signs of the author's apprehension at being forced after all to publish something that was less than comprehensive. The *Origin* also contained gestures of piety, or at least gestures offered to those who would see an argument about transmutation as intrinsically impious. In fact, Darwin prefaced the book with a quotation from just such a person, William Whewell. The *Origin*'s epigraph from Whewell's Bridgewater Treatise read, "But with regard to the material world, we can at least go so far as this—we can perceive that events are brought about not by insulated interpositions of Divine power, exerted in each particular case, but by the establishment of general laws."[66] If one assumes that Darwin most sought to forestall criticism of his position on transmutation, then this epigraph serves the clear purpose of hinting that descent with modification might have been the Creator's own method for creating a diversity of living forms. No doubt Darwin intended as much, for he made the point explicitly in the *Origin*'s closing lines, arguing that "it accords better with what we know of the laws impressed on matter by the Creator, that the production and extinction of the past and present inhabitants of the world should have been due to secondary causes [than that they were a consequence of innumerable direct acts of creation]."[67] But using the quotation from Whewell served the second purpose of alluding to Whewell's position that the pursuit of these "general laws"—that is to say, the act of "generalizing" or theorizing—was the highest objective of science itself. As I discussed earlier, Whewell was noted for his statements about when and how theorizing could safely be undertaken: subsequent to amassing a requisite body of *facts* through the work of descriptive science. By quoting Whewell Darwin signified his own philosophical caution about generalizing.

Nevertheless, the main text of the *Origin* opened with a three-paragraph statement of contrition for the inadequacy of the book's content as a foundation for generalizing.[68] The evidence I presented earlier in this chapter makes it possible to recognize that Darwin's words at the beginning of the *Origin* were not just a pro forma expression of humility but rather were a manifestation of his own private anxiety about when and how to publish a theory. His words conjured a version of his own past, however, that had been fashioned to resemble the Whewellian scheme for developing valid generalizations.

"When on board H.M.S. 'Beagle,' as naturalist," he began, "I was much struck with certain facts in the distribution of the inhabitants of South America, and in the geological relations of the present to the past inhabitants of that continent. These facts seemed to me to throw some light on the origin of species—that mystery of mysteries, as it has been called by one of our greatest philosophers." Here, of course, Darwin was alluding to the period's other great authority on proper scientific methods of inquiry, John Herschel. And then, less than a hundred words into the *Origin*, Darwin switched from a truly autobiographical vignette to something like a parable.

> On my return home, it occurred to me, in 1837, that something might perhaps be made out on this question [of the origin of species] by patiently accumulating and reflecting on all sorts of facts which could possibly have any bearing on it. After five years' work I allowed myself to speculate on the

subject, and drew up some short notes; these I enlarged in 1844 into a sketch of the conclusions, which then seemed to me probable: from that period to the present day I have steadily pursued the same object.[69]

As the many scholars of Darwin's notebook period have made amply clear, he did not in fact proceed in this stepwise manner from empiricism to hypothesis. Rather, he considered multiple causes of transmutation between 1837 and the occasion in 1842 when he drew up his "pencil sketch" of the theory of natural selection after completing *The Structure and Distribution of Coral Reefs*. And as I have argued, far from resisting his desire to engage in private speculations on species in this period, he relished the activity as a stimulating diversion from the increasingly burdensome task of formulating public speculations in geology. There is, it seems to me, just one interpretation in which the passage quoted above could be literally true: if Darwin was tacitly defining "speculation" not as an act of thinking but as a specific genre of prose writing exemplified by the unpublished 1842 sketch and 1844 essay he alluded to. But this semantic distinction is incompatible with his usage in a query he sent to Huxley the month before the *Origin* was published: "Can you tell me of any good & <u>speculative</u> foreigners to whom it would be worth while to send copies of my Book 'on origin of species'[?]"[70]

In what manner did Darwin intend his readers to interpret this story of methodical progress from facts to theory? On this point he was explicit: "I hope that I may be excused for entering on these personal details, *as I give them to show that I have not been hasty in coming to a decision.*"[71] While the phrasing leaves it somewhat unclear precisely which "decision" Darwin was talking about, the thrust of the entire first paragraph leads to the conclusion that he was referring to the decision to abandon species fixity in favor of transmutation, that is, he was referring to the theory itself.

The other decisions Darwin had to explain, to publish *now* and *in this form*, were the topic of the second paragraph of the *Origin*. He declared that his two decades' work would still take "two or three more years" and that his health was "far from strong."[72] For this reason, he said, "I have been urged to publish this Abstract." Hence the nature of the book. And he carried on to explain the scenario of the previous year, in which Wallace's paper had arrived and "Sir C. Lyell and Dr. Hooker, who both knew of my work . . . honoured me by thinking it advisable to publish, with Mr. Wallace's excellent memoir, some brief extracts from my manuscripts." Hence the abandonment of his long-held plan for publication and the need instead to act rapidly by following through with the present volume.

The third paragraph of Darwin's sustained apologia addressed the consequent shortcomings of the text. "This Abstract, which I now publish, must necessarily be imperfect." He explained that it lacked references, which demanded an added layer of faith between reader and author. Darwin felt he "must trust to the reader reposing some confidence in my accuracy."[73] References to other authorities were a way to signal one's trustworthiness and scholarly discernment.[74] Darwin, in his urgency to expand on the material presented to the Linnean Society, would have to do without this credential, which weakened the "substratum of facts" he had once told Morlot "ought surely to be first given" before laying out a theory.[75]

Darwin tried to minimize this flaw in the *Origin*'s foundation by promising to buttress it with further books. *"No one can feel more sensible than I do,"* he attested, "of the necessity of hereafter publishing in detail all the facts, with references, on which my conclusions have been grounded."[76] He offered no explanation for why he, *among all authors*, ought to be especially sensitive to the need to follow statements of theory with supplemental facts, but we can notice here the echo of his own private confessions of mortification: at having burnt his fingers by committing the sin of speculation in short geological papers before failing to produce the book he had originally promised would detail, as it were, the facts on which his conclusions had been grounded. All he said by way of explanation within the text of the *Origin* was that "[a] fair result can be obtained only by fully stating and balancing the facts and arguments on both sides of each question: and this cannot possibly be here done."[77]

It is natural to ask how effective Darwin's self-fashioning turned out to be in shaping the responses to the *Origin*.[78] This cannot possibly be done here either—at least not in any detail. Legions of scholars have analyzed the responses to the book by allies, opponents, and everyday readers; I can only hope that the arguments I have laid out will be useful to those who pursue the topic further.[79] One persistent theme in current scholarship on the reception of the *Origin* is that the reputation of the book's author shaped the way it was read. Readers judged the work in part based on its having been written by Darwin, a gentleman scholar who had then spent almost two decades publishing sober and noncontroversial work, and not by a known controversialist or indeed someone who wrote under the veil of anonymity.[80] As one reviewer put it in the weeks after the *Origin*'s publication, "There are forms of speculation so wild and improbable, or, at any rate, so alien to our ordinary habits of thought, that they can only obtain a fair consideration under the protection of some illustrious name." The theory of natural selection would inspire "incredulity" if presented anonymously or by an amateur, "but the case is widely different when this theory is put forward by Mr. Darwin, a man confessedly in the foremost ranks of natural philosophy . . . [and] honoured among his scientific peers." This reviewer, the Oxford-educated (and aptly named) cleric Richard Church, explained that a theory by Darwin deserved "respectful attention," particularly because the *Origin* was "the result of twenty-three years' patient investigation and is . . . only the condensed argument of a larger work which, if the author's health permit, will follow it in two or three years, and supply the vast substratum of detailed facts which alone can give security to the airy and tottering fabric he is labouring to erect."[81]

This is just one review among the hundreds that were written, but it illustrates some of the continuities that linked 1839 and 1859.[82] Once again Darwin was getting attention for the speculations in his just published book. Readers of the *Origin*, in common with the readers of young Darwin's *Journal of Researches*, had to balance the audacity of the author's claims against his promise that future work would provide more evidence to support them. Compared with readers of Darwin's earliest forays into scientific authorship, however, those who opened the green cloth covers of the *Origin* could add his quarter-century of dogged productivity to the scales as well.

Dealing with Darwin's "Recollections"

As public attention after 1859 became focused on Darwin's species theory, the coral reef theory came to seem far less speculative than it once had appeared. Such a shift occurred in Darwin's perception as well as in the broader scientific discussion of his career. And whereas privately theorizing about species may at one time have felt less fraught to Darwin than the challenge of actually publishing on geology, the topic of the *Origin* did indeed make it much more controversial than any theory about coral reefs was ever likely to be. This contrast between the two theories was exemplified by the way Darwin was awarded the Copley Medal of the Royal Society in 1864.[83] His name had been put forward by younger allies in each of the preceding two years, and both times he had been blocked by members of the society's council who were hostile to, or at least ambivalent about, the *Origin*—even though his supporters had tactfully avoided citing natural selection as one of the accomplishments that made him worthy of the Copley Medal. In council discussion of the 1863 nominees, Darwin's champion William Carpenter tried to combat other members' criticism of the *Origin* by reading from a footnote that John Stuart Mill had just added to a new edition of his *System of Logic.* "Mr. Darwin's remarkable speculation on the Origin of Species is another unimpeachable example of a legitimate hypothesis. . . . It is unreasonable to accuse Mr. Darwin (as has been done) of violating the rules of Induction."[84] Rather pointedly, however, that year's medal was awarded instead to the person who had leveled this very accusation: Darwin's old geology instructor, Adam Sedgwick. After writing to Darwin the month the *Origin* was published to tell him, "You have de- serted . . . the true method of induction," Sedgwick had gone on to savage the book in an unsigned review that made the same point: "I must in the first place observe that Darwin's theory is not inductive,—not based on a series of acknowledged facts pointing to a general conclusion."[85]

Darwin's Copley Medal nominators were finally successful on their third try, but in awarding the medal to (an absent) Darwin the society's president, Edward Sabine, went out of his way to specify that the award had been given not for the *Origin* but rather for other contributions to science. The one he enumerated first and at greatest length was Darwin's "preeminently successful" solution to the problem of coral reefs and his "applying [that solution] to the explanation of geological phenomena." After summarizing the 1842 book and explaining the broader implications of the theory for understanding stratigraphy and crustal movement, Sabine declared it "one of the most important illustrations which geology [has] received since it [was] shaped into a science."[86] To the outrage of some of Darwin's allies—especially Thomas Henry Huxley—Sabine raised the *Origin* in his award citation in order to inject his own opinion on the species theory, saying, "Some amongst us may perhaps incline to accept the theory indicated by the title of [the *Origin*], while others may perhaps incline to refuse, or at least to remit it to a future time, when increased knowledge shall afford stronger grounds for its ultimate acceptance or rejection."

By the mid-1870s the coral reef theory was arguably Darwin's most uni- versally admired scientific accomplishment. James Dwight Dana's work during and after the United States Exploring Expedition had lent con- siderable additional weight to the subsidence theory, and no significant challenge had materialized (though one would emerge later in the decade when the results from the 1872–76 voyage of HMS *Challenger* were published).[87] The level of sustained interest in the theory was reflected by the fact that Darwin's 1842 book was almost impossible to acquire. His friends had been unable to buy copies, and his son Horace had learned that a Cambridge bookseller spent three years trying to locate a copy before finding one in Berlin.[88] In December 1873, the sixty-four-year-old Darwin wrote to his former publisher, Smith and Elder, to inquire about reprinting it with light revisions. In a brief new preface to the second edition (dated February 1874) he declared, perhaps ungenerously but not unreasonably, that in the thirty-two years since the first edition was published, "only one important work on the same subject has appeared, namely, in 1872, by Professor Dana, on Corals and Coral-Reefs."[89] Meanwhile the *Origin* was in its sixth edition. In two of Darwin's newer books, *The Descent of Man* (1871) and *The Expression of the Emotions in Man and Animals* (1872), he had finally applied his theories of natural selection and sexual selection to the evolution of human bodies and behaviors. These works were receiving far more attention in the 1870s than Darwin's coral reef theory ever had, but debate over the species theory was hardly settled. By contrast, Darwin could state comfortably in an autobiographical sketch he began writing in this period that the coral reef theory "was thought highly of by scientific men, and . . . is, I think, now well established."[90]

Being cognizant of this contrast between the time-tested coral reef theory and the still contentious evolutionary theories is essential when evaluating these autobiographical statements, which Darwin began to write for his family in 1876 and continued adding to until the year before he died. Though they are nowadays most often known by the title of a subsequently published version, *The Autobiography of Charles Darwin*, Darwin himself called them "recollections of the development of [my] mind and character."[91] As James Secord points out, it is suggestive that Darwin omitted the word my from the title in his first draft, for he may indeed have intended the piece to serve as a guide for others on developing mind and character in addition to being (or under the pretext of being) a narrative of his own development.[92]

It has become increasingly clear to scholars that the claims in "Recollections" often fail to correspond to the facts of Darwin's life as we understand them from other evidence.[93] How, then, should we read this text? Could it have been intended as something other than a documentary account of what Darwin was thinking and feeling at key moments in his life? The title suggests as much: these were *recollections*. And many of the episodes he wrote about were worth recalling and recounting precisely because they served as parables on how to live and work as a man of science. In what follows I will argue that Darwin's self-reflections and his assessments of fellow savants were intended as a set of prescriptions for such living and working. By 1876 his sons George and Francis had embarked on scientific careers. Therefore, even if the text was not written with publication in mind (and ostensibly it was not, though his family did publish an emended version of it five years after his death), these parables would have a readership that could benefit from its object lessons.

Consistent with the idea that Darwin intended to offer *general* insights on the development of mind and character, he supplemented the autobiographical material with assessments of the manner of thinking (and degree of philosophical caution) exhibited by a number of others. He wrote, for example, that while as a young man he had admired his grandfather Erasmus Darwin's transmutationist tract *Zoonomia*, he subsequently became "much disappointed" in the work on account of "the proportion of speculation being so large to the facts given."[94] Here, then, was the publishing advice he had given to Morlot and Woodd, repackaged as a family story. Darwin reported in turn that his father had a "mind [that] was not scientific" in Darwin's view. "He did not try to generalize his knowledge under general laws; yet he formed a theory for almost everything."[95] Here again the implication was that, to be properly scientific, theorizing must emerge from the proper ordering and explicating of evidence. And finally, writing about himself, Darwin mused, "My mind seems to have become a kind of machine for grinding general laws out of large collections of facts."[96]

Lyell had died just a year before Darwin began writing the "Recollections." Darwin commemorated him with affection, reminiscing, "One of his chief characteristics was his sympathy with the work of others; and I was as much astonished as delighted at the interest he showed when on my return to England I explained to him my views on coral reefs. This encouraged me greatly, and his advice and example had much influence on me."[97] Darwin wrote more about Lyell in a section he added in 1881, just twelve months before his own death. Once again he gave special attention to the moment in 1837 when he had told Lyell about the coral reef theory. This time he emphasized that "my views on coral-reefs . . . differed from his" before reiterating, "I was greatly surprised and encouraged by the vivid interest which he showed." Darwin concluded that the science of geology was indebted to Lyell "more so, as I believe, than to any other man who ever lived," citing the "powerful effects of Lyell's works" in combating the "wild hypotheses" of the Frenchman Élie de Beaumont. Darwin did not address Lyell's own reputation for speculation, but it is noteworthy to see that he characterized Lyell's impact on the science of geology not in terms of his work on specific rocks or localities (as he did, for example, in describing the "services rendered to geology" by Murchison) but rather in terms of his ability to improve on others' speculations.[98]

Darwin rated others in his scientific circle more directly according to their approaches to description and theorizing. Some fell short because their interests were parochial or they had not shared his own zeal for the value of generalizing. Of his original Cambridge mentor, John Stevens Henslow, he wrote, "His strongest taste was to draw conclusions from long-continued minute observations. His judgment was excellent, and his whole mind well-balanced; but I do not suppose that anyone would say that he possessed much original genius." The venerable botanist Robert Brown, whom Darwin had become close to in the years immediately after the voyage, was capable of "pour[ing] forth a rich treasure of curious observations and acute remarks, but they almost always related to minute points, and he never with me discussed large and general questions of science." At the other end of the spectrum, younger than Darwin and altogether too much given to speculation, lay Herbert Spencer. "His deductive manner of treating every subject is wholly opposed to my frame of mind," Darwin wrote in an extended and unflattering portrait. "His conclusions never convince me: and over and over again I have said to myself, after reading one of his discussions,—'Here would be a fine subject for half-a-dozen years' work.'"

It might come as a surprise, therefore, to learn that Darwin described his own coral reef theory as a product of the same kind of reasoning for which he criticized Spencer. "No other work of mine," he wrote, "was begun in so deductive a spirit as this; for the whole theory was thought out on the west coast of S[outh] America before I had seen a true coral reef." He claimed that when he had first encountered "living reefs" (in the Pacific) he "had therefore only to verify and extend [his] views." According to these reflections, Darwin had "been incessantly attending to the effects on the shores of S. America of the intermittent elevation of the land, together with the denudation and the deposition of sediment. This necessarily led me to reflect much," he continued, "on the effects of subsidence, and it was easy to replace in imagination the continued deposition of sediment by the upward growth of coral. To do this was to form my theory of the formation of barrier-reefs and coral atolls."[99]

As I have foreshadowed throughout this book, this claim is at first glance contradictory to several of my arguments, and particularly to my emphasis on the eureka moment at Tahiti. Notice, though, just how divergent this recollection of the coral reef theory was from every version of the theory that Darwin himself had previously published. The 1837 paper and the 1842 book both suggested that the theory was prompted by contemplating the phenomenon of encircling reefs that stood off from the shore of an island. The 1835 essay he wrote en route from Tahiti to New Zealand indicated much the same, except that it specified a precise moment when he pondered one particular encircling reef from a singular vantage point. My claim that the coral reef theory existed in many permutations over time suggests that our question need not be whether the late-life autobiographical account reflected how he "really" came up with "the theory." Rather, the question is why it made sense, in Darwin's memory or in his effort at self-fashioning, to retroactively equate work in South America with the insights from Tahiti.

One answer is that theories themselves can change the narratives by which individuals make sense of their scientific careers. I have already shown how Darwin's idea at Tahiti defined a plan of writing and fieldwork through the rest of the voyage, and it is hardly surprising that developing a new theory would alter the course of any scientist's future work. But the eureka moment in Tahiti imposed *retrospective* coherence on Darwin's activities as well. He had discovered unexpectedly that certain ideas dating to his time in South America were relevant to solving the well-known question of the formation of coral reefs.[100] Once Darwin "had" the theory, he forevermore remembered his investigations into South American elevation and subsidence as *inherently* relevant to the theory (indeed, as constituting the theory), and it became all but impossible to remember that he had ever pursued this line of study for other reasons, independent of the theory he would eventually produce. The closing sentence of that reminiscence ("To do this [analogize upward coral growth to accumulation of sediment] was to form my theory of the formation of barrier-reefs and coral atolls") would stand as a decent description of the insight that "forcibly struck" Darwin at Tahiti if not for his insinuation that this moment had already occurred in South America. Without the teleology Darwin's memory had imposed, he might instead have written, of his insights about sedimentation and subsidence in South America, "*all that remained* was to form the theory."

This perspective on theories and memory applies to Darwin's evolutionary theorizing as well as to the coral reef theory. Frank Sulloway, the historian who debunked the myth that Darwin became an evolutionist during his time at the Galápagos, wrote, "Contrary to the legend, Darwin's

[Galápagos] finches do not appear to have inspired his earliest theoretical views on evolution, even after he finally became an evolutionist in 1837; *rather it was his evolutionary views that allowed him, retrospectively, to understand the complex case of the finches.*"[101] Sulloway went on to rebuke the authors of bad "textbook" histories who had oversimplified Darwin's biography to the point of distortion by "telescoping history around one dramatic moment of insight in the Galapagos Archipelago."[102] But Sulloway failed to chastise Darwin for being a bad historian of his own work, in the sense that Darwin too "telescoped" history around certain dramatic moments of insight. His subsequent evolutionary theorizing made it *seem* to him as though the distribution of certain animals encountered during the voyage had always determined the theory he would eventually develop. Likewise, to paraphrase Sulloway, geologizing in South America appears not to have sparked Darwin's thinking about the shape of coral reefs; rather, it was his eventual explanation for reef formation that allowed Darwin, retrospectively, to take it for granted that the key to understanding coral reefs had always lain in studying the geology of South America.

A second way to make sense of Darwin's recollection about the coral reef theory becomes apparent in contradistinction to his description of the species theory in the same text. Unlike the ostensibly deductive process by which he derived the coral reef theory, Darwin's "Recollections" credited the species theory to his rigorous adherence to the inductive method. "My first note-book [on species] was opened in July 1837," he wrote, citing observations he made during the *Beagle* voyage that had caused him to doubt species fixity. He continued with a phrase that is now well known. "I worked on true Baconian principles & *without any theory* collected facts on a wholesale scale." It was only after fifteen months of "systematic inquiry," and after reading "for amusement" Thomas Robert Malthus's *Essay on the Principle of Population*, that Darwin—in this account—lit on an explanation for descent with modification: the idea of natural selection. "Here then I had *at last* got a theory by which to work."[103] Scholars have subsequently picked apart virtually all of those claims. Darwin's transmutation notebooks reveal that he was hypothesizing enthusiastically in 1837, well before he (re)read Malthus.[104] Natural selection, it turns out, wasn't his first theory of descent with modification, it was his last one—or at least one that was useful enough to keep working with.

Now is the moment when it becomes useful to remember that Darwin composed his "Recollections" at a time when the coral reef theory was a secure and noncontroversial accomplishment and when many condemnations of his ultracontroversial species theory were couched in critiques of the "method" he had used to develop it. With this perspective it becomes possible to appreciate simultaneously how Darwin could *afford* to caricature his coral reef theory as "most deductive" and why he *needed* to describe his species theory as "pure[ly] Baconian" in origin. Both of these caricatures fit precisely with the authorial persona Darwin sought to fashion from 1839 onward, that of a dogged empiricist who was constantly vigilant because even he had formerly been susceptible to being carried away by flights of speculation. In the "Recollections" Darwin reinforced this point by recalling his attempt to incorporate Glen Roy into his grand geological synthesis. "This paper [on Glen Roy] was a great failure, and I am ashamed of it." His mistake had been faulty reasoning, and his "error," he concluded, "has been a good lesson to me." This was a key axis of the "development of [Darwin's] mind and character." He portrayed himself—and I believe genuinely understood himself—to have grown from a clever but immature young man given to deductive speculation into a mature man of science whose insights were the product of methodical work.

The opportunities and challenges Darwin had faced as a young author were shaped by details of place, field, and time: in London, in geology, in the 1830s. Some of this specificity may be illustrated with reference to the most famous opinion Darwin ever expressed about the relation of his work to Lyell's. He expressed the sentiment in an 1844 letter to Lyell's father-in-law, Leonard Horner. "I always feel as if my books came half out of Lyell's brains & that I never acknowledge this sufficiently. I have always thought that the great merit of the Principles, was that it altered the whole tone of one's mind & therefore that when seeing a thing never seen by Lyell, one yet saw it partially through his eyes."[105] This suggestive comment has been quoted widely since 1903, when Francis Darwin included it in an edition of his father's letters, and it has helped to ensure that Lyell's indirect influence on Darwin, through the *Principles,* is more famous than the crucial face-to-face coaching and behind-the-scenes advocacy I have documented. But Darwin continued the famous sentence with a qualification that is rarely quoted. It offers a glimpse of the previously unrecognized second-guessing that I have shown he endured after trying to emulate Lyell's self-assured style of presenting theories. In fact, his sentiment sounds quite different when his full reflection is included: "When seeing a thing never seen by Lyell, one yet partially saw it through his eyes—*it would have been in some respects better if I had done this less.*"[106]

Darwin's reference to seeing through Lyell's eyes brings to mind Henry

De la Beche's cartoon parody of Lyell's zeal for theorizing (fig. 22): "Take a view, my dear sir, through these glasses, and you will see that the whole face of nature is as blue as indigo." The cartoon, in turn, offers a way to understand why Darwin might have come to regret how far he became prone to seeing the world as Lyell saw it. I have argued that during the *Beagle* voyage itself Darwin was, if anything, less apt to interpret landscapes on Lyell's terms than we might have expected from the subsequent record. After the voyage, what Darwin came to see through Lyell's eyes was not just the face of nature but the social world of science. Yet the De la Beche caricature reminds us that the perspective Darwin learned was not shared by many, let alone most, of the men he desired to impress within the geological community. Lyell's enthusiasm for theoretical authorship, and his project's resemblance to discredited "theories of the earth," created resistance that other approaches might have avoided.

Here then it was highly significant that Darwin was trying to establish himself and, more important, his theories in the *English* geological community, and in the first half of the nineteenth century. Whereas theorizing had been a prime characteristic of German *Naturphilosophie* and Wernerian geognosy, and of the geological and zoological programs of Francophone savants such as Jean-André Deluc, Georges Cuvier, and Étienne Geoffroy Saint-Hilaire, the specialist sciences of zoology and geology had been established in Britain as explicitly antitheoretical pursuits. Veneration for ostensibly descriptive science by the founders of the Geological Society (established in 1807) was intended to counteract the tradition of exuberant theorizing that characterized both Continental natural history and those earlier British "theories of the earth" by authors such as Thomas Burnet and William Whiston. The new generation of specialists sought to distinguish themselves both from their speculative forebears and from contemporaries who grubbed in practical matters such as mining.[107]

Meanwhile, of course, Lyell, Darwin, and others did seek causal explanations as part of their work in geology and natural history, and their contemporaries John Herschel and William Whewell devoted themselves to giving accounts of how generalizing might responsibly be pursued (though Whewell claimed that geology had yet to advance to the stage when a "general system of geology" might be established). But as Sandra Herbert has phrased it, the establishment in Britain of "specialized societies devoted to the study of botany, geology, and zoology . . . altered social forms for expression within science."

For those in natural history who were theoretically inclined, the accepted scientific posture in the 1830s was not that of the theorist. . . . Those responsible for the situation . . . were those very philosophers of science and specialists who, looking backward to what they saw as a period of excessive speculation, saw their own cautionary attitude toward theory as salutary and conducive to orderly growth in science.[108]

The way Darwin retrospectively characterized his development as a geologist, then, actually recapitulated his colleagues' mythology about the development of the field itself, as an immature and speculative science transmuting into a disciplined program of collective investigation.

Conclusion

As I declared at the outset, this book is intended not as a biography but as a study in the history and sociology of science. Now, in closing, I want to emphasize the ways my approach and my arguments bear on broader themes in science studies.

It may seem incongruous for a book in the social studies of science to be focused on a single person, but my main interests relate to the tension between individuality and collectivity *within* a scholarly community. I have investigated how Darwin came to possess the knowledge and access necessary to be inducted into such a community. I have tried to understand the challenges his scholarly community posed to him (how, for example, to be innovative and intelligible at the same time). And I have been concerned with the problem, for a community, of allocating credit to individual people who contribute to a body of knowledge that is meaningful only to the extent that it is shared.

Lyell, Darwin, and Authorship

In analyzing how Darwin came to be a credible author of scientific theories I did not take it for granted that my focus should be—or that Darwin's focus at any given time necessarily was—the theory of evolution by natural selection. I proceeded by paying attention to his everyday activities—what many historians would term his "practices"—and observing how and where he spent his time. And Darwin's life looks very different from such a perspective. It is not just that he devoted a higher proportion of his most creative years to laboring over a theory of coral reef formation than he did to any other topic or task, it is that he thought his reputation was tied inexorably to that work and to the larger geological program to which it belonged. I have become convinced that Darwin's anxiety during the "notebook" years was caused not by his private activities but by his public obligation to write a synthetic geology book that lived up to the audacious theorizing of his early geological papers. Only in retrospect could it seem obvious (to Darwin or to us) that he *should* have been devoting attention to his epoch-making species theory, let alone that he had more reason to be anxious about the reaction to the *topic* than to the seeming scientific rigor of his work.

Many of the events that loomed largest to me as a consequence of examining Darwin's career in this way could aptly be described in terms of the "reception" of Lyell's theoretical system rather than the development of Darwin's.[1] There is nothing new about claiming that Darwin was influenced by Lyell's work, but I have showed that Lyell took considerably more overt action on Darwin's behalf than we are normally led to believe. In addition to giving help behind the scenes and in public, Lyell also imposed himself on Darwin and took substantial liberties with his unpublished work. This should perhaps not be surprising. Only because the junior figure in this story turned out to be the more famous member of the duo does it seem at all counterintuitive to find that the mentor often had more agency than the pupil.

As we have seen, though, Lyell certainly did not simply cast theories out into the world, and Darwin was not a passive recipient or even an independent one. Rather, the pair worked *together*, and rapidly, once Darwin was back in England. They positioned Darwin as an independent worker who came to disagree with Lyell on a significant geological question *as a result* of following the methods modeled in Lyell's publications. Even while Darwin was disputing Lyell's earlier theory of coral reef formation he was using his own new theory to advocate the master's broader claims about geology and geologizing. That it was even possible to be "more Lyellian" than Lyell himself is revealing of the degree to which Lyell had staked a claim on a particular style of geological theorizing. But indeed, as William Whewell reported to John Herschel (in an exchange between two men who contemplated the nature of theorizing as intensely as anyone ever has), Darwin was recognizably "an extreme Lyellist" within just a few months of returning from the *Beagle* voyage.[2]

Here let me emphasize three points I made in passing earlier in the text. First, that Lyell's patronage, and particularly his premeditated concession to Darwin's coral reef theory, helped to verify Darwin's candidacy as a credible author of geological theories. Second, that as the two worked to establish Darwin's bona fides they were simultaneously changing the *kind* of author Lyell was. The *Principles* went from being a text written by Lyell to being the foundation of a system that had produced valuable work by someone else. It is easy to forget that Lyell was himself still in his thirties when Darwin returned from the *Beagle* voyage, a provocateur as much as a scientific statesman. And third, I reiterate that this early episode established a pattern of backstage planning and coordinated action between Lyell and Darwin that was a central feature of Darwin's preparation to publish his species theory in its originally intended long form. When the arrival of Alfred Russel Wallace's essay threw Darwin's plans into confusion, Lyell once again masterminded a public presentation of Darwin's work. This time he had the aid of Joseph Dalton Hooker, whom Darwin had groomed as an ally and—for lack of a better term—research assistant in the very fashion Lyell had done with Darwin. As with the unveiling of Darwin's 1837 coral reef paper, the 1858 announcement of natural selection by Lyell and Hooker at the Linnean Society worked to the mutual benefit of a senior and a junior figure. This time it was Wallace's stature being raised even as his insights from far-flung research sites lent independent corroboration to Darwin's grander ambitions.

Darwin's relationships—with those who helped him during the voyage and with Lyell—bear on questions of scientific "authorship." I refer here not only to publishing per se but to the successful claiming of something like what today is called "intellectual property."[3] We have seen how authority, or proprietorship of ideas, was context dependent. During the voyage, Darwin was treated by those on board as the ship's "philosopher." The officers he was friendly with were only too eager to contribute specimens and observations as fodder for Darwin's expansive ruminations. On returning to England, however, Darwin himself had the status of a mere collector and observer in relation to the elite members of the scientific community (even as his own collections and observations subsumed some of the officers').[4] Lyell worked first to enroll Darwin in the science of geology in just such a subordinate capacity, then very rapidly helped to establish him as a legitimately philosophical geologist, someone who might be taken seriously as an author of theories. As FitzRoy complained, Lyell's influence pulling Darwin up to his exalted position overshadowed the credit owed to the hydrographers who first "held the ladder by which you mounted to a position where your industry, enterprise, and talent could be thoroughly demonstrated."[5] In terms of social class and formal rank in his field, FitzRoy himself considerably surpassed Darwin. He was both metaphorically and, upon return, literally the author of the voyage's achievements, and he was greatly esteemed as a surveyor and observer by Lyell and his ilk. But when it came to the realm of geological *theory*, a topic he did pursue in a chapter of his Narrative titled "A very few Remarks with reference to the Deluge," he was taken no more seriously by Lyell (and Darwin) than the missionary John Williams had been in 1837, and I think not just because FitzRoy and Williams hewed toward "scriptural geology."[6]

Neither Lyell (as adviser) nor the *Beagle*'s officers (as contributors) were ever listed as Darwin's formal coauthors in the fashion that is common in the sciences today.[7] This fact might lead some to underestimate the extent of his collaborations and the depth of his indebtedness. The surveyors, their crewmen, and Darwin's servant Syms Covington all resemble the "invisible technicians" who, as Steven Shapin has shown, contributed to Robert Boyle's chemical experiments but were rarely acknowledged by name in his publications. In describing why even Boyle's most trusted (and fully acknowledged) assistant was never accorded coauthorship, let alone sole authorship, of the experiments he conducted and wrote up, Shapin wrote that "Boyle was the *author* because Boyle possessed *authority*. It was he who presided over the scientific workplace—indeed, it was his house; it was he who possessed the acknowledged right to set the agenda of work, who could effectively command the skilled labor of others, who could define the boundaries between skill and knowledge."[8] With respect to HMS *Beagle* and the labors conducted during the voyage, the person who answers to this description is FitzRoy. When it came to publishing charts and the formal narrative, FitzRoy was indeed the authority and the named author. One of the remarkable things about Darwin's career is that he received such active assistance, such leeway to work, and such control over his own specimens, especially in comparison with many of his contemporaries naturalizing on other naval voyages.[9]

In many ways Darwin's debts to Lyell resembled those of a modern research student or postdoctoral fellow to her supervisor (and vice versa). Sociologists of science have long been studying the benefits and costs, to each party, of students and advisers sharing formal authorship. Most famously, Robert Merton addressed the topic in his paper on what he called the "Matthew Effect," the disproportionate accrual of credit to individuals who were already the most renowned among a given group of collaborators or among simultaneous claimants to a discovery.[10] Merton offered an extended analysis of the deeply felt dilemma faced by eminent scientists, whether to add their names to students' publications. To do so would draw more attention to the work but diminish the proportion of credit accorded to the student who had conducted it. We might imagine Lyell having faced an analogous decision: not whether to append his name to Darwin's publications but whether to advertise his role as an active participant or to have Darwin emphasize his debt to Lyell's *publications*. As we have seen, the pair followed the latter course and, not coincidentally, Darwin has perennially been described as a student of the *Principles* rather than a student of Lyell's.[11]

There are countless historical examples of the kind of mentorship Lyell provided to Darwin, but we are more accustomed to seeing them between people with a formal pedagogical relationship. But I have illustrated that in addition to his well-known training in field research methods by his university instructors Grant and Sedgwick before the *Beagle* voyage, Darwin received an even more rigorous and personalized training in the social and expository skills of a geologist when he came to reside in London. Under Lyell he benefited from patronage much like the support offered to students who trained in university laboratories in the nineteenth century. Lyell provided for Darwin all the cajoling, advising, reading, rewriting, and backstage advocacy that laboratory leaders provided for their students in nineteenth-century chemistry and physiology laboratories.[12] And when we recognize that the bulk of the *research* Darwin did to develop and support his coral reef theory occurred not during the *Beagle* voyage but in the libraries of London, and we recall Lyell's role as Darwin's main theoretical interlocutor on coral reefs, their relationship takes on all the more significance. Little surprise that in a letter recalling his coral reef book, Darwin closed by bidding Lyell "Farewell My dear old Master."[13] I have illustrated that when Lyell and Darwin spoke of their relationship as that of master and student they meant it not as a casual metaphor but in a way that drew on established social roles and implied serious obligations on both sides.

My point is not to argue that we need to devote even more attention to the giants of the history of science, but rather to show how much more Darwin's success had in common with the smaller successes of contemporaries who came through academic research schools in Glasgow or Giessen than we might have imagined based on the strength of his posthumous reputation. Likewise, I do not mean to discount the significance of the cumulative advantage Darwin received from his wealth, his gender, and his family name. Instead I mean to call attention to a factor—his intensive coaching in the art of scientific self-fashioning—that set him apart from many who, despite sharing some of those inborn advantages, nevertheless found their speculations taken less seriously than Darwin's. But the tendency to overlook Lyell's active interventions has indeed been compounded by the fame Darwin eventually achieved. As Merton himself pointed out in his discussion of the Matthew Effect, "Should the younger scientist move ahead to do autonomous and significant work, this work *retroactively* affects the appraisals of his role in earlier collaboration."[14] Perhaps most important, I want to emphasize that neither a "top-down" (Lyell focused) nor a "bottom-up" (Darwin focused) perspective on this collaboration between unequal partners could be entirely satisfactory. What my study illustrates is the significance of the transactions *between* mentor and pupil (and those undertaken on one another's behalf) in shaping their work for audiences and audiences for their work.

Studying Practices, Learning about Theories

Over the past thirty years or so, a strong tradition of research on locally specific practices of experiment, observation, and collecting has helped to portray the history of science as something more than just a succession of ideas. These works provide rich descriptions of what scientists actually did from day to day and often illustrate the interpenetration between activities taken to constitute "science" and a range of "social" factors. The most ambitious of them explain how various spaces, methods, or instruments came to be deemed properly scientific at all.

When these "practice" studies began appearing, the authors often framed them explicitly as counterreactions to a narrowly defined, seemingly elitist "history of ideas," and to idealized models of the scientific method that were based on philosophical reconstructions rather than on empirical research by historians or ethnographers.[15] One protagonist of this trend, Andrew Pickering made a case that in some instances "what scientists do is just as important as the knowledge they produce." Significantly, however, he also urged the need to apply insights gleaned through studying "science-as-practice" back toward studies of "science-as-knowledge."[16] Gradually, though, studying scientists' everyday activities has come for many scholars to constitute an end in itself. Robert Kohler, for example, emphasized his book's "lack of attention to issues of epistemology and 'social construction'" in *Lords of the Fly*, arguing instead for a "pragmatic conception of credibility and truth [that] locates the causes of scientists' behavior in the production process rather than in the realm of theoretical beliefs or professional and political ideologies."[17] In a much more recent work that uses Joseph Hooker's career to ask some of the same science-studies questions I have asked about Darwin, Jim Endersby takes the step of organizing an entire book according to a set of distinct "practices"—loosely defined—including traveling, collecting, and classifying but also such things as "seeing," "settling," "associating," and "governing." He makes it clear that he did so with polemical intent, hoping "to sharpen the contrast with a history of ideas in which disembodied concepts wander vaguely across an intellectual landscape." In a decision consistent with this perspective, Endersby explains, "I have chosen to emphasize 'writing' as a practice rather than the more abstract notion of 'theorizing.'"[18] Endersby closes his book by discussing theory making, however, arguing that this act was remote, both temporally and spatially, from a botanist's craft activities. "The practices of collecting and classifying," he concludes, "were prior to those of . . . philosophizing [theorizing] in two related senses: chronologically . . . and physically."[19] Whether as a consequence of Endersby's method, his choice of biographical subject, or both, the result, it seems to me, is that theories come to appear even more disembodied from the everyday work of science.

Instead of studying scientific "practices" as a goal in itself, as Kohler and Endersby in their own ways have done, I have aimed to apply the results of doing so to answering the kinds of questions that practice-oriented histories of science now so often marginalize. Rejecting histories of "disembodied" ideas is one thing, but after putting the "body" into our accounts of the history of science, it is worth asking whether the result offers us the chance to write new kinds of history of ideas.[20]

I modeled this tactic, of turning the *method* of studying everyday scientific activities back toward *questions* more characteristic of intellectual history, on a particular body of work in the history of the physical sciences. It is important, therefore, to show the ways my approach and my conclusions diverge from that important work on "cultures of theory" in physics.[21] The authors of those studies used a set of approaches familiar from research on laboratory life and the history of experiment to demonstrate the degree to which even math-intensive physics also depended on material culture, training, and technique. Taking a similar approach to the history of theorizing in natural history, however, raises distinct challenges, because natural history involved no equivalent explicit vocational division between experimentalists and theorists of the sort that emerged in the physical sciences.[22] For the aforementioned historians, studying theorizing meant studying whatever it was that "theoretical physicists" actually spent their time doing. In contrast, naturalists like Lyell and Darwin may have "had" theories, but none defined themselves exclusively as "theorists."[23]

For these historians of theoretical physics, it stood as a significant accomplishment to show that a seemingly abstract science had "practices" just like any other. Given that I am studying specimen- and fieldwork-intensive sciences, I have had roughly the opposite challenge: to examine whether "practices" had theories. I have come to a conclusion about the relation between practice and theory in natural history very different from the one Endersby arrived at in his study of Hooker's career. Neither during the ostensibly fieldwork-oriented phase of Darwin's career nor in the ensuing years when he was confined to England was there either a unidirectional relation or, necessarily, a chronological lag between acts of data gathering and analysis. I have shown the significance of Darwin's specific locales, including not just the Andes or Tahiti but also the *Beagle*'s library and the scientific institutions of London, as providing certain resources for Darwin's theories. I have shown how specific tools and techniques, such as those related to hydrographic surveying, reshaped *both* Darwin's collecting *and* his emerging theories. The sounding lead supplied Darwin with a mode of "seeing" and collecting underwater; those opportunities shaped his speculations, which in turn guided the questions he tried to answer through continued hydrographic work. The general idea that theory can guide observation, and vice versa, was as familiar to Darwin as it is to science-studies scholars today. I have sought to argue that particular *ways* of observing and collecting, rather than just the observations and specimens themselves, can fuel certain *types* of theorizing, as for instance when Darwin reasoned about species distribution as a consequence of his work with the sounding lead.

I have wrestled throughout this book with the contradiction that Darwin's theories were ever-changing yet retained, in his mind and in his speech, their *identity*. Thus, while Darwin rarely spoke of theorizing as a discrete act, he did talk of theories as though they were independent, almost physical, things. He spoke of "building theories," of "putting together the frame of a geological theory." In describing other men of science he saw one "rid[ing] his theory very hard" and another "attack[ing] my theory." He believed that if Lyell were to reject the species theory, it would be "my fault & not the theory's fault."

Darwin's statement that "a man is as tender of his theories as of his child<<ren>>" provides one of the best illustrations of his attitude.[24] In Darwin's view both children and theories could grow and change without losing their essential quality. Just as having children could imbue everyday life with deeper meaning and purpose, "having" theories could give purpose and significance to the workaday aspects of studying nature. And just as Darwin yearned for his children—or to be more precise, his *sons*—to go out and find independent success, he likewise envisioned his theories eventually having lives of their own that would enrich the scientific community and, crucially, define the legacy of their parent. Just as he strove to offer his sons every possible advantage *even though* he maintained great confidence in their personal merit, he was convinced that even "true" theories had to be "promoted." This way of talking about *Darwin's* perspective on theories can lead us astray if it permeates our analytical language as well. Likening theories to children can make it seem as though the "mature" theory was the single possible outcome of an earlier theory that shares its name, as though some permutation could be described as an "early" or "incomplete" statement of a theory. It is important, from the historian's perspective, not to surrender to that sense of inevitability. If we want to understand what and how scientists were thinking *while they worked*, we must avoid deeming one version of a theory authoritative and using it as the baseline for analyzing "how far" they had come at a given point.

As I argued at the end of the previous chapter, though, Darwin was a "bad historian" of his own theories. He did often talk as though "the theory" (as he knew it at any given time) was the inevitable product of a developmental process. And, of course, the eventual outcome of (successful) theorizing could come to feel inevitable precisely because it came to reflect how nature actually was. Once a theory had been adopted (but only then) it could appear to have been drawn forward by the very physical reality it now explained, rather than being pushed forward by a set of individual and collective choices *to pursue* and *not to pursue* particular research questions in particular ways. Likewise, when theories were discarded it could come to seem that they had been exclusively a product of contingency, uncertainty, and merely provisional knowledge. Worse still, Darwin was prone to recalling that the *purpose* of his earlier work had always been to produce not just any mature theory but *the* eventual theory-as-he-knew-it. My accounts, both of the shift from zoology to geology as the interest that governed his study of corals and reefs and of the unexpected uses to which he put his ideas about South American crustal movement, show that his motivations and the uses for his work changed continually, as he also changed his intended audiences and those potential audiences themselves changed over time.

One of the hazards attending the study of a scientific career as success-ful as Darwin's is the risk of something like historical confirmation bias. We know in retrospect that certain observations, ideas, or experiments will prove momentous, so it is difficult to overcome the tendency to assume that those instances felt especially noteworthy to our historical actors as they happened, or that whatever significance they were accorded at the time was given for the same reason we know of them today. It is also easy to imagine that it would be self-evident to colleagues, rivals, teachers, and students that the young Darwin's work would be worth their attention. As I have argued elsewhere, we must pay special attention to when and why such individuals' ideas *initially* came to be seen as important and credible, before their authors had acquired the fame that makes it all too easy to take it for granted that these ideas would be given a serious hearing.[25] The role of Darwin's own well-known retrospections in shaping the histories that others have written simply magnifies the problem.

As the reputation of the *Origin* has grown while memories of other nineteenth-century theorists and their publications have dimmed, it seems increasingly self-evident to many that Darwin authored a revolution in science and social thought. In the face of such oversimplification it can be difficult to distinguish the man from the movements he may or may not have inspired: to distinguish Darwin from the many kinds of Darwinism. Nevertheless one thing that has remained constant is a respect, whether grudging or reverent, for the power of Darwin's ideas. His long and com-plex career has so often been boiled down to a single product, a single phrase: "Darwin's theory."

This book has illustrated the implications for Darwin himself of the fact that theories cannot be isolated from the identities of their authors and their audiences. Even Darwin, whose theories seem so powerful that they should have been able to stand on their own, whose confidence never wa-vered that his theories of coral reef formation and the origin of species were right, nevertheless treated them as things to be nurtured and supported. He considered the grooming of theories and audiences for one another to be skills *intrinsic* to the practice of science. And he learned that—like any other—these scientific skills could be taught and developed, if never quite perfected.

Acknowledgments

I imagine most scholars can empathize with Darwin's lament, while attempting to keep up the pace of his geological writing, that "life is now measured by volume, chapters, & sheets." In the foregoing pages I have described Darwin's struggle to become an author, his efforts to balance claims to originality with acknowledgment of the intellectual traditions that gave his new ideas significance, and the pressure he faced to demonstrate his scholarly maturity by completing a reputable book. He did not face these challenges alone. He received extraordinary support from a series of deeply committed mentors and advocates, his wife, Emma, and his extended family. In his travels and at home he benefited from the dedication and generosity of innumerable people inside and outside the scholarly community. As I bring this book to a close I am reminded, as ever, that I have been the beneficiary of all these kinds of generosity myself.

In her role as an author, Janet Browne sparked my original interest in many of the questions that led me to study voyaging, natural history, and Darwin's early career. As a mentor, from the time when as an undergraduate I first corresponded with her, to my postdoctoral fellowship, and to this day, she has shaped my way of thinking and helped ensure that I had the opportunity to continue studying our shared interests. And as a colleague and friend she continues to teach me, encourage me, and set an example. I am so grateful to her.

My outlook on the history of science—and my desire to spend a career belonging to this scholarly community—were shaped above all by Andrew Warwick. Andy taught me to see that the production and transmission of scientific knowledge were puzzles to be explained, and he introduced me to texts (and, in so many valuable cases, people) offering exhilarating ways to answer these questions. His own scholarship illustrates with extraordinary clarity how knowledge has been—and can only be—produced as a collective activity, and his "coaching" in how to read, think, and talk like a historian of science is what gave me the desire and the skills to try to contribute to the field myself.

The field Andy introduced me to contains many wonderful and remarkable characters. A handful of them have been not just my main supporters but also my most cherished interlocutors over the years, and their generosity with time, effort, and wisdom humbles me. Simon Schaffer started a conversation with me when I was a beginning student in the UK, and his willingness to continue it for almost two decades since has been one of the great pleasures and privileges of my life. From the same early days and ever since, Rob Iliffe has been teaching me how (and why) to think about scientific self-fashioning and the relations between authors and audiences. His influence on the arguments I make in this book is immeasurable. But I would never have thought of writing about coral reefs, let alone found it so exciting, if it hadn't been for Graham Burnett and his extraordinary breadth of knowledge and depth of enthusiasm. Angela Creager guided me through my PhD, pushing and pulling me when I needed it most, helping me to learn what it meant to be a professional scholar. Moreover, Angela's teaching, her advice on many dissertation chapter drafts, and her utterly eye-opening report on the finished product provided me with the intellectual terms in which I have thought about my work ever since. Jim Secord, meanwhile, is the reader I often imagined when I was writing, and he is the person whose comments on various pieces of my work over the past dozen years had the greatest impact on what has emerged. More recently, Lukas Rieppel has become my most indispensable friend-reader-disputant; his ideas enrich much of what I have written in these pages. My greatest intellectual debt is to Laura Stark, who is continually teaching me new ways of thinking, in addition to being the person who knows me best.

So much has been written about Darwin that it can feel daunting to try to contribute something original to the field. Such anxieties are more than offset, however, by the pleasure and intellectual excitement I have experienced by joining the community of Darwin scholars. No acknowledgment of my debt to Sandra Herbert's work could ever be sufficient. She is the main authority on both of this book's overarching topics, Darwin's career as a geologist and his identity as an author of theories. Her 2005 book came out shortly before I traveled to Cambridge to study Darwin's manuscripts for the first time, and it is the foundation of much of what I have done here. I am just as deeply indebted to Martin Rudwick for his immensely valuable work on the history of geology in general and on Lyell in particular, for framing (in his article "Darwin in London") many of the questions I was most eager to try to answer by writing this book, and for being such an energetic conversationalist, host, and friend for so many years.

I am grateful as well to Leonard Wilson for his invaluable work on Lyell and for generously giving me access to his duplicates of Lyell's notebooks from 1836 to 1840 before I had a chance to see them at Kinnordy House; to Jon Hodge for his exacting writing, spectacular generosity, and boundless enthusiasm for discussing Darwin, Lyell, and Grant; to Jim Moore for a decade's worth of sincere, challenging, and fantastically enjoyable conversations about Darwin; to the late David Stoddart, whose writing introduced me to the history of coral reef science and whose personal munificence and dynamism exceeded even the written version; to David Kohn for the resources so many Darwin scholars depend on; to Simon Keynes for sharing proofs of his book while it was in press. These people are all formidable scholars whose work provided the point of departure for what I have tried to accomplish here. Every one of them showed me unfailing intellectual and personal generosity as I endeavored to make my small mark on their field.

I am grateful to my fellow members of the Darwin Correspondence Project, particularly Janet Browne, Jim Secord, and Alison Pearn, for their consistent support and encouragement. I am deeply indebted to my UK-based colleagues for their knowledge and advice, especially Rosy Clarkson, Samantha Evans, Shelley Innes, Francis Neary, Anne Secord, Liz Smith, and Paul White, and I am very grateful to my US-based colleagues Kelly Buttermore, Myrna Perez, Jenna Tonn, and Rebecca Woods for their friendship and collaboration. I thank John van Wyhe for his doggedness and generosity in assembling the life-changing *The Complete Work of Charles Darwin Online* and express my admiration to Gordon Chancellor for his exquisite interpretive work in essays published on that site. My heartfelt thanks to my teachers at Indiana University for introducing me to the history of science, and to Richard Sorrenson in particular for giving me the enthusiasm and the opportunity to go to graduate school. *Dankon* to Michael Gordin for teaching me so much and, even more, for insisting that I work so hard. Finally, as I look back I can pinpoint several moments when I hit a roadblock of some kind. I was revived and given a way to move forward at different moments by Hasok Chang's resolve, Ken Alder's generosity, Pam Henson's wisdom, Ole Molvig's advice, and Suman Seth's compassion and clear-headedness. Helen Rozwadowski, Katey Anderson, and Michael Reidy have buoyed me time and again, sometimes because I needed it and usually just by making bright times even brighter.

Several people read a full draft of the manuscript at one or more of the stages it has passed through. For their generosity and expertise I am extremely grateful to Katey Anderson, Patrick Anthony, Sandra Herbert, Ole Molvig, Jim Secord, Carolyn Taratko, J'Nese Williams, and Tyler Wren. I made many important changes and additions to my argument after most of them read it (indeed, because their readings yielded such valuable suggestions), so my expressions of gratitude for their help do not indicate that they have seen or endorsed the final product. All mistakes and misjudgments are my responsibility alone.

I have benefited from discussing particular topics or approaches in this book with Megan Barford, Andrew Berry, Mark Borrello, Gordon Chancellor, Joyce Chaplin, Miguel Chavez, Jamie Cohen-Cole, Henry Cowles, Alex Csiszar, Helen Curry, Jill Darrell, Andy Daum, Sarah Dry, Peter Galison, Michael Gordin, Hampton Howell, Catherine Jackson, Iain McCalman, Carla Nappi, Naomi Oreskes, Danielle Picard, Greg Radick, Joanna Radin, the late (and acutely missed) Ron Rainger, Bill Rankin, Michael Reidy, Madison Renner, Brian Rosen, Helen Rozwadowski, David Sepkoski, Suman Seth, Sam Schweber, Anne Secord, Steven Shapin, and Paul White. Others gave me opportunities to visit and experience some of the field locations I write about in this book; for that I am grateful to Frank and Hinano Murphy in Moorea, Kylie and Ash James at Cocos, Boban in Takapoto, and my friends aboard *Aranui 5*. I thank Christopher Mapes for his assistance with my bibliography, Oscar Martinez for sharing his picture of the Rodados Patagónicos, Alan Franks for his technical advice on my photographs, and Michael Barton for a particularly timely e-mail. It is a pleasure to acknowledge the wonderful Ellen Guarente for her many generous deeds on my behalf. I am profoundly and eternally indebted to Christopher Rigling for his skill, compassion, wit, and wisdom. I am grateful to Daniela Bleichmar, Tom Boeve, James Byrne, Andy Graybill, Ole Molvig, Tania Munz,

Joe November, Nick Popper, Suman Seth, Jen Stiens, and Matt Wisnioski for setting inspirational examples of scholarship and friendship for me in graduate school and beyond.

I have had the tremendous pleasure of being a member of faculty at Vanderbilt University while this book was taking shape. I am so grateful to my colleagues for their advice and friendship. I offer special thanks to Celia Applegate, David Blackbourn, Jeff Cowie, Marshall Eakin, Jim Epstein, Joel Harrington, Paul Kramer, Peter Lake, Jonathan Lamb, Matt Ramsey, Hanna Roman, Samira Sheikh, and Rachel Teukolsky for the roles they individually played in helping me with this book, and extend further heartfelt thanks to these and so many other colleagues who have taught, advised, and entertained me over lunches, dinners, and walks, notably Patrick Abbot, Larisa DeSantis, and Antonis Rokas. Thank you to history department staff including Tiffany Giese, Christen Harper, Susan Hilderbrand, Chris Lindsey, and Heidi Welch. And a very enthusiastic thank-you to my science-studies colleagues Michael Bess, Sarah Igo, Ruth Rogaski, and Arleen Tuchman, and to the wonderful group of graduate students doing history of science, for making this such an exciting and rewarding place to be. Thanks as well to Vanderbilt library staff, and to Vanderbilt's Robert Penn Warren Center for hosting our science-studies reading group.

It is especially nice to be able to thank my closest family members for their enthusiasm and encouragement over so many years and in so many places. Thank you, Mum, Dad, Heather, Zac, Roger, and Abi! And thank you, Laura! Thank you, UJ, for sharing adventures in England, Wales, Scotland, Tahiti, and beyond. Thank you, Ricardo, for saving my life! Thank you to the rest of my Frydman and Sponsel family members (especially to AD for taking care of me so often while I'm in the UK), and to my Morris, Wright, and Gooch family too, for all the love and support you have given me. And, for many years' worth of inspiration and kindness, thank you to my friends from the UK (especially the Shepherds and the Wyatts), San Antonio (especially the Hamlins and Kanes), Bloomington (especially Andy Carlson, Keith and Anne Leonard, Greg Mazu, Gina Ronzello, Dan and Katie Stark, and Rich Stazinski), Princeton (especially Jon Beyer and Tyler Wren), and Nashville (especially Patrick Harkins, Shannon Williams, and Rob Jackson).

Many individuals helped me strengthen this work when I presented my research at their institutions. I am grateful to colleagues at the Bermuda Institute of Ocean Sciences, the University of Cambridge Department of History and Philosophy of Science and the Sedgwick Museum of Earth

Sciences, the Franklin & Marshall College Department of Earth and Environment, the Harvard University Department of the History of Science, the Indiana University Department of History and Philosophy of Science, the (US) National Evolutionary Synthesis Center, the Northwestern University Program in Science in Human Culture, the Princeton University Program in History of Science, the Scripps Institution of Oceanography, the University of Minnesota, the University of California San Diego Science Studies Program, the University of South Carolina Department of History, the University of Sydney, and the University of Wollongong, and at meetings of the American Historical Association, the International Society for the History, Philosophy, and Social Studies of Biology, the Lone Star Historians of Science meeting, the History of Science Society, and at a workshop in Halifax organized by Katey Anderson and Helen Rozwadowski. I am also grateful to students in my courses at Harvard and Vanderbilt for helpful suggestions and thought-provoking questions about the history of science and exploration.

I have been extremely fortunate in receiving expert help from archivists and librarians. I offer my enthusiastic thanks to Godfrey Waller and so many others for such knowledgeable and kind assistance in the manuscripts room at Cambridge University Library, to Guy Hannaford and Ann Browne at the UK Hydrographic Office, to Tori Reeve at Down House, and to the archivists who have supported my research at the American Philosophical Society, the Natural History Museum in London, and Edinburgh University Library among many others. I offer particular thanks to the late Lord Lyell for permission to quote from the Kinnordy House manuscripts, and to the Gifford family (especially Mary Gifford) for such kind hospitality in allowing me to study them in person. I am also grateful to have had access to a range of extraordinary resources for Darwin scholars, including the Darwin Correspondence Project (whose members have been immensely helpful before and since I joined their ranks), *The Complete Work of Charles Darwin Online*, the Darwin Manuscripts Project, Nick Gill's index to the Darwin manuscripts at Cambridge University Library, and the notebook transcriptions published in the 1980s and in the past decade. I have benefited from the decisions of many libraries and other collections to make versions of their holdings available online, either directly or through outlets such as Google Books and the Biodiversity Heritage Library. And I am grateful to the institutions and individuals, acknowledged throughout the book, who granted the University of Chicago Press and me permission to reproduce images in their possession.

I am delighted to have the opportunity to acknowledge support for my research and writing from the (US) National Science Foundation for grant SES 05–22664, to the Max Planck Institute for the History of Science, the Mrs. Giles Whiting Foundation, the Smithsonian Institution, and the Scripps Institution of Oceanography for fellowships, to the Andrew W. Mellon Foundation and the American Philosophical Society for a library research grant, to Princeton University, and to Vanderbilt University. I have benefited from publishing an essay in the journal *Isis* that forms the basis for some passages of this book, and I am grateful to the journal's referees and its editor, Floris Cohen, for helping me strengthen my arguments. Thank you to the many people at the University of Chicago Press who helped to bring this book into print, including Mary Corrado and Evan White, and to Alice Bennett for her copyediting. Finally, I have had the incredibly good fortune to receive the support, kindness, and wisdom of my editor, Karen Merikangas Darling. Thank you, Karen.

Notes

Introduction

1. Darwin to W. D. Fox, May [1832], in *Correspondence of Charles Darwin*, ed. Burkhardt et al. (hereafter cited as *CCD*), 1:232–33.

2. Charles Darwin to Adolph von Morlot, 10 October [1844], *CCD*, 3:64–66. Rebecca Stott offers a valuable analysis of "speculation" in Darwin's work in chapter 7 of Stott, *Darwin and the Barnacle*. On broader contexts for, and meanings of, speculation in British geology at this time see O'Connor, *Earth on Show*.

3. I am using the term "identity" here, and in the book's title, in an anachronistic fashion. I have mixed feelings about doing so. This is not a book about "identity politics" in the modern sense, though as I point out at several junctures, Darwin's wealth and gentlemanly status were a sine qua non for almost everything he did. Because this is a book about science as a communal activity and as a collectively held body of knowledge, I am particularly concerned with credibility and authority within scientific communities. Yet I will show, as other scholars have memorably done in other cases, that the way one fashioned or assessed a "scientific" reputation was inextricable from norms operating more broadly than just in scientific discourse.

In addition to the risk of distracting readers by using a word that has such potency in present-day discourse, I am also hesitant because, as Rogers Brubaker and Frederick Cooper (among others) have argued, the term has considerable drawbacks as an analytic category. As they write, "'Identity' ... tends to mean too much (when understood in a strong sense), too little (when understood in a weak sense), or nothing at all (because of its sheer ambiguity)" (Brubaker and Cooper, "Beyond Identity," 1). For example, the term (when used by itself) entirely elides any distinction between self-understanding and reputation. One of the main themes of this book is the tension between those two things. For this reason I do not write much about "identity" per se within the text of the book. In the limited space available in a book title, however, the vagueness of the term seems a reasonable price to pay for the benefit of its versatility. In the space of a few letters it lets me allude to the fact that this book is about Darwin's self-understanding *and* his reputation—indeed, my recurrent focus is Darwin's concern *about* his reputation and his efforts to reshape it. I am equally interested in these issues with respect to Charles Lyell as well, and I argue that Lyell and Darwin worked consciously to link their reputations in a mutually beneficial manner.

On Darwin's self-understanding and reputation as a geologist see Herbert, "Darwin the Young Geologist"; Herbert, "Charles Darwin as a Prospective Geological Author"; Herbert, *Charles Darwin, Geologist*; Rhodes, "Darwin's Search for a Theory of the Earth"; Rudwick, "Charles Darwin in London"; Rudwick, "Darwin and the World of Geology"; Rudwick, *Worlds before Adam*; Secord, "Discovery of a Vocation"; Chancellor, "Darwin's Geological Diary"; Brinkman, "Charles Darwin's *Beagle* Voyage." On the *Beagle* voyage in general see Keynes, *"Beagle" Record*; Browne and Neve, *Voyage of the "Beagle"*; Keynes, *Fossils, Finches, and Fuegians*; McCalman, *Darwin's Armada*; Pearn, *Voyage round the World*.

4. On Darwin's theory of coral reef formation see Stoddart, "*Coral Islands* by Charles Darwin"; Stoddart, "Darwin, Lyell, and the Geological Significance of Coral Reefs"; Stoddart, "Grandeur in This View of Life"; Stoddart, "Theory and Reality"; Stoddart, "'This Coral Episode'"; Stoddart, "Darwin and the Seeing Eye"; Montgomery, "Charles Darwin's Theory of Coral Reefs and the Problem of the Chalk"; Burkhardt, "Darwin's Early Notes on Coral Reef Formation"; Ghiselin, "Introduction"; Herbert, *Charles Darwin, Geologist*; Dobbs, *Reef Madness*.

5. I see a parallel between Darwin's careful crafting of his argument and evidence and the preparations that, according to Bruno Latour, Louis Pasteur made before unveiling his anthrax vaccine. Latour argues that in Pasteur's case (and in general) laboratories are such powerful spaces because they provide opportunities outside public view for scientists to try, err, revise, and prepare before extending their laboratory findings to wider publics. This argument of Latour's also stimulated my work on another case of cultivating evidence and audiences in advance, that of Arthur Stanley Eddington's announcement of the results of the 1919 eclipse expedition to test a prediction made by Einstein's general relativity theory. See Sponsel, "Constructing a 'Revolution in Science.'"

6. Colp, *To Be an Invalid*; Desmond and Moore, *Darwin*; Colp, *Darwin's Illness*.

7. In pointing out that there is inadequate evidence for the species theory's being Darwin's main source of distress I am echoing (for different reasons) a reaction to Colp's work made by the psychologist John Bowlby in his biography of Darwin. Bowlby's own argument (building on his research as a pioneer of attachment theory) was that Darwin's troubles were a product of his mother's death and his desire to please male authority figures. Bowlby wrote, "Colp has advanced

his hypothesis that Darwin's long years of ill health are to be attributed mainly to anxiety engendered by his ideas of evolution. While the evidence suggests this may have played some part, I believe it should be seen within the context of a much more general problem, namely his deep desire to earn the approval of his father and other father-figures and, at all costs, to avoid arousing their criticism, which he was always expecting." Bowlby, *Charles Darwin*, 216. I do identify Darwin's desire to earn Lyell's approval as a main goal of his geological work.

8. On the absence of a role for the dedicated theorist in Darwin's communities of geologists and zoologists, see Herbert, "Place of Man." I have discussed this myself, building always on Herbert's work, in Sponsel, "Amphibious Being," and Sponsel, "Pacific Islands and the Problem of Theorizing." Rudwick disagrees slightly with Herbert in "Charles Darwin in London."

9. Charles Darwin to Charles Lyell, [14] September [1838], *CCD*, 2:104–8; emphasis added. This is one of the pieces of evidence Bowlby pointed to in his critique of Colp: "Reading and speculating about species, he makes clear in his letter to Lyell . . . he regarded at this time as no more than a private hobby which, because of its far-reaching and exciting prospects, tempted him to neglect his proper duties." Bowlby, *Charles Darwin*, 216.

10. Some of these questions date to early work in the sociology of science by Robert Merton and others. However, my interest in examining how engagements within a scientific community shaped the very *content* of scientific claims strives to mirror more directly some of the claims made by the so-called Strong Programme in the sociology of scientific knowledge as well as various reactions to and by its architects. Many readers will already know this literature intimately; for others its origins and diversity can be judged by a sample including such works as Bloor, *Knowledge and Social Imagery*; Shapin, "History of Science and Its Sociological Reconstructions"; Collins, *Changing Order*; Knorr-Cetina, *Manufacture of Knowledge*; Latour and Woolgar, *Laboratory Life*.

11. I am, of course, far from being the first historian to draw on the insights and provocations of the sociology of scientific knowledge (SSK). A justifiably famous early example is Shapin and Schaffer, *Leviathan and the Air-Pump*. Some of my concerns echo Harwood, *Styles of Scientific Thought*. My views of the British geological community in particular draw from Martin Rudwick's SSK-inflected studies: Rudwick, "Charles Darwin in London"; Rudwick, *Great Devonian Controversy*. Above all, though, I am asking questions taught to me by Andrew Warwick and Rob Iliffe in the years just before they published Warwick, *Masters of Theory*, and Iliffe, "Butter for Parsnips."

By choosing this scientific community as a historical case I am also taking advantage of a uniquely detailed and multidimensional base of sources. Martin Rudwick has pointed out, and exploited to great effect in his own work, the richness of manuscript material from British geologists in this period. These sources, he argues, provide the opportunity to study developments at the optimal "'graininess' or degree of temporal 'resolution.'" Rudwick, *Great Devonian Controversy*, 8–9. Rudwick was himself drawing on an observation in Cannon, "History in Depth."

Even compared with that of his most prolific scientific contemporaries, however, Darwin's work is exceptionally well documented. He obsessively made—and saved—notes about his research, and he carried out investigations of all sorts: observation in the field, dissection under a microscope, comparison of preserved specimens, compilation in libraries and map rooms, experiments, and interrogation in person and by mail. He was (with the help of his wife, Emma, and later his daughter Henrietta, along with a series of paid assistants) a remarkably prolific letter writer at a time and place when written correspondence was a flourishing medium of scientific communication. His reputation ensured that much of his outgoing correspondence was saved by the recipients and that the notes and letters in his possession were preserved after his death. More than fifteen thousand letters written by or to Charles Darwin survive, having been located, transcribed, and analyzed by the scholars working on the ongoing Darwin Correspondence Project. Also indispensable (and ever-growing), *The Complete Work of Charles Darwin Online* has made Darwin's publications, and increasingly his manuscripts, readily available to modern readers around the world. Thanks to these and many other scholarly initiatives, a historian studying Darwin can expect that the documentary record will contain answers to detailed and broad questions alike and that there is a good chance of actually locating those answers within the vast corpus of sources.

12. The historically oriented work of another sociologist, Augustine Brannigan, helped shape my views in this respect. See Brannigan, *Social Basis of Scientific Discoveries*.

13. This concern is applicable across many fields of history. Matthew Eddy makes the point specifically with respect to social studies of pre-twentieth-century science in Eddy, "Fallible or Inerrant?," 97.

14. Bruno Latour recommends "follow[ing] scientists and engineers through society" in *Science in Action*. The term society implies a rather macro-scale analysis, but Latour's very purpose in formulating this approach was to bridge the gap he diagnosed between micro- and macrosocial studies of science. I invoke macro-scale phenomena to explain why the *Beagle* voyage occurred in the first place and why there were many constituencies who desired an explanation for the formation of coral reefs, but most of my attention is trained on microsocial interactions, first between shipmates and then between metropolitan geologists. My argument about Darwin's approach to publishing his species theory is based in part on a claim about the *narrowness* of the audience he was most concerned to convince: scientific specialists (*as* specialists), rather than as members of the social class to which they belonged, let alone the whole reading public. Therefore broad social and political contexts for Darwin's theorizing play a smaller role in my book than they do in Desmond, *Politics of Evolution*; Desmond and Moore, *Darwin*; and Secord, *Victorian Sensation*. On Latour's purpose in "following" scientists see also "Give me a Laboratory," especially 141–44.

15. The key steps in Darwin's career are inevitably described in terms of his interaction with other scholars' *books* rather than the authors themselves. Alexander von Humboldt's *Personal Narrative*, which described his 1799–1804 journey to the Americas, fueled the young Darwin's desire to pursue scientific travel. Lyell's

Principles of Geology, we are led to believe, taught him the potential of small changes to produce major transformations in the course of a sufficiently deep Earth history. And Darwin's famous eureka moment about natural selection happened not during a successful experiment or while making a surprising observation but while reading a book on political economy, Thomas Robert Malthus's *Essay on the Principle of Population*. I will argue, on the contrary, that Darwin's apparent debts to books by Humboldt and Lyell conceal face-to-face interactions (with surveyors and with Lyell himself) that endowed those works with particular significance.

16. Science studies scholars (including historians of science) often end up using the term symmetry nowadays, it seems to me, to stake a position that would more closely correspond to the "impartiality" tenet articulated by the founders of the Strong Programme. Bloor, for example, argues that a satisfactory explanation in SSK should be *both* "impartial" (with respect to our current understanding of the truth or falsity of a given scientific claim or tradition) and "symmetrical" (in the sense that "the same *types* of causes would explain, say, true and false beliefs" [emphasis added]). Bloor, *Knowledge and Social Imagery*, 7.

17. I use this term gritting my teeth and at risk of implying that ecology has some essential quality toward which Darwin's work inevitably pointed. I will offer a more subtle explanation of what I mean in chapter 2.

18. In doing so I will be expanding on Sponsel, "An Amphibious Being." For the broader history of British hydrography in the period see Blewitt, *Surveys of the Seas*; Ritchie, *Admiralty Chart*; Day, *Admiralty Hydrographic Service*; Dawson, *Memoirs of Hydrography*; Cock, "Sir Francis Beaufort and the Co-ordination of British Scientific Activity"; Cock, "Rear-Admiral Sir Francis Beaufort"; Cock, "Scientific Servicemen in the Royal Navy"; Barford, "Fugitive Hydrography"; Barford, "Naval Hydrography, Charismatic Bureaucracy, and the British Military State."

19. My approach has much in common with that of the late Dov Ospovat, who made a detailed study of how Darwin's theory of evolution by natural selection changed in the years between 1838 and 1859 (Ospovat, *Development of Darwin's Theory*). Ospovat showed that the considerable revisions Darwin made to his explanation of the origin of species were largely responses to contemporaneous developments in practical and theoretical natural history. Although Ospovat's attention was often trained on the most intimate details of Darwin's private notes and manuscript drafts, he concluded his book by arguing that "the development of Darwin's theory . . . can best be described as a social process." By this he meant that Darwin's private ideas and modifications to his theory were shaped by the larger world of specialist science to which he aimed to contribute, and therefore that "the formation and transformation of Darwin's theory represent not so much the results of an interaction between the creative scientist and nature as between the creative scientist and socially constructed conceptions of nature." Ospovat claimed that his case study on Darwin's evolutionary theory "suggests how the work of an individual scientist in the privacy of his study or laboratory—even a scientist who is engaged in creating a revolutionary theory designed to supplant existing explanations—is molded by the ideas and attitudes of his professional colleagues" (quotations on 229 and 233).

I agree wholeheartedly with Ospovat's perspective on the way scientific work must, almost by definition, emerge from and be shaped by existing questions, definitions, and theories. In this book I aim not only to illustrate that point with respect to Darwin's geological and geographical work, but to go a step beyond that. For while Ospovat was sensitive to the ways Darwin engaged and wrestled with the ideas of his contemporaries, he offered no account of how Darwin developed his techniques or strategies of engagement with other men of science. Ospovat provided a portrait of the man of science expertly navigating the social and intellectual currents of his chosen discipline, but he treated these skills as though they were themselves fixed parts of Darwin's character—as though they were inborn traits. By contrast, I have striven to demonstrate how Darwin originally acquired skills relevant to authoring theories and the attitudes toward practices of speculating and generalizing that Ospovat found already present by the time Darwin was refining his species theory in the 1840s and 1850s. Ospovat, *Development of Darwin's Theory*.

20. Darwin's intellectual debts to Lyell, via his *Principles of Geology*, have been most strongly elaborated by Jon Hodge, to whose work I am myself indebted. Hodge, "Structure and Strategy of Darwin's 'Long Argument'"; Hodge, "Darwin and the Laws of the Animate Part of the Terrestrial System"; Hodge, "Development of Darwin's General Biological Theorizing"; Hodge, "Notebook Programmes and Projects of Darwin's London Years." See also Wilson, *Charles Lyell*; Cannon, "Charles Lyell, Radical Actualism, and Theory"; Stoddart, "Darwin, Lyell, and the Geological Significance of Coral Reefs."

21. As I argue in chapter 10, however, the descriptive terminology Darwin used in his reef catalog was calculated to support his theory implicitly.

Chapter One

1. Lyell, *Principles of Geology* (1830), 1:82. Lyell went on to point out that an amphibious being who could *also* go underground would be in an even more advantageous position to study geology.

2. Bougainville, *Voyage autour du monde*, 182–83. He went on to explain, "Navigation is extremely perilous among these low islands riddled with breakers and strewn with reefs, where it is advisable, especially at night, to use the greatest caution."

3. Cook, *"Endeavour" Journal*, 1:344.

4. Beaglehole, *Life of Captain James Cook*, 236–46.

5. On the length of lines carried by eighteenth-century navigators see Rozwadowski, *Fathoming the Ocean*, 32.

6. Forster, *Observations Made during a Voyage round the World*, 150–52.

7. Ibid., 150–51.

8. Flinders, quoted in Jameson, "On the Growth of Coral Islands," 382.

9. Lamouroux, *Histoire des polypiers coralligènes flexibles*. (All translations are my own.) Baudin himself did not survive the voyage. On the distribution of François Péron's specimens from the voyage to Lamouroux and others, see Burkhardt, "Unpacking Baudin."

10. Lamouroux, *Histoire des polypiers coralligènes flexibles*, lix.

11. [Eschscholtz], "On the Coral Islands."

12. Ibid., 331.

13. Although the appendix's authorship was anonymous in the original publication, it was specifically credited to Chamisso when it was published in Britain as a stand-alone article, Chamisso, "On the Coral Islands of the Pacific Ocean." Eschscholtz's authorship was asserted in Sluiter, "Eine geschichtliche Berichtigung." See Stoddart, "Darwin, Lyell, and the Geological Significance of Coral Reefs," 214.

14. Appel, "Jean-René-Constant Quoy"; Coleman, "Joseph Paul Gaimard."

15. Quoy and Gaimard, "Mémoire sur l'accroissement des polypes lithophytes considéré géologiquement," 284.

16. Ibid., 289. Eschscholtz believed that ring-shaped reefs existed because the largest corals inhabited the circumference of a reef and were first to reach the surface, where debris from their erosion choked off growth in the lagoon. For him the shape of a submarine mountain accounted only for the outside dimensions of the reef atop it. [Eschscholtz], "On the Coral Islands," 334.

17. See commentary by Beechey, *Narrative of a Voyage to the Pacific and Beering's Strait* (1832), 192. Quoy and Gaimard's conclusions were widely noticed by geologists in Britain, including Robert Jameson, Henry De la Beche, and (as discussed below) Charles Lyell. Jameson, "On the Growth of Coral Islands"; De la Beche, *Geological Manual*, 141–42.

18. Lyell referred to "my volcanic crater theory" in Lyell to J. F. W. Herschel, 24 and 26 May 1837, RS HS 11.422 and 11.450. This letter is partially transcribed in Lyell, *Letters and Journals of Sir Charles Lyell*, vol. 2. See also Wilson, *Charles Lyell*, 447–49.

19. Beechey, *Narrative of a Voyage to the Pacific and Beering's Strait* (1831). The section on "Peculiarities of the Coral Islands" is 1:186–95. Beechey's descriptions of individual coral islands are found throughout his first volume in the order he encountered them.

20. Ibid.

21. Ibid., 1:193.

22. Ibid., 1:192.

23. Lyell argued that, despite their large size, the features Buch and Élie de Beaumont called "craters of elevation" could be explained by reference to "actual" (meaning "contemporary") causes. He described an active submarine volcano undergoing successive eruptions: "In hot countries coral reefs . . . must often, during long intervals of quiescence, obstruct the vent, and thus increase the repressive force and augment the violence of eruptions. The probabilities, therefore, in a submarine volcano, of the destruction of a larger part of the cone and the formation of a more extensive crater, are obvious." Lyell, *Principles of Geology* (1830), 1:392.

24. Ibid., 1:167–68.

25. Ibid., 2:296.

26. Ibid., 2:295.

27. Francis Beaufort, "Memoranda for Commander Fitzroy's orders," 11 November 1831, UKHO MB 2, 2–24; reprinted in Browne and Neve, *Voyage of the*

"Beagle," 384–99; quotation on 391. Many of the specifics of Beaufort's orders address concerns raised by Rear Admiral R. W. Otway from his post at Rio de Janeiro in August 1828, where his flag lieutenant was the twenty-three-year-old Robert FitzRoy. See minute of 14 August 1828, UKHO MB 1, 206. On FitzRoy's first post in South America see Dawson, *Memoirs of Hydrography.*

28. For a brief discussion of these issues see the editors' introduction to Browne and Neve, *Voyage of the Beagle.* Jon Hodge has argued, building on the work of Cain and Hopkins, that the *Beagle* voyage exemplified a patrician form of "gentlemanly capitalism." Hodge, "Notebook Programmes and Projects of Darwin's London Years," 65–67; Cain and Hopkins, "Gentlemanly Capitalism and British Expansion Overseas, I"; Cain and Hopkins, "Gentlemanly Capitalism and British Expansion Overseas, II"; Cain and Hopkins, *British Imperialism.*

29. Beaufort, "Memoranda for Commander Fitzroy's orders," in Browne and Neve, *Voyage of the Beagle,* 392.

30. Ibid., 397.

31. Ibid. Here I have followed the original wording and punctuation of the manuscript version, UKHO MB2, 21.

32. Cock, "Rear-Admiral Sir Francis Beaufort"; Cock, "Sir Francis Beaufort and the Co-ordination of British Scientific Activity." Cock's thesis was an indispensable guide to the complex organization of the UKHO archive.

33. Beechey, *Narrative of a Voyage to the Pacific and Beering's Strait* (1831), xii.

34. Beaufort to FitzRoy, 14 November 1831, UKHO LB 3, 280.

35. "Scientific Voyage." The topic of coral reef formation had received wide coverage that year in reviews of Beechey's narrative.

36. Sponsel, "Coral Reef Formation," 65–68; Cock, "Sir Francis Beaufort and the Co-ordination of British Scientific Activity."

37. Lyell to Gideon Mantell, 15 February 1830, in Lyell, *Life, Letters and Journals of Sir Charles Lyell,* 1:262.

38. Lyell, "Journal to Miss Horner," 4 February 1831, in ibid., 1:368–69.

39. Lyell, "Journal to Miss Horner," 13 February 1832, in ibid., 1:371.

40. Ibid.

41. "Proposed orders for Captain Blackwood," UKHO MB 3, September 1837–May 1842, 409–16; Francis Beaufort to Francis Blackwood, UKHO LB 12, quoted in Goodman, *"Rattlesnake."*

42. Beaufort reported these concerns from Lord Auckland, then first lord of the Admiralty, to Owen Stanley on dispatching him to survey the perilous Torres Straits. Francis Beaufort to Owen Stanley, 28 November 1846, UKHO LB 14, 214–15. Cited in Cock, "Sir Francis Beaufort and the Co-ordination of British Scientific Activity," n. 420.

43. On Beaufort's administrative roles and support for the projects to investigate magnetic variation and the tides, see Cock, "Sir Francis Beaufort and the Co-ordination of British Scientific Activity"; Reidy, *Tides of History*; Barford, "Fugitive Hydrography"; Barford, "Naval Hydrography, Charismatic Bureaucracy, and the British Military State."

44. Hodge, "Darwin as a Lifelong Generation Theorist"; Secord, "Discovery of

a Vocation"; Sloan, "Darwin's Invertebrate Program"; Sloan, "Making of a Philosophical Naturalist."

45. Secord, "Discovery of a Vocation"; Herbert, *Charles Darwin, Geologist,* chap. 1; Browne, *Charles Darwin: Voyaging,* 65–72.

46. Desmond, "Grant, Robert Edmond." See also Secord, "Edinburgh Lamarckians."

47. In his study of Darwin's marine zoology, Phillip Sloan explained the contemporary taxonomic understanding of coral and coral-like organisms thus: "The discussions in the 1820s and 1830s about the possibility of a true 'zoophytal' creature had involved workers in invertebrate zoology in a search for defining criteria of plants and animals that particularly concerned the status of the colonial invertebrates and infusoria. Another taxonomic group drawn into this problem was the coralline algae, curious forms occurring primarily in warm waters, and often involved in coral-reef formation. Currently these are placed unambiguously among the plants, but they were a subject of substantial debate in the early nineteenth century. On the one hand, many of these forms occurred in branching tufts with a calcareous skeleton, or even in spreading fungus-like forms, strongly reminiscent of some of the corals, and in this were unlike any known plant. On the other hand, they lacked evident polyps, and showed none of the other animal functions. Generally Darwin's authorities placed the 'Corallina' (as distinguished from the animal 'corallines') among the colonial animals, but there was clearly much uncertainty on this." Sloan, "Darwin's Invertebrate Program," 98.

48. Desmond and Parker, "Bibliography of Robert Edmond Grant."

49. Rehbock, "Early Dredgers"; Rozwadowski, *Fathoming the Ocean,* chap. 4; Egerton, "History of the Ecological Sciences, Part 35: Beginnings of British Marine Biology."

50. These notes are reprinted in Barrett, *Collected Papers of Charles Darwin,* 285–91.

51. Browne, *Charles Darwin: Voyaging,* 80–88.

52. Sloan, "Darwin's Invertebrate Program"; Hodge, "Darwin as a Lifelong Generation Theorist."

53. On the relationship between Darwin and Henslow see Walters and Stow, *Darwin's Mentor.*

54. Herbert, *Charles Darwin, Geologist,* 39–47. Darwin and Sedgwick set off together on 5 August 1831 and separated on, or shortly after, 11 August.

55. Secord, "Discovery of a Vocation," 144–50.

56. These lessons were not the only ones Darwin's success as a producer of knowledge depended on. As I argue below, Charles Lyell also coached him in practices of authorship and presentation.

57. Henslow also had field experience as a geologist and was professor of mineralogy for three years before taking the botany chair in 1825. On Henslow's geology, see Herbert, *Charles Darwin, Geologist,* 36.

58. The edition Henslow and Darwin read was a translation by Helen Maria Williams: Humboldt and Bonpland, *Personal Narrative of Travels to the Equinoctial Regions of the New Continent.*

59. Darwin to Susan Darwin, 9 September 1831, *CCD*, 1:145.

60. Darwin recalled this book in his autobiographical recollections. See Darwin, *Evolutionary Writings*, 368.

61. Darwin, *Evolutionary Writings*, 373. Darwin undoubtedly paid more attention to Jameson's lectures and museum specimens than the *Autobiography* implies, as illustrated in Secord, "Discovery of a Vocation," 134–42; Herbert, *Charles Darwin, Geologist*, 32–36.

62. Jameson lecture notes, EUL GEN 122. As this reference to the receding universal ocean indicates, Jameson was (to Darwin's retrospective disgust) a noted devotee of the "Neptunist" geology of Abraham Gottlob Werner.

63. Jameson's lecture notes, EUL GEN 122. This lecture is not in Jameson's handwriting and may have been copied or translated by a secretary from another author's published work (as was the case with several other sets of notes on the same distinctive blue paper).

64. Jameson, "On the Growth of Coral Islands." Jameson also contributed shorter appendixes on coral island formation to the first (1813) through fourth British editions of the book, which were translated by Robert Kerr. The 1827 edition was "translated from the last French edition" by Jameson (although it in fact contained many passages identical to those in Kerr's translation).

65. Browne, *Charles Darwin: Voyaging*, 72.

66. Jameson's lecture notes, EUL GEN 122. The proof sheets in this folder appear to be from one of the first three editions, as they contain no mention of Chamisso or Quoy and Gaimard.

67. Sedgwick, "Syllabus of a Course of Lectures on Geology," quoted in Clark and Hughes, *Life and Letters of the Reverend Adam Sedgwick*. Darwin claimed in his autobiography that he had not attended Sedgwick's lectures, though other students recalled differently. See Secord, "Discovery of a Vocation," 143. I would add that Darwin's statement implies only that he did not attend them *on first arriving at Cambridge*, while his vivid description rather implies that he had eventually attended them: "Public lectures on several branches were given in the University, attendance being quite voluntary; but I was so sickened with lectures at Edinburgh that I did not even attend Sedgwicks eloquent & interesting lectures. Had I done so I sh[oul]d probably have become a geologist earlier than I did." Secord, *Evolutionary Writings*, 377.

68. In a letter written to Darwin after they had separated (in response to a letter from Darwin that is not extant), Sedgwick explained, "Your information did not however surprise me, as madrepores [stony corals] are quite as likely to be met with as terebratulae, which seem to occur here and there thro' ye Snowdonian chain." Adam Sedgwick to Charles Darwin, 4 September 1831, *CCD*, 1:137–39.

69. Quotation from Darwin to FitzRoy, [10 October 1831], *CCD*, 1:174–75. One of the instructions from Grant was to "kill [zoophytes] by gradual additions of fresh water so that polypi hang out[, and] Actineae by pouring boiling water in their interiors." DAR 29.3:78.

70. Coldstream to Darwin, 13 September 1831, *CCD*, 1:151–53.

Chapter Two

1. Cannon, *Science in Culture*, chap. 3.

2. On Darwin's relation to Humboldt, see ibid.; Camerini, "Darwin, Wallace, and Maps"; Richards, *Romantic Conception of Life*, chap. 14; Egerton, "Humboldt, Darwin, and Population"; Chancellor, "Humboldt's Personal Narrative"; Egerton, "History of Ecological Sciences, Part 37"; Herbert, *Charles Darwin, Geologist*, 12–17. For analysis of the similarities of Darwin's texts to Humboldt's, see Tallmadge, "From Chronicle to Quest"; Leask, "Darwin's Second Sun."

3. "Instruction manuals" refers to scholarship on the importance of tacit knowledge in various types of scientific replication, as well as to the work of historians who have emphasized the inadequacy of textbooks as a means of inculcating a trainee with the values and competencies required to contribute to a scientific community. Of the former studies, my thinking has been shaped most heavily by face-to-face training with Andrew Warwick and by Collins, *Changing Order*, chap. 3; Shapin and Schaffer, *Leviathan and the Air-Pump*, chap. 6: "No one built a pump from written instructions alone" (281). My strongest debts to the latter (allied) tradition of scholarship are to Warwick, *Masters of Theory*; Geison and Holmes, "Research Schools."

4. Entry for 17 January 1832, Keynes, *Charles Darwin's "Beagle" Diary*, 24.

5. Entry for 17 January 1832, ibid. For other mentions of his coral collecting at the Cape Verde Islands, see the diary entries for 28 January and 3 February 1832.

6. Entry for 6 February 1832, ibid., 34.

7. Ibid., 33.

8. On the link between the methodology of Darwin's invertebrate researches at Edinburgh and those during the voyage, see Porter, "*Beagle* Collector and His Collections," 974–75.

9. In an important recent essay, Paul Pearson and Chris Nicholas argue that Darwin may not have read Lyell's *Principles* before arriving at St. Jago and that his immediate interpretation of the geology of the Cape Verde Islands posited extreme flooding inconsistent with the main thrust of Lyell's work. Pearson and Nicholas, "'Marks of Extreme Violence.'" On the geology of the Cape Verde Islands as Darwin's inspiration to pursue geology as a vocation, see Secord, "Discovery of a Vocation." For a detailed account of Darwin's collection of specimens and study of the landscape there, along with his considerations of elevatory movements and the origin of "diluvium," see Herbert, *Charles Darwin, Geologist*, 141–58.

10. Lyell, *Principles of Geology* (1830). See also Herbert, *Charles Darwin, Geologist*, 152–56.

11. Darwin to Henslow, 18 May-16 June 1832, *CCD*, 1:236–39, quotation on 236.

12. Darwin to Henslow, 18 May-16 June 1832, *CCD*, 1:236–39, quotation on 237; Keynes, *Charles Darwin's Zoology Notes*, 14–15.

13. Darwin to Henslow, 18 May-16 June 1832, *CCD*, 1:236–39, quotation on 237.

14. Entry for 10 January 1832, Keynes, *Charles Darwin's "Beagle" Diary*, 21. Keynes remarked that this appears to have been among the first uses of such a net

for scientific collecting at the water's surface (rather than by dredging along the bottom as Coldstream had recommended).

15. Beaufort, "Memoranda for Commander Fitzroy's orders," in Browne and Neve, *Voyage of the "Beagle,"* 386.

16. Entry for 27–28 March 1832, Keynes, *Charles Darwin's "Beagle" Diary*, 48.

17. These entries were made on 26–28 March 1832; Keynes, *Charles Darwin's Zoology Notes*, 31–32.

18. Hutton, *Philosophical and Mathematical Dictionary*, sec. "Sounding."

19. Descriptions of the bottom were important parts of navigators' accounts of their explorations. Bougainville, for example, wrote in the narrative of his circumnavigation, "I shall add, for the use of those who may be plying here in thick weather, that a gravelly bottom shews that they are nearer the coast of Terra del Fuego than to the continent; where they will find a fine sand, and sometimes oozy bottom." Bougainville, *Voyage round the World*, 132.

20. This was recorded as specimen "392 not [in] spirits." Keynes, *Charles Darwin's Zoology Notes*, 33. Nearby notes show that Darwin was likely drawing his taxonomic information from what he called the "Dic Class," the seventeen-volume French natural history reference edited by Bory de Saint-Vincent. Bacillaria and Arthrodia, or "Bacillareès & Anthrodieès [*sic*]," as Darwin called them here, were unicellular algae of the sort now called diatoms. See Bory de Saint-Vincent, *Dictionnaire classique d'histoire naturelle*.

21. Cannon, *Science in Culture*; Dettelbach, "Humboldtian Science"; Browne, *Secular Ark*.

22. In my "Amphibious Being" essay (2016) I speculated that Humboldt's training and practical work as a mining engineer might have played a role in generating his "geographical sensibility" much as hydrography later did for Darwin, particularly since Freiberg's subterranean mines also introduced the factor of depth (negative elevation) to the study of distribution. New work by Patrick Anthony has since shed considerable light on the connections between Humboldt's Freiberg mining career and his phytogeography. Anthony, "Underground Enlightenment."

23. On Darwin's study of corallines at Botafogo Bay, see the *Diary* entries for 8, 15, 20–22, 27 June 1832, Keynes, *Charles Darwin's "Beagle" Diary*, 73–78.

24. Darwin to Henslow, [c. 26 October]-24 November [1832], *CCD*, 1:279–82.

25. Entry for 4 April 1833, Keynes, *Charles Darwin's "Beagle" Diary*, 149.

26. Keynes, *Charles Darwin's Zoology Notes*, xiii.

27. Darwin and his contemporaries had no conception that hermatypic corals have symbiotic algae-like zooxanthellae living within their polyps.

28. See, e.g., Down House Notebook 1.18, "Santiago Book" (CUL MS microfilm 532), 6; DAR 37.2:791v; DAR 38.2:893v; Stoddart, "*Coral Islands* by Charles Darwin," 9.

29. Keynes, *Charles Darwin's Zoology Notes*, xiv, 187–88.

30. Keynes, *Charles Darwin's "Beagle" Diary*, 232–33.

31. Keynes, *Charles Darwin's Zoology Notes*, 232–33; see also the discussion on xv.

32. Darwin to Catherine Darwin, 9–20 July 1834, *CCD*, 1:391–94.

33. Lamouroux, *Histoire des polypiers coralligènes flexibles*, lxxxiv. He continued, "It is therefore to draw the attention of men instructed in these new objects that I have published this work, in which I have sought to gather all that has been said on the Polypiers by previous authors, and to expand the domain of science through some new observations."

34. Lyell, *Principles of Geology*, 2:182.

35. Darwin to Henslow, 24 July-7 November 1834, *CCD*, 1:397–402. From internal context, this part of the letter appears to have been written on, or shortly after, 24 July.

36. Darwin to Henslow, 11 April 1833, *CCD*, 1:306–8.

37. Darwin to Henslow, 18 July 1833, *CCD*, 1:321–23. On the weather, see the *Beagle* captain's log 1831–34, NAK, ADM 51/3054, 42v. The weather was recorded several times each day via a series of code letters that referred to a corresponding key found on page 79 of the log.

38. Darwin to Henslow, March 1834, *CCD*, 1:368–71. The BAAS was divided into "sections" relating to different areas of science, hence Darwin's use and emphasis of the word to describe the work he considered fundamentally zoological. On the early history of the BAAS see Morrell and Thackray, *Gentlemen of Science*.

39. I offer here a mild adjustment to Sandra Herbert's observation that Beaufort's coral reef instruction was "Darwin's only direct assignment as a geological author." Herbert, "Charles Darwin as a Prospective Geological Author," 189. My argument in part II also stands somewhat opposed to this statement because I show that Lyell imposed a number of obligations on Darwin not as a field geologist but as a geological author.

40. The "Transactions of the Sections" in the BAAS Report for 1832 show geography and geology grouped into one section and zoology, anatomy, and physiology in another. *Report of the First and Second Meetings of the British Association for the Advancement of Science*, viii.

41. Here I am amplifying the compelling arguments about Grant's significance to Darwin's thought that have been advanced by Sloan and Hodge. For recent statements see Hodge, "Notebook Programmes and Projects of Darwin's London Years," 45; Sloan, "Making of a Philosophical Naturalist," 27.

Chapter Three

1. See especially Love, "Darwin and Cirripedia prior to 1846"; Herbert, *Charles Darwin, Geologist*.

2. Herbert, "From Charles Darwin's Portfolio, 30 (Darwin's n. 2). For a recent perspective on these pebbles, the Rodados Patagónicos, see Martinez, Rabassa, and Coronato, "Charles Darwin and the First Scientific Observations on the Patagonian Shingle Formation (Rodados Patagónicos)."

3. Darwin to Henslow, March 1834, *CCD*, 1:368–72.

4. Herbert, "From Charles Darwin's Portfolio: An Early Essay on South American Geology and Species," 31 (Darwin's nn. 5–6). On the date of the manuscript, see 25–27.

5. Ibid. (Darwin's n. 6).

6. Darwin to Henslow, 10 March 1834, *CCD*, 1:368–72.

7. This style of reasoning from present to past was fundamental to the work of all serious geologists at this time, though it has been widely associated with Charles Lyell because of his assertion that present processes *operating at their observed intensities* were sufficient to explain past geological events. On this point more broadly see chapter 7 of Rudwick, *Worlds before Adam*. On Lyell's own mythmaking role in establishing himself as a pioneer of geological method, see also Porter, "Charles Lyell and the Principles of the History of Geology."

8. Charles Darwin to Caroline Darwin, 29 April 1836, *CCD*, 1:494–97.

9. Herbert, "From Charles Darwin's Portfolio," 32 (Darwin's n. 8).

10. DAR 34.1:45.

11. DAR 34.2:131–52, and "Valley of S. Cruz," DAR 34.2:104–10.

12. DAR 34.1:40–60. On dating the essay, see Herbert, *Charles Darwin, Geologist*, 399.

13. In May 1834 he wrote, "NB When I say concentric. I mean not truly so.—but an enlargement of the curve of the world." DAR 34.2:110v.

14. DAR 34.2:199.

15. On Darwin's attention to earthquakes see Herbert, *Charles Darwin, Geologist*, 217–32; White, "Darwin, Concepción, and the Geological Sublime."

16. The alternative would be that the whole globe was expanding when horizontal elevation occurred, which Darwin found untenable in itself, and which would have diverged wildly from the conventional wisdom of continental geologists who understood the vertical relief of the earth's crust to be wrinkling caused by the ongoing *shrinking* of a cooling globe. See Greene, *Geology in the Nineteenth Century*, chap. 2.

17. Down House notebook 1.15, "Santiago Book," CUL MS microfilm 532. These quotations from Darwin's field notebooks are drawn from my own transcriptions. Full transcriptions have since been published in *Charles Darwin's Notebooks from the Voyage of the Beagle*, ed. Chancellor and van Wyhe.

18. Ibid. A corresponding entry in Darwin's geological diary clarifies his meaning in this cryptic note, which is rendered confusing by the absence of the word "our" in the original: "When the reason refuses to admit, the vast, the almost incomprehensible powers of destruction in water, it is well to recollect the shingle plains of Patagonia: perhaps if the bottom of Pacific was exposed to view, our wonder would be reversed, we should only marvel, that part of the Andes yet remain" (DAR 36:433). I thank Michael Reidy for pointing me to this passage and correcting my interpretation (in Sponsel, "Amphibious Being," 266) of the Valparaiso Notebook entry, which I took to refer to the elevation of the seafloor itself rather than to valleys caused by erosion of such uplifted land, and I read "wonder would be reversed" as a question, namely that Darwin wondered whether the process of uplift was reversed elsewhere.

19. Misdated entry for 5 April [actually 4 April] 1835, Keynes, *Charles Darwin's "Beagle" Diary*, 321.

20. Darwin to John Stevens Henslow, 18 April 1835. *CCD*, 1:440–45.

21. Darwin to W. D. Fox, May [1832], *CCD*, 1:232–33.

22. R. Alison to Darwin, 25 June 1835, *CCD*, 1:452.

23. Darwin to Catherine Darwin, 31 May 1835, *CCD*, 1:449–50.

24. DAR 34.1:87–92.

25. Down House notebook 1.18, "Santiago Book," CUL MS microfilm 532 (Darwin's pp. 7, 12).

26. DAR 34.2:151–52v.

27. Down House notebook 1.18, "Santiago Book," CUL MS microfilm 532 (Darwin's pp. 6–8).

28. Ibid. (Darwin's p. 14).

29. Ibid. (Darwin's p. 15).

30. See, for example, Burkhardt, "Appendix V: Darwin's Early Notes on Coral Reef Formation," 567; Sulloway, "Further Remarks," 369; Herbert, *Charles Darwin, Geologist*, 169. For readers who are not already familiar with such claims, their basis will become more evident when I analyze Darwin's "Recollections" in the final section of chapter 11. For those who are interested in this issue: we see here that Darwin shifted from discussing the accumulation of sedimentary beds that contain broken corals to discussing the preservation of intact coral reefs. This passage in the Santiago Book does correspond exactly to the thought process he recounted in the "Recollections," whereby he developed his coral reef theory by "replac[ing] in imagination the continued deposition of sediment by the upward growth of coral." But these notes, which so forcefully corroborate that part of Darwin's reminiscences, do not support his conclusion that simply "to do this was to form my theory of the formation of barrier-reefs and coral atolls." As I argue in chapter 11, it would have been more apt for Darwin to write "all that remained was to form the theory."

31. Down House notebook 1.18, "Santiago Book," CUL MS microfilm 532 (Darwin's p. 21).

32. Darwin to Catherine Darwin, 9–20 July 1834, *CCD*, 1:391–94.

33. Darwin to Henslow, 11 April 1833, *CCD*, 1:306–8. Darwin to Catherine Darwin, 9–20 July 1834, *CCD*, 1:391–94.

Chapter Four

1. On Darwin's activities at the Galápagos see Herbert, "'Universal Collector,'" and other entries in the same volume. On the delay between his visit to the Galápagos and his development of an evolutionary theory see Sulloway, "Darwin and His Finches."

2. Young, "Names of the Paumotu Islands."

3. Entry for 9 November 1835, Keynes, *Charles Darwin's "Beagle" Diary*, 364.

4. FitzRoy, *Narrative*, 2:506.

5. Entry for 9 November 1835, Keynes, *Charles Darwin's "Beagle" Diary*, 364.

6. FitzRoy, *Narrative*, 2:507–8.

7. Entry for 13 November 1835, Keynes, *Charles Darwin's "Beagle" Diary*, 365.

8. FitzRoy, *Narrative*, 2:508.

9. Entry for 17 November 1835, Keynes, *Charles Darwin's "Beagle" Diary*, 368.

10. Entry for 17 November 1835, ibid.

11. Humboldt and Bonpland, *Personal Narrative of Travels to the Equinoctial Regions of the New Continent,*" xxvi.

12. Volume 6 of the English translation of Humboldt's *Personal Narrative* contained a plate titled "Journey towards the summit of Chimborazo attempted on the 23rd June 1802" with the (partial) subtitle "A sketch of the Geography of the Plants in the Andes of Quito." Humboldt and Bonpland, *Personal Narrative of Travels to the Equinoctial Regions of the New Continent.* Quotation from Humboldt, "New Inquiries," 279. On precedents for Humboldt's depictions see Ebach, *Origins of Biogeography,* chap. 3. For an argument about the emergence of a "vertical consciousness" in the sciences around the turn of the nineteenth century see Reidy, *Tides of History,* 274–81; Reidy, "Oceans through Islands to Mountains."

13. This was, incidentally, the period when using the cartographic convention of the contour line, or isohypse, was becoming common. Using isobaths (contours of a given depth) in hydrographic charts, on the other hand, was already well established. See Robinson, "Genealogy of the Isopleth."

14. Entry for 17 November 1835, Keynes, *Charles Darwin's "Beagle" Diary,* 369.

15. Stoddart, "*Coral Islands* by Charles Darwin," 7 (Darwin's p. 5). The passage in full ended "& there remains a Lagoon Island." To the modern reader it is likely confusing to encounter Darwin's use of the now antiquated term lagoon island to refer to an atoll (a ring-shaped reef that encircles an *empty* lagoon). Eimeo was quite literally an island inside a lagoon, namely the lagoon created by its barrier reef. Paradoxically, however, in early nineteenth-century usage the *removal* (by subsidence) of that high island from within Eimeo's barrier reef would leave behind the phenomenon (annular reef and empty lagoon) known as a "lagoon island."

16. Entry for 22 November 1835, Keynes, *Charles Darwin's "Beagle" Diary,* 378.

17. Stoddart, "*Coral Islands* by Charles Darwin," 10 (Darwin's p. 11).

18. Ibid., 11 (Darwin's p. 13).

19. Ibid., 10 (Darwin's pp. 11–12).

20. The deleterious effect of sediment on the growth of corals is one reason deforestation (and its attendant increase in erosion) is now considered a chief threat to the health of coral reefs. I have not, however, found any naturalist predating Darwin who argued that sediments checked reef growth. Of course, neither Darwin nor any of his European contemporaries thought of coral reefs as needing protection. On the contrary, such voyagers desired protection from the navigational hazards posed by the growth of reefs. See Sponsel, "From Cook to Cousteau."

21. Stoddart, "*Coral Islands* by Charles Darwin," 11 (Darwin's p. 13).

22. Entry for 22 November 1835, Keynes, *Charles Darwin's "Beagle" Diary,* 378.

23. "Fungia," specimen 1334, Keynes, *Charles Darwin's Zoology Notes,* 301.

24. Stoddart, "*Coral Islands* by Charles Darwin," 10 (Darwin's pp. 11–12).

25. Ibid., 11 (Darwin's p. 14).

26. There is a vast literature on the history of scientific research on and in the Pacific. Traditionally these studies emphasized the Pacific as a "laboratory" for generating and testing Western ideas, while more recent scholarship drawing implicitly or explicitly on postcolonial studies has attended to the (power) relations

between visiting scientists and residents of the Pacific. For the former see, for example, MacLeod and Rehbock, *Darwin's Laboratory*. For the latter see, for example, Anderson, "Hybridity, Race, and Science" and Radin, *Life on Ice*. For discussion of islanders' systems of knowledge in Polynesia and in the Pacific more generally see Sivadundaram, "Science" and Turnbull, "Masons, Tricksters, and Cartographers." On Cook and Tupaia see Turnbull, "Cook and Tupaia." On problems caused by the testing at Bikini Atoll see Weisgall, *Operation Crossroads*.

27. For example, Darwin learned from Patagonian gauchos to recognize a second species of "ostrich" (now rhea) and from residents of Bahia Blanca, Argentina, to notice the holes dug by the "*casarita*" bird. Throughout the voyage, and in retrospect most famously in the Galápagos, Darwin also learned a great deal from colonial administrators.

28. Stoddart, "*Coral Islands* by Charles Darwin," 11 (Darwin's p. 14).

29. Entry for 23 November 1835, Keynes, *Charles Darwin's "Beagle" Diary*, 378.

30. The original manuscript is at DAR 41:1–12 (Darwin's pp. 1–22). DAR 41:13–22 is a fair copy probably made by Darwin's servant Syms Covington, which contains a small number of annotations by Darwin and FitzRoy not transcribed by Stoddart.

31. Stoddart, "*Coral Islands* by Charles Darwin," 5 (Darwin's p. 1).

32. In addition to Tahiti and Eimeo, the *Beagle* passed within sight of two further islands of the Society group, Ulitea (Raiatea) and Huaheine (Huahine). See *Beagle* captain's log 1835–36, NAK, ADM 51/3055, 34.

33. Stoddart, "*Coral Islands* by Charles Darwin," 6 (Darwin's p. 4).

34. Ibid., 8 (Darwin's p. 8).

35. Ibid., 14 (Darwin's p. 16).

36. Ibid., 14 (Darwin's p. 17).

37. Ibid., 16 (Darwin's p. 20).

38. Ibid., 15 (Darwin's p. 19).

39. I echo the title of Humboldt's "Essay on the Geography of Plants," Humboldt, *Essai sur la géographie des plantes*.

40. Stoddart, "*Coral Islands* by Charles Darwin," 13 (Darwin's p. 15).

41. Ibid., 17 (Darwin's p. 22a).

42. Ibid. (Darwin's pp. 22–22a).

43. Darwin's French was poor, and he may have been guided to this passage in Humboldt's "Fragmens asiatiques" by an extended summary of the work by William Conybeare from the second meeting of the BAAS. In a separate memoir Conybeare discussed "that part of the theory of M. Elie de Beaumont, which asserts that the lines of disturbance of the strata, assignable to the same age, are parallel." *Report of the First and Second Meetings of the British Association for the Advancement of Science*, 393–94, 587–91.

44. I have done this myself in chapter 2 of Sponsel, "Coral Reef Formation."

45. DAR 37.2:787 (Darwin's p. 41).

46. DAR 37.2:791v.

47. DAR 37.2:791 (Darwin's p. 42). See Herbert, *Charles Darwin, Geologist*, 170.

48. DAR 37.2:791 (Darwin's p. 42). Sandra Herbert discusses these notes in the

context of Darwin's views while at the Galápagos. I can offer several kinds of evidence to support my view that they were added to the longer run of notes only after the insight at Tahiti. These comments are written on the front and back of page 42 of Darwin's Galápagos notes, which he numbered 1–46 (with two pages numbered 1 and four pages numbered 41). The material on page 44 could not have been written anytime immediately after the Galápagos visit, because it quotes information from the *Beagle* weather journal about the temperatures two days after the departure from Tahiti. The question therefore is whether the writing was interrupted by the visit to Tahiti before Darwin wrote page 42, or between pages 42 and 44. Two kinds of clues aside from the intellectual content suggest that Darwin had been to Tahiti before he wrote page 42. The first is the page headings. Whereas the pages from 1 to 6 are headed "1835 October Galapagos Id," and those up to the second page 41 are headed "1835 Galapagos Isd," the page in question is headed only "Galapagos Id." The second indication is in Darwin's habitual misspellings. In the zoological notes of a dissection he performed on a Galápagos specimen (which must have been written there because the polyp was still alive), Darwin still used his characteristic South American spelling of "corall." The crater notes use the "coral" spelling that he picked up while writing the "Coral Islands" essay, and also use the spelling "Pacific" (as opposed to "Pacifick") that he otherwise failed to use for several months after leaving South America. In one of Frank Sulloway's original articles on the diagnostic study of Darwin's misspellings, he specifically mentioned that this page of the Galápagos notes had usage inconsistent with the Galápagos period, but he seems not to have considered that it was written later. However, given that the spellings and the ideas are both characteristic of the post-Tahiti period, I believe there is good reason to believe these notes were written after Darwin was turned into an active opponent of the crater theory by his view of Eimeo. See "note p" in Sulloway, "Further Remarks," 364.

49. DAR 37.2:792 (Darwin's p. 43).

50. Ibid.

51. DAR 37.2:793 (Darwin's p. 44).

52. DAR 37.2:793v.

53. Ibid.

54. I made an argument along these lines in my paper "An Amphibious Being," but I develop it further at the end of chapter 11 in the present book.

Chapter Five

1. The *Beagle* touched Australia well south of the tropics and never came close to the barrier reefs on the northeast coast.

2. Beaufort's instructions mentioned stopping at the Keeling Islands to "settle their position," but only if a winter crossing forced FitzRoy to pass north of Australia through the Torres Strait. FitzRoy decided sometime early in 1836 to make the stop at Keeling despite taking the southerly route. Beaufort, "Memoranda for Commander Fitzroy's orders," in Browne and Neve, *Voyage of the "Beagle,"* 393. Armstrong and Herbert think it plausible that Darwin's interest in seeing a coral island

may have encouraged FitzRoy to deviate from the path in the itinerary. Armstrong, *Darwin's Other Islands*, 196; Herbert, *Charles Darwin, Geologist*, 234–35.

3. John Clements Wickham to Phillip Parker King, 9 March 1836 (sent from King George's Sound), King papers, SLNSW MSS 7048/1/3. I am very grateful to Megan Barford for directing me to this quotation. On the *Nautical Magazine* see Barford, "Fugitive Hydrography."

4. The most detailed accounts of Darwin's sojourn at South Keeling are Armstrong, *Under the Blue Vault of Heaven*, and Armstrong, *Darwin's Other Islands*.

5. Entry for 1 April 1836, Keynes, *Charles Darwin's "Beagle" Diary*, 413.

6. My full transcription of Darwin's South Keeling field notes is available online. Sponsel, "Charles Darwin's Notes on the Geology and Corals of the Keeling Islands."

7. Stoddart, "*Coral Islands* by Charles Darwin," 11 (Darwin's p. 14). The statement of Quoy and Gaimard's views quoted by Darwin comes from De la Beche, *Geological Manual*, 141.

8. Stoddart, "*Coral Islands* by Charles Darwin," 13–14 (Darwin's pp. 15–16).

9. Ibid., 13 (Darwin's p. X15[c]).

10. Ibid., 12 (Darwin's p. X15[b]).

11. DAR 41.41 (Darwin's p. 3).

12. [Walsh], *Manual of British Rural Sports*, 443–44.

13. DAR 41.41 (Darwin's pp. 3 and 4). Though he referred to these masses as *Astrea* throughout his field notes, Darwin subsequently identified them as *Porites*. In his 1842 book he described the outer margin of Keeling as "almost entirely composed of a living Porites, which forms great irregularly rounded masses (like those of an Astraea, but larger) from four to eight feet broad, and little less in thickness." Darwin, *Structure and Distribution of Coral Reefs*, 6.

14. DAR 41:41 (Darwin's p. 4).

15. Ibid.

16. DAR 41:42 (Darwin's p. 5).

17. Ibid.

18. DAR 41:41–42 (Darwin's pp. 4–5).

19. DAR 41:48 (Darwin's p. 15v).

20. DAR 41:44 (Darwin's p. 9v). He was unable to name one of the two types of *Corallina* and identified it only by a small icon that he first used on DAR 41:42.

21. Keynes, *Charles Darwin's Zoology Notes*, 418–19.

22. DAR 41:47 (Darwin's pp. 14–14v).

23. DAR 41:42 (Darwin's p. 6).

24. Ibid.

25. DAR 41:40–41 (Darwin's pp. 2–3).

26. DAR 41:40 (Darwin's p. 2).

27. Ibid.

28. DAR 41:45 (Darwin's p. 11).

29. DAR 41:47 (Darwin's p. 14v).

30. FitzRoy, *Narrative*, 2:634.

31. Keynes, *Charles Darwin's Zoology Notes*, 418–19.

32. DAR 41:40v (Darwin's p. 1v).

33. DAR 41:44 (Darwin's p. 10).

34. Ibid.

35. Ibid.

36. Ibid.

37. DAR 41:45 (Darwin's p. 11).

38. DAR 41:45 (Darwin's pp. 11–12).

39. DAR 41:45–46 (Darwin's pp. 12v-13).

40. DAR 39.2:135v. Despite his seeming to imply it, I cannot find any explicit statement in Darwin's Keeling notes to suggest that he thought the lagoon's branching forms would be more resistant to mortality from settling sediment.

41. DAR 39.2:135.

42. DAR 41:45 (Darwin's p. 12v).

43. DAR 41:43 (Darwin's p. 7v). On Darwin's use of the term greenstone see Herbert, *Charles Darwin, Geologist*, 115–16. See also "Glossary," in Lyell, *Principles of Geology*, vol. 3.

44. I diverge here from David Stoddart's analysis of these diagrams, in which he takes the "3 ft" to refer to an emergence of the "corals in situ" three feet above sea level. The note scribbled below (not mentioned by Stoddart) reads, "NB. If tides had been very small. when such a mass had accumulated, there would be very little Breccia." My figure also revises several small details of Stoddart's transcription of the writing on the diagrams and offers readings of a handful of words he considered illegible. See Stoddart, "Darwin and the Seeing Eye," 11–15.

45. Sandra Herbert has used the term symbiotic to describe the collaboration between FitzRoy and Darwin at South Keeling. This is an apt description that overlooks only the complementary activities of the *Beagle*'s other surveyors (and the seamen who supported them). Herbert, "Doing and Knowing," 318.

46. Sulivan's log book quoted in Keynes, *Charles Darwin, Robert FitzRoy and the Voyage of HMS "Beagle,"* 89.

47. DAR 41:53 (Darwin's p. 20v).

48. FitzRoy, *Narrative*, 2:634.

49. DAR 41:53 (Darwin's pp. 20–20v).

50. DAR 41:56 (Darwin's p. 25).

51. DAR 41:55 (Darwin's p. 24).

52. DAR 41:54, 41:56, and 41:57 (Darwin's pp. 22, 25, and 26).

53. DAR 41:51 (Darwin's p. 18).

54. Entry for 7–11 April 1836. Keynes, *Charles Darwin's "Beagle" Diary*, 417.

55. DAR 41:55 (Darwin's p. 24v). On the difficulties of keeping the lead line "dead up and down," see Rozwadowski, *Fathoming the Ocean*, chap. 3.

56. FitzRoy, *Narrative*, 2:630.

57. See, for example, Darwin's list of "Sulivans. outside deep soundings" on DAR 41:53 (Darwin's p. 20).

58. FitzRoy, *Narrative*, 2:630.

59. DAR 41:56 (Darwin's p. 25v).

60. DAR 41:54 (Darwin's p. 21).

61. DAR 41:51 (Darwin's p. 18). Darwin tended to labor over arithmetic, and in many cases I cannot see how his slopes could have been produced from the measurements he was working with. It is worth noting that when he lived in London after the voyage, Darwin commissioned his brother Erasmus to calculate angles of inclination for him (see the notes by Erasmus in DAR 39.1:28–30, e.g., "14 ½ miles base with 200 ft gives 0°7′48‴"). Given horizontal displacements measured in miles, and elevation changes measured in feet, Erasmus returned angles indicating that he figured 6,080 feet to the mile. This shows that Darwin's measurements were in nautical miles (the distance of one-sixtieth of a degree of longitude at the equator) rather than statute miles.

62. DAR 41:56 (Darwin's p. 25v). Beechey's diagram was published in *Narrative of a Voyage to the Pacific and Beering's Strait* (1831), 1:193.

63. DAR 41:57 (Darwin's p. 26).

64. DAR 41:52 (Darwin's p. 19).

65. For a concise and useful discussion of Humboldt's expectation that knowledge of underlying laws would emerge from such activity, see Dettelbach, "Humboldtian Science."

66. DAR 41:56 (Darwin's p. 25).

67. DAR 41:49 (Darwin's p. [16]).

68. In the 1842 book Darwin wrote, "I at first concluded that the whole [of South Keeling atoll] consisted of a vast conical pile of calcareous sand, but the sudden increase of depth at some points, and the circumstance of the line having been cut, as if rubbed, when between 500 and 600 fathoms were out, indicate the probable existence of submarine cliffs." Several pages later, after discussing the inclination of several atolls studied by Beechey, Moresby, and others, he added, "Here then occurs a difficulty;—can sand accumulate on a slope, which, in some cases, appears to exceed fifty-five degrees? . . . M. Élie de Beaumont has argued, and there is no higher authority on the subject, from the inclination at which snow slides down in avalanches, that a bed of sand or mud cannot be formed at a greater angle than thirty degrees. . . . I must conclude that the adhesive property of wet sand counteracts its gravity, in a much greater ration than has been allowed for by M. Élie de Beaumont. From the facility with which calcareous sand becomes agglutinated, it is not necessary to suppose that the bed of loose sand is thick." Darwin, *Structure and Distribution of Coral Reefs*, 9, 23.

69. DAR 41:56 (Darwin's p. 25v).

70. On Darwin's Beagle diary, or "Log Book," see Browne, *Charles Darwin: Voyaging*, 194.

71. Entry for 12 April 1836, Keynes, *Charles Darwin's "Beagle" Diary*, 418.

72. Ibid.

73. Darwin to William Darwin Fox, 15 February 1836. *CCD* 1:491–93.

74. Saint-Pierre, *Paul et Virginie*, 239 (my translation). The novel was translated into English by Helen Maria Williams, who went on to translate the edition of Humboldt's *Personal Narrative* that Darwin owned. Williams's translation of *Paul et Virginie* replaced Saint-Pierre's description of the reefs (*rescifs*, an antiquated spell-

ing) with references to a "chain of breakers" or "sand banks mingled with breakers." See Saint-Pierre, *Paul and Virginia*, 180–81.

75. [Saint-Pierre], *Voyage à l'Isle de France*. On corals and reefs see, for example, 1:156–60, 277, 285–86, 321–23.

76. Beaufort, "Memoranda for Commander Fitzroy's orders," in Browne and Neve, *Voyage of the "Beagle*," 392. FitzRoy's lone sentence: "We anchored in Port Louis, at the Mauritius, on the 29th of April: sailed thence on the 9th of May: passed near Madagascar—thence along the African shore—and anchored in Simon's Bay, at the Cape of Good Hope, on the 31st." FitzRoy, *Narrative*, 2:638.

77. DAR 38.2:894. On Darwin's fieldwork in Mauritius see Armstrong, "Charles Darwin's Geological Notes on Mauritius," and Armstrong, *Darwin's Other Islands*, chap. 17.

78. Ibid.

79. DAR 38.2:895.

80. Ibid.

81. DAR 38.2:899.

82. DAR 38.2:900.

83. Ibid.

84. Ibid.

85. DAR 38.2:901.

86. DAR 38.2:900–901.

87. Darwin, "Geology," 190–92.

88. Darwin, "Geology," 175–76; emphasis added.

Chapter Six

1. Charles Darwin to Caroline Darwin, 29 April 1836, *CCD*, 1:494–97.

2. Stoddart, "Darwin, Lyell, and the Geological Significance of Coral Reefs," 206–7; Herbert, *Charles Darwin, Geologist*, 242–43; Rudwick, *Worlds before Adam*, 492.

3. In his biography of Darwin, John Bowlby was highly attentive to Lyell's seniority, but (in a book whose express purpose was to apply insights from psychoanalysis and child psychology to the study of Darwin's life) he saw this as one in a series of father-son relationships rather than as a relationship of master and research student. Bowlby, *Charles Darwin*.

4. Charles Darwin to Caroline Darwin, 29 April 1836, *CCD*, 1:494–97.

5. On the disciplinary commitments of the geological community to which Darwin aspired see Herbert, *Charles Darwin, Geologist*, and Rudwick, "Darwin in London."

6. Secord, *Controversy in Victorian Geology*, chap. 1.

7. On the territoriality of British geological practice at this time, see Secord's extended discussion of the importance of, and ultimate controversy over, the "collaborative boundary" between the Welsh research sites of Sedgwick and Murchison. Secord, *Controversy in Victorian Geology*, especially 90–92.

8. Lyell, *Principles of Geology*, 1:56–57. See also Secord, *Controversy in Victorian Geology*, 25.

9. *CCD*, appendix 4, 2:443. Darwin added, "If I don't travel.—Work at transmission of Species."

10. Secord, *Controversy in Victorian Geology*.

11. Ibid., 30.

12. Like Humboldt, Buch had studied with Werner at Freiberg. For English-language studies of the European geological context at this moment, and particularly on theories of mountain building, see Greene, *Geology in the Nineteenth Century*, as well as Rudwick, *Worlds before Adam*. I here borrow the term dynamics from William Whewell's 1837 usage, as discussed in the next chapter.

13. Barrett et al., *Charles Darwin's Notebooks*, 17–81. On Darwin's preparations for authorship see Herbert, "Charles Darwin as a Prospective Geological Author"; Herbert, *Charles Darwin, Geologist*, 164–73; Barrett et al., *Charles Darwin's Notebooks*, 18.

14. Barrett et al., *Charles Darwin's Notebooks*, 25 (Red Notebook, p. 18).

15. Ibid., 42–44 (Red Notebook, pp. 68e–73). On Darwin's visions of a "simple" geology see Rhodes, "Darwin's Search for a Theory of the Earth"; Herbert, *Charles Darwin, Geologist*, chap. 7.

16. See Browne, *Charles Darwin: Voyaging*, 334–39; Herbert, *Charles Darwin, Geologist*, chap. 5, especially 173–76.

17. Herschel's diary entry for 15 June 1836 reads, "Raining in showers all day and night. Captain FitzRoy, Mr Darwin, Capt Alexander, Mr C Bell & Mr & Mrs Hamilton dined here at 6. Capt F and M^r D came at 4 & we walked together to Newlands." Herschel W0019, "Diary 1836," HRC. I have found no evidence to suggest that Darwin told Herschel his coral reef theory at this time. Herschel received at least two letters the following year describing the theory, one each from Lyell and Murchison, and he referred to it in a letter of his own in December 1837. See the last section of this chapter as well as chapter 7.

18. See Charles Darwin to W. D. Fox, 15 February 1836, *CCD*, 1:491–92.

19. Darwin reported that Sedgwick had told him before the voyage that he would be willing to propose Darwin as a member of the Geological Society. Charles Darwin to J. S. Henslow, 9 July 1836, *CCD*, 1:499–501.

20. Darwin, *For Private Distribution*.

21. J. S. Henslow to R. W. Darwin, quoted in the letter from Caroline Darwin to Charles Darwin, 29 December [1835], *CCD*, 1:473–75.

22. Darwin, "Geological Notes." Note that Sedgwick's extracts are not reprinted in Darwin and van Wyhe, *Charles Darwin's Shorter Publications*, They do appear in Barrett, *Collected Papers of Charles Darwin*, 1:16–19.

23. Sedgwick to Samuel Butler, as quoted by Susan Darwin in her letter to Charles Darwin, 22 November 1835, *CCD*, 1:469–70.

24. William Buckland to J. S. Henslow, 13 December 1835, APS Scientists Collection 509.L56.

25. Susan Darwin to Charles Darwin, 12 February 1836, *CCD*, 1:488–89.

26. "Geological Society." On Erasmus Darwin, see Susan Darwin to Charles Darwin, 22 November 1835, *CCD*, 1:469–70.

27. Beaufort to F. W. Beechey, 14 October 1835, UKHO LB6, 1834–36.

28. In a recent article, John van Wyhe has argued (to quote the subtitle of the piece) that "Charles Darwin really was the naturalist on HMS *Beagle*." See van Wyhe, "My Appointment Received the Sanction of the Admiralty." The alternative van Wyhe objects to is the claim that Darwin's primary role on the voyage was as a gentlemanly "companion" to FitzRoy rather than as the official ship's naturalist. In my view van Wyhe goes further than necessary when, after acknowledging that "[the role of] 'companion' and [the title of] official naturalist were not contradictory or mutually exclusive," he goes on to argue in the next sentence that "it is more accurate to refer to Darwin as the 'naturalist' because 'companion' has become loaded with misleading baggage." I do not see the advantage to replacing one artificially narrow description of Darwin's role with another one. More important, much of the apparent tension would be resolved if vocation were not conflated with formal status. There is no evidence that Darwin was formally appointed at the outset of the voyage as a member of the crew whose *job* was to carry out natural history collecting or research. There is, on the other hand, ample evidence that Darwin's primary reputation and self-identity during the voyage was as a naturalist. Quotations from Beaufort and FitzRoy (on van Wyhe's pp. 323–24) illustrate that both went out of their way to emphasize that Darwin's industrious naturalizing was done *despite* his being a volunteer. It seems equally clear that Darwin's status on the voyage came over time to seem more "official" to all involved, particularly after the voyage, once his accomplishments were becoming apparent. (He was, for example, named as "Naturalist" in FitzRoy's *Narrative* and other government-sponsored publications.) Such a title would have lent validity to Darwin's solicitation of government funds to assist in publication and presumably also helped to ensure that his success would reflect well on the Admiralty. For those who want to consider the question for themselves, van Wyhe's article presents an admirable breadth of source material. Much of what he cites as evidence that Darwin was something other than a purely private companion may also be read (against the grain of van Wyhe's most polemical statements) as evidence that he was something other than a straightforward member of the crew. See van Wyhe, "My Appointment Received the Sanction of the Admiralty."

29. Huxley, *Life and Letters of Sir Joseph Dalton Hooker*, 1:41.

30. DAR 41:13–22. For a physical description of the manuscript, see Stoddart, "*Coral Islands* by Charles Darwin," 4.

31. DAR 41:18.

32. DAR 41:22. The original sentence is in the diary entry for 12 April 1836. Keynes, *Charles Darwin's "Beagle" Diary*, 418.

33. DAR 41:22.

34. Charles Darwin to Caroline Darwin, 29 April 1836, *CCD*, 1:494–97.

35. DAR 5:B98–99. These notes are discussed in Sloan, "Darwin's Invertebrate Program"; Keynes, *Charles Darwin's Zoology Notes*.

36. Lyell, "Address to the Geological Society, Delivered at the Anniversary, 1836," 367.

37. On Lyell's dispute with Greenough over elevation, see Herbert, *Charles Darwin, Geologist*, 217–18.

38. On Lyell's correspondence with Robert Alison in South America, see his letter to Francis Beaufort, 28 November 1835. UKHO Incoming Letters pre-1857, box 1, L242.

39. Darwin, *For Private Distribution*.

40. Lyell, "Address to the Geological Society, Delivered at the Anniversary, 1836," 367.

41. On De la Beche's caricatures and his rivalry with Lyell see Rudwick, "Emergence of a Visual Language for Geological Science"; Rudwick, *Worlds before Adam*, 397–99; McCartney and Bassett, *Henry De la Beche*.

42. De la Beche, *Sections and Views*, iii–iv. On the likelihood that De la Beche wrote this preface after the first volume of Lyell's *Principles* appeared, see Rudwick, *Worlds before Adam*, 325.

43. On Lyell's essays in the *Quarterly* see Rudwick, *Worlds before Adam*, 202–16, 253–57. On Lyell's ambitions as an author and his relationship with the Murrays see Secord, "Introduction [to Lyell's *Principles*]," especially xiii–xv.

44. Charles Lyell to John Murray II, 18 April 1827. "Letters to John Murray, publishers, of Charles Lyell," MS.40726, NLS.

45. Charles Lyell to John Murray II, 6 April 1829. "Letters to John Murray, publishers, of Charles Lyell," MS.40726, NLS.

46. Charles Lyell to John Murray II, 24 November 1829. "Letters to John Murray, publishers, of Charles Lyell," MS.40726, NLS.

47. Lyell to Adam Sedgwick, December 6, 1835, Lyell, *Life, Letters and Journals of Sir Charles Lyell*, vol. 1.

48. Lyell to Francis Beaufort, 15 December 1835, UKHO Incoming Letters pre-1857, box 1, L244.

49. Valuable discussions of these widely shared attitudes include Larsen, "Not since Noah"; Endersby, *Imperial Nature*; Knell, *Culture of English Geology*; Rudwick, "Charles Darwin in London"; Herbert, "Place of Man."

50. Darwin to Henslow, 6 October [1836], *CCD*, 1:507–8.

51. Darwin to Caroline Darwin, 24 October [1836], *CCD*, 1:509–10.

52. Ibid.

53. Though Darwin met Grant while both were at the University of Edinburgh, Grant had held the combined chairs in comparative anatomy and zoology at the newly founded University of London. Desmond, "Grant, Robert Edmond."

54. Desmond and Moore, *Darwin*, 203.

55. Darwin to Henslow, [30–31 October 1836], *CCD*, 1:512–15.

56. Ibid.

57. Lyell to Darwin, 26 December 1836, *CCD*, 1:532–33.

58. William Whewell to John Herschel, 4 December 1836, in Todhunter, "William Whewell," 250.

59. Darwin to W. D. Fox, 6 November 1836, *CCD*, 1:516–17.

60. For a description of Erasmus Darwin's efforts to translate "a certain German pamphlet on the corals of the red sea," see Elizabeth Wedgwood to Hensleigh Wedgwood, [16] November [1836], *CCD*, 1:519–21. The pamphlet was probably Ehrenberg, *Über die Natur und Bildung der Coralleninseln und Corallenbänke im Rothen Meere*. On Hensleigh Wedgwood's comment, see his letter to Charles Darwin, [20 December 1836], *CCD*, 1:530. Wedgwood also annotated the diary manuscript. See Keynes, *Charles Darwin's "Beagle" Diary*, 419. See also Charles Darwin to Caroline Darwin, [7 December 1836], on his expectation of "detailed criticisms" from Hensleigh and Fanny Wedgwood *CCD*, 1:524.

61. This paper became Darwin, "Observations of Proofs of Recent Elevation on the Coast of Chili."

62. Lyell to Darwin, 26 December 1836, *CCD*, 1:532–33.

63. KHM, Lyell notebook 63, 63.

64. Secord, *Evolutionary Writings*, 399–400.

65. On Lyell's "wild excitement," see Judd, "Darwin and Geology," 358. In his autobiographical recollections Darwin recalled his surprise at Lyell's encouraging reaction to a view that differed from his own. For my discussion of this, see the final section of chapter 11 below.

66. Lyell to Babbage, 6 January 1837, Correspondence of Charles Babbage, BL Add. MS 37190.8, quoted in Wilson, *Charles Lyell*, 435.

67. Lyell to Darwin, 13 February 1837, *CCD*, 2:4–5.

68. I will discuss this further below. The quotation comes from Lyell to Leonard Horner, 12 March 1838, Lyell, *Life, Letters and Journals of Sir Charles Lyell*, 2:39–41.

69. Lyell to Darwin, 13 February 1837, *CCD*, 2:4–5. For a useful perspective on Lyell's public roles in these years see Morrell, "London Institutions and Lyell's Career."

70. Lyell, "Address to the Geological Society, Delivered at the Anniversary, 1837," 510–11.

71. See Darwin to Leonard Jenyns, 10 April [1837], *CCD*, 2:15–17.

72. Darwin to W. D. Fox, [12 March 1837], *CCD*, 2:10–12.

73. See Darwin to Henslow, 18 [May 1837], *CCD*, 2:17–19, and Darwin, *Journal of Researches*, 554. The abstracts were themselves eventually abstracted as Charles Darwin, "A Sketch of the Deposits Containing Extinct Mammalia in the Neighbourhood of the Plata," and Darwin, "On Certain Areas of Elevation and Subsidence."

74. Kinnordy House manuscripts, quoted by permission of Lord Lyell and the Gifford family. The notebooks were cited in the 1972 volume of Leonard Wilson's biography of Lyell, and his chapter on Lyell and Darwin planted the seed for many of the ideas I have developed in this book. I only recently had the opportunity to study the Kinnordy House manuscripts in person, but several years ago Professor Wilson was kind enough to show me extracts photocopied from Lyell's notebooks 63 and 64, covering late 1836 and 1837, and subsequently I was able to view his copies of additional notebooks up to the end of 1840. I am extremely grateful to Professor Wilson for sharing this material with me.

75. KHM, Lyell notebook 63, especially 61, 69, 89, 101–2, 105–13, and Lyell notebook 64, especially 6–9, 16, 32, 38–39.

76. KHM, Lyell notebook 63, 108–9.

77. Caroline Darwin to Charles Darwin, [21 February 1837], *CCD*, 2:7–8. Henslow tried to contribute to Darwin's springtime research on reefs by sending his former pupil a "chart and account of Diego Garcia," which Darwin considered "a beautiful instance of a Lagoon Island." It is acknowledged in Darwin's letter to Henslow, [28 May 1837], *CCD*, 2:21–22.

78. Charles Darwin to Caroline Darwin, 27 February 1837, *CCD*, 2:8–9.

79. Lyell, "Address to the Geological Society, Delivered at the Anniversary, 1836," 365. On the relation of Darwin's coral reef observations to Lyell's later writing on the origin of the Chalk formation of Europe, see below.

80. Noteworthy discussions of the content and style of the Red Notebook entries are in Darwin, *Red Notebook of Charles Darwin*, and Gross, *Starring the Text*, chap. 6.

81. Barrett et al., *Notebooks*, 62–63 (Red Notebook, pp. 130–33).

82. These phrases come from ibid., 56–57 (Red Notebook, pp. 111–12).

83. Ibid., 58 (Red Notebook, p. 117). See also Sandra Herbert's interpretations of these passages in Herbert, *Charles Darwin, Geologist*, 175, 243.

84. Lyell to J. F. W. Herschel, 24 and 26 May 1837, RS HS 11.422 and 11.450. This letter is partially transcribed in Lyell, *Life, Letters and Journals of Sir Charles Lyell*, vol. 2. See also Wilson, *Charles Lyell*, 447–49.

85. For an overview of Herschel's philosophy of science, especially as it related to Darwin's work in the 1830s, see Hull, "Darwin's Science and Victorian Philosophy of Science"; Ruse, "Darwin's Debt to Philosophy." For similar issues in connection with Lyell's work see the many works of Jon Hodge already cited and also Ruse, "Charles Lyell and the Philosophers of Science." I am grateful to Charles Pence for sharing his work in progress on this topic: Pence, "Sir John F. W. Herschel and Charles Darwin."

86. Herschel did in fact heed this advice. In December of that year, in a letter to astronomer George Gipps, who was at the Paramatta Observatory near Sydney, he wrote, "There is one <u>Geological</u> datum which it would be desirable to possess in every seaport . . . viz: the mean level of the sea as referred to some absolutely permanent & identifiable mark on the land—the recent discoveries of Mr Darwin relative to the ~~formation of~~Coral formations of the Pacific render this a point of some considerable interest on a sea coast so peculiarly beset with these formations as that of Australia." Herschel to Gipps, 26 December 1837, RS HS 19.72. I thank Simon Schaffer for pointing this letter out to me.

Chapter Seven

1. At the beginning of the section on coral reef formation in the *Journal of Researches*, Darwin included a footnote (p. 554) saying, "This sketch was read before the Geological Society, May, 1837." I consider the text that followed to indeed be the paper Darwin read at the Geological Society on 31 May 1837, for three reasons:

first, because the content follows so closely the summarized versions of the paper that were published immediately in the *Athenaeum* and the *Proceedings of the Geological Society of London*; second, because (although the actual release of Darwin's *Journal* was delayed until 1839) these pages of the book were printed so soon after the paper was delivered that it is unlikely the content was revised; and finally, because the content corresponds perfectly with the detailed firsthand account of the talk's contents published in 1837 by John Phillips (see below). The published summaries of the paper are Darwin, "Geological Society [Report of Darwin's 'Areas of Elevation and Subsidence' Paper]," and Darwin, "On Certain Areas of Elevation and Subsidence."

2. The society's curator and librarian, William Lonsdale, was an expert in the taxonomy of fossil corals, but he did not have Darwin's experience studying modern reef-building corals in the field.

3. Darwin, *Journal of Researches*, 554.

4. Ibid., 556.

5. Darwin, "On Certain Areas of Elevation and Subsidence," 553.

6. Down House Notebook 1–3. Darwin also recorded his struggles with this question in DAR 38.1:891, DAR 38.2:892, and DAR 29.3:48b.

7. Darwin, *Journal of Researches*, 557.

8. Ibid., 561.

9. Ibid., 557–58.

10. On Darwin's reasoning using the "principle of exclusion," see Darwin, *Evolutionary Writings*, 391; Rudwick, "Darwin and Glen Roy."

11. Darwin, "On Certain Areas of Elevation and Subsidence," 553.

12. Darwin, *Journal of Researches*, 559.

13. Lyell to John Pye-Smith, 1 or 2 June 1837, published on 17 September 1935 in a letter to the *Times* of London by A. S. Pye-Smith. A copy may be found in DAR Add. 8904.3:138.

14. Rosen and Darrell, "Generalized Historical Trajectory."

15. Darwin, "On Certain Areas of Elevation and Subsidence," 553.

16. Darwin, *Journal of Researches*, 564.

17. They would be classed as fringing reefs because, as Darwin said, those reefs described by Quoy and Gaimard "do not require a foundation at any greater depth than that from which the coral-building polypi can spring." Ibid., 561.

18. Ibid. Note that throughout his analysis of Quoy and Gaimard's joint publication, Darwin referred by name solely to Quoy.

19. Ibid., 562.

20. In his presidential address of 1838 Whewell reminded the society, "We have had placed before us the map, in which Mr. Darwin has . . . divided the surface of the Southern Pacific and Indian oceans into vast bands." Whewell, "Address to the Geological Society, 1838," 644.

21. For a brief discussion of "maps as instruments of thought" for Darwin, see Camerini, "Evolution, Biogeography, and Maps."

22. Darwin, *Journal of Researches*, 566.

23. I discuss Darwin's geographical imagination below in analyzing his 1842 book.

24. Darwin, *Journal of Researches*, 565. Darwin claimed this anomaly "has never been attempted to be solved." In fact, this was Eschscholtz's stated reason for his argument (against Forster) that lagoon islands must be underlain by submarine mountains.

25. Darwin, "On Certain Areas of Elevation and Subsidence," 554.

26. [Whewell], Review of *Principles*, vol. 2 (1832) by C. Lyell.

27. On various operative meanings of "uniformity" in geology at this time see Rudwick, "Uniformity and Progression."

28. Lyell, *Principles of Geology* (1830), vol. 1.

29. Darwin, *Journal of Researches*, 566–67.

30. Whewell coined the term consilience, which literally meant a "jumping together" of facts from different classes of knowledge. Whewell, "Philosophy of the Inductive Sciences," 5:67–68. On Darwin's use of Whewell's ideas in the philosophy of science, see Hull, "Darwin's Science and Victorian Philosophy of Science"; Ruse, "Darwin's Debt to Philosophy."

31. Lyell, *Principles of Geology* (1830), 1:1–4.

32. Darwin, *Journal of Researches*, 567.

33. For brief summaries of the significance of this work by their contemporaries, see Lyell, "Address to the Geological Society, Delivered at the Anniversary, 1837," 487; Whewell, "Address to the Geological Society, Delivered at the Anniversary," 1838," 646–47. For Darwin's reaction to Herschel and his reliance on Hopkins's publications of the late 1830s, see Herbert, *Charles Darwin, Geologist*, 210–15.

34. I am not certain Darwin had this paleontological implication in mind.

35. Lyell, "Address to the Geological Society, Delivered at the Anniversary, 1837," 506.

36. Darwin, *Journal of Researches*, 568.

37. Darwin's undated reading notes on Lesson's *Voyage autour du monde*, at DAR 29.3:48b, include, "May we think the continent of which Tahiti was a peak had a peculiar vegetation & that the trade wind drifted seeds to E. Indian Isd. ??" These notes were probably written at or near St. Helena, on the leg of the *Beagle* voyage between the Cape of Good Hope and the final return to Britain. Darwin's Red Notebook contains a reference to Lesson at p. 62, shortly before the first notes mentioning St. Helena.

38. Darwin, *Journal of Researches*, 568.

39. Ibid., 569.

40. John Herschel to Lyell, 20 February 1836. Quoted in Wilson, *Charles Lyell: The Years to 1841*, 438–39.

41. Barrett et al., *Charles Darwin's Notebooks*, 191 (B Notebook's p. 82). He had a specific location and pair of species in mind in this example, namely the rhinoceroses of Java and Sumatra. As his coral reef paper revealed, he believed these islands were in the process of elevation. On Darwin's studies of the distribution of East Indian fauna, see Camerini, "Evolution, Biogeography, and Maps."

42. On this topic see Kottler, "Charles Darwin's Biological Species Concept and Theory of Geographic Speciation"; Richardson, "Biogeography and the Genesis of Darwin's Ideas on Transmutation"; Kohn, "Theories to Work By."

43. Barrett et al., *Charles Darwin's Notebooks*, 227 (B Notebook's pp. 224–25).

44. Ibid., 227, 247 (B Notebook's pp. 227 and 246). Lyell had posited that changes of fauna in the geological record were directly proportional to the time series represented by the strata in question. See Rudwick, "Charles Lyell's Dream of a Statistical Palaeontology."

45. Darwin, *Journal of Researches*, 569.

46. Darwin, "On Certain Areas of Elevation and Subsidence," 554.

47. The sources Darwin named in the full text printed in the *Journal of Researches* were FitzRoy (twice), Kotzebue, Beechey (twice), De la Beche (twice), Flinders (twice), Dampier, Lyell (three times), Dillon (twice), Quoy (three times), Forster, Lesson (twice), Labillardière, Bennett, Bligh, Bougainville, Cook, Ehrenberg, and William Owen. (He also mentioned La Peyrouse in connection with his shipwreck at Vanikoro.) De la Beche, though mainly remembered for his geological work on British soil, had seen coral banks when he lived in Jamaica and discussed them in the source Darwin cited. See De la Beche, *Geological Manual*, 141–42.

48. Quoted in Wilson, *Charles Lyell: The Years to 1841*, 435.

49. Darwin, *Journal of Researches*, 568.

50. On compensatory movements hypothesized by Lyell, see Lyell, *Principles of Geology* (1830), 1:113–17 and 473–77.

51. Stoddart, "*Coral Islands* by Charles Darwin," 17 (Darwin's pp. 22–22a).

52. Diary entry for 12 April 1836, Keynes, *Charles Darwin's "Beagle" Diary*, 418.

53. See the discussion above of Darwin's reading of Lesson's *Voyage*.

54. For Lyell, "one complete revolution" would have taken place in the cycle of geological changes when "various causes of change both igneous and aqueous [had] remodel[ed] . . . the entire crust of the earth." Lyell, *Principles of Geology* (1830), 2:271.

55. Lyell to J. F. W. Herschel, 24 and 26 May 1837, RS HS 11.450; emphasis added.

56. On the tradition of discussion at the Geological Society in this period, see Rudwick, *Great Devonian Controversy*, 25–26; Secord, *Controversy in Victorian Geology*, 14–24; Thackray, *To See the Fellows Fight*.

57. Murchison, "Address to the Geological Society, Delivered at the Anniversary, 1833," quoted in Morrell, "London Institutions and Lyell's Career," 138.

58. A flavor of the (sometimes friendly) antagonism between Lyell and Murchison can be had in Page, "Rivalry between Charles Lyell and Roderick Murchison."

59. J. S. Henslow to Darwin, 24 August 1831, *CCD* 1:128.

60. This paragraph describes a sentiment held by some nineteenth-century geologists. As I hope my arguments in Part I made clear, I do not myself subscribe to the idea that the acts of observing and collecting could be devoid of theory.

61. B. J. Sulivan to Darwin, 29 November 1881, DAR 177. This letter was prompted by the publication of Katherine Lyell's *Life, Letters and Journals of Sir Charles Lyell*. Sulivan was eager to find out "if [Lyell] described in any way the Meeting at the Geological S[ociety] when you read your 'Coral Island' paper . . . the only meeting of that Society I ever attended."

62. Emphasis added.

63. Lyell to J. F. W. Herschel, 24 and 26 May 1837, RS HS 11.422 and 11.450. This letter is partially transcribed in Lyell, *Life, Letters and Journals of Sir Charles Lyell*, vol. 2. See also Wilson, *Charles Lyell: The Years to 1841*, 447–49.

64. The 31 May meeting was the penultimate Geological Society meeting of the season. The final one was held on 14 June, after Lyell's departure. The society reconvened on 1 November, when Darwin read his paper on the formation of mold. For the schedule, see announcements in the 1836 issues of the *Athenaeum*, e.g., 515 (9 September 1837): 660; 526 (25 November 1837): 866.

65. Darwin to Henslow, [28 May 1837], *CCD*, 2:21–22.

66. Darwin to W. D. Fox, 7 July [1837], *CCD*, 2:29–30; emphasis added. On the relationship between Darwin and Fox see Larkum, *Natural Calling*.

67. Phillips, *Treatise on Geology*, 1:306.

68. Ibid., 1:310–11.

69. Ibid., 1:313.

70. Ibid.

71. Desmond and Moore suggest that Whewell's high praise for Darwin was intended, at least in part, to persuade him to accept a labor-intensive position as one of the secretaries of the society. Desmond and Moore, *Darwin*, 235–36.

72. Whewell, "Address to the Geological Society, Delivered at the Anniversary, 1838," 632–33. For a brief discussion of his first use of the term geological dynamics, in response to the first volume of Lyell's *Principles*, see Wilson, *Charles Lyell: The Years to 1841*, 305.

73. Whewell, "Address to the Geological Society, Delivered at the Anniversary, 1838," 632.

74. Ibid., 647–48.

75. Ibid., 643.

76. Ibid., 624.

77. Ibid., 643–45.

78. Ibid., 644.

79. On Whewell's reviews of *Principles*, see Wilson, *Charles Lyell: The Years to 1841*, chaps. 9–10; Rudwick, *Worlds before Adam*, secs. 23.1 and 24.4.

80. Whewell, "Address to the Geological Society, Delivered at the Anniversary, 1838," 644–45.

81. Ibid., 645. It must be noted that when he gave this speech, Whewell had read an advance copy of Darwin's *Journal of Researches*. See Darwin to Henslow, [21 January 1838], *CCD*, 2:69–70: "I have sent a copy of my journal, as far as complete, to Mr Whewell, for him to review the Geolog: part, in his anniversary speech." On the "Cambridge Network" of social interaction among men of science in which Whewell and Darwin participated, see Cannon, *Science in Culture*.

Chapter Eight

1. Charles Darwin to William Lonsdale, 3 August [1837], *CCD*, 2:35. Darwin requested that "the abstract which I read on my views concerning coral Forma-tions" be withdrawn because "Capt. FitzRoy . . . wishes the principal results to be

given" in their joint "account of the Voyage of the Beagle." Darwin's use of the word "abstract" in this letter creates some confusion as to his intentions: I believe it refers not to the abbreviated abstract of the paper he read (which would have been written by Lonsdale in his capacity as secretary and was indeed published in the society's *Proceedings*). Rather, Darwin is referring to the full paper he read as a mere abstract of the fuller views on coral formations that he intended to publish in a future book. The full paper was indeed never published in the society's organ for transmitting the full text of papers, its *Transactions*.

2. Charles Darwin to Charles Lyell, [14] September [1838], *CCD*, 2:104–8.

3. Bowlby, *Charles Darwin*, 244.

4. Lyell to Darwin, 29 August and 5 September 1837, *CCD*, 2:41–43.

5. For the arrangement and the information about Darwin's specimens and diagrams, see Lyell to John Pye-Smith, 1 or 2 June 1837, published on 17 September 1935 in a letter to the *Times* of London by A. S. Pye-Smith. Clippings of this published letter turn out to have been placed both in the Darwin archive and in Williams's papers (DAR Add. 8904.3:138 and LMS CWM South Seas Personal, box 2, folder 11), but I have not found any further reference confirming whether this specific meeting took place as scheduled.

6. Williams, *Narrative of Missionary Enterprises in the South Sea Islands*. On the date of publication, see Gutch, *Beyond the Reefs*, 110–11.

7. In this case, of course, the subsidence must have been followed by elevation in order for the coralline strata to have emerged several hundred feet above sea level. This was precisely how Darwin interpreted the history of the island of Mangaia, Williams's prime example, in his 1842 book. See Darwin, *Structure and Distribution of Coral Reefs* (1842), 132, 139.

8. See Lyell's letter to Pye-Smith.

9. Rudwick, "Charles Darwin in London," 190–91.

10. On Whewell's induction as president of the Geological Society, Murchison referred obsequiously to these two as "philosophers in higher walks, who did not disdain to look upon the earth." Murchison reported this toast to Herschel in a letter of 2 February 1837, RS HS 12.390.

11. See the entry under June 1837 in "List of Donations to the Geological Society."

12. See, for example, Darwin's descriptions of the Cook and Austral Islands. Darwin, *Structure and Distribution of Coral Reefs*, 132, 155.

13. Ellis, *Polynesian Researches*.

14. Sivasundaram, *Nature and the Godly Empire*, 211–12.

15. In the same week as Lyell's departure, the *Athenaeum* published a review of Williams's book that revealed a more general secular disdain for "the habitual unconscious arrogance of one who fancies himself in the immediate guidance of heaven." Review of *A Narrative of Missionary Enterprises* by John Williams.

16. I borrow the metaphor of social topography from Rudwick, "Charles Darwin in London."

17. Darwin to W. D. Fox, 7 July [1837], *CCD*, 2:29–30.

18. Robert FitzRoy to Darwin, 16 November 1837. *CCD*, 2:57–59.

19. Ibid.

20. In the preface to the *Zoology* of the voyage as it was ultimately published, Darwin wrote, "I must here, as on all other occasions, take the opportunity of publicly acknowledging with gratitude, the obligation under which I lie to Captain FitzRoy, and to all the Officers on board the Beagle, for their constant assistance in my scientific pursuits, and for their uniform kindness to me throughout the voyage." On the preface, see *CCD*, 2:59n1.

21. Darwin to W. D. Fox, 7 July [1837], *CCD*, 2:29–30.

22. Darwin to W. D. Fox, May [1832], *CCD*. 1:232–33.

23. Darwin to Henslow, [21 January 1838], *CCD*, 2:69–70.

24. See *CCD*, 2:70n3.

25. Darwin to Caroline Wedgwood, [May 1838], *CCD*, 2:84–85.

26. Lyell to Darwin, 6 and 8 September 1838, *CCD*, 2:99–102.

27. Darwin to Babbage, [1838], *CCD*, 2:67.

28. Darwin to C. Whitley, [8 May 1838], *CCD*, 7:468–69.

29. See, e.g., Darwin to Caroline Wedgwood, [May 1838]; Darwin to Lyell, [14] September [1838]; Darwin to W. D. Fox, 24 October [1839], *CCD*, 2:84–85, 2:104–8, 2:234–35.

30. Darwin to Lyell, [14] September [1838], *CCD*, 2:104–8.

31. Lyell, *Elements of Geology*.

32. Ibid., 96–97.

33. Wilson, *Charles Lyell: The Years to 1841*, 506.

34. For useful tables illustrating Lyell's 1838 nomenclature for the fossiliferous rocks, see Lyell, *Elements of Geology*, 280–81.

35. William Montgomery has examined the relation of Darwin's coral reef theory to ongoing geological debates over the Chalk. He argues that Lyell's interest in coral reefs was driven by his desire to find an alternative to the Wernerian notion that the Chalk had formed as a chemical precipitate and by claims that Lyell's theory of the organic deposition of chalk (derived from corals) remained operative alongside Darwin's theory of reef formation. Montgomery is right to point out that the problem of the Chalk was one of the arenas in which Darwin's theory was disputed, but I believe he has overstated his case. In contending that the debates over reef formation were essentially arguments about the origin of the Chalk, Montgomery ignores the long-standing concern (dating from J. R. Forster) over the potential role of corals in building new continents and thereby changing the proportions of land and sea. The issue of the Chalk was important to Lyell and others, but it should not obscure the fact that the study of reefs was brought to bear on even more fundamental questions about the remodeling of the earth's crust. See Montgomery, "Charles Darwin's Theory of Coral Reefs and the Problem of the Chalk."

36. Lyell, *Elements of Geology*, 329.

37. Darwin, "Geological Society [Report of Darwin's Paper 'On the Formation of Mould']." This comment was incorporated into the version printed in the Geological Society's *Proceedings*, but not the one in the *Transactions*: Darwin, "On the Formation of Mould" (1838); Darwin, "On the Formation of Mould" (1840). See

also Barrett, *Collected Papers of Charles Darwin*, 1:53n4. Buckland's referee's report (9 March 1838) is transcribed in *CCD*, 2:76.

38. Lyell, *Elements of Geology*, 320–21.

39. Ibid., 324.

40. Lyell, *Principles of Geology* (1830), 2:298.

41. Darwin to Lyell, [19 December 1837], *CCD*, 2:65–66.

42. Compare Darwin's letter of 19 December 1837 to Lyell, *Elements of Geology*, 329–30.

43. Ibid., 329.

44. Ibid., 362–63.

45. Ibid., 402.

46. Ibid., 403.

47. Darwin to Lyell, 9 August [1838], *CCD*, 2:95–99.

48. Lyell to Darwin, 6 and 8 September 1838, *CCD*, 2:99–102.

49. Ibid.

50. Darwin to Lyell, [14] September [1838], *CCD*, 2:104–8.

51. Ibid.

52. Vorzimmer, "Darwin Reading Notebooks," 120. The books in question were Herschel, *Preliminary Discourse*, and Whewell, *History of the Inductive Sciences*.

53. Darwin, "On the Connexion of Certain Volcanic Phaenomena."

54. Herbert, *Charles Darwin, Geologist*, chap. 7, especially 225–30; Rhodes, "Darwin's Search for a Theory of the Earth."

55. Lyell to Leonard Horner, 12 March 1838, Lyell, *Life, Letters and Journals of Sir Charles Lyell*, 2:39–41.

56. In a letter to John Herschel of February 1838, Murchison described the critics of Lyell's theory of the metamorphosis of rocks as "the practical geologists of the highest rank, including Sedgwick, Buckland, &c. & even some theoretical writers such as [George Poulett] Scrope." RS HS 12.390.

57. For the insight that Lyell "never speaks above his breath, so that everybody keeps lowering their tone to his," see Emma Darwin to Elizabeth Wedgwood, 2 April 1839. In *Emma Darwin: A Century of Family Letters*, 1:40. Wilson, *Charles Lyell: The Years to 1841*, 459. On Lyell's training and career in law, see Wilson's chapters 5 and 6.

58. Lyell to Horner, 12 March 1838; postscript quoted in ibid., 456; emphasis added.

59. Darwin to John Phillips, [November 1840], *CCD*, 2:273–74.

60. He contended that the preservation of fossils and of old sea beaches should be considered the exception rather than the rule. For instance, he pointed to a number of locations, ranging from his home county of Shropshire to the coasts of Scandinavia, where exposed deposits of undoubted marine origin had been found not to contain any marine shells, presumably because they had been dissolved by acid rain. Likewise, he pointed out that durable terraces like the roads might have been formed only where a special combination of currents and tides were acting on a coastline of a particular geological composition.

61. Darwin, "Observations on the Parallel Roads of Glen Roy," 78. On Darwin's reasoning in the Glen Roy paper, see Rudwick, "Darwin and Glen Roy."

62. DAR 210.8:1. Darwin's notes on career and marriage were written on the back of a letter from Leonard Horner, 7 April [1838], presumably sometime between that date and 11 November 1838, when Emma Wedgwood accepted his marriage proposal. The notes are published in full in *CCD*, 2, appendix IV.

63. Charles Lyell to John Murray II, 22 May 1832. "Letters to John Murray, publishers, of Charles Lyell," MS. 40726, NLS.

Chapter Nine

1. There have been many efforts to diagnose Darwin retrospectively with various ailments. For perspectives I have found most valuable in considering the effects of Darwin's illnesses see Colp, *To Be an Invalid*; Colp, *Darwin's Illness*; Browne, "I Could Have Retched All Night." Both authors review the various retrospective diagnoses that have been offered; Browne offers a useful taxonomy of these in Browne, "I Could Have Retched All Night," 241.

2. Let me emphasize this time range. I am here analyzing the pre-1842 period, before Darwin had completed his book on coral reefs. In chapter 11 I address Darwin's later attitudes toward the act of speculation in general and toward his geological and species theories in particular.

3. Darwin to Leonard Jenyns, 15 July [1839], *CCD*, 2:206–7.

4. Darwin, *Structure and Distribution of Coral Reefs*, 119.

5. See Darwin to J. Shillinglaw (Secretary, Royal Geographical Society), [1839-May 1842], Darwin to Librarian of the Royal Geographical Society, [19 March 1839], Darwin to John Washington, 14 October [1839], and Darwin to John Washington, [14 October 1839], *CCD*, 2:153, 2:177–78, 2:229–30, 2:230.

6. See preface to Darwin, *Structure and Distribution of Coral Reefs*.

7. On Darwin's reliance on correspondents, see Browne, *Charles Darwin: The Power of Place*, 10–11.

8. Darwin to William Henry Smyth, 7 August [1839], *CCD*, 2:212–13. Smyth's reply was inconclusive. See Darwin, *Structure and Distribution of Coral Reefs*, 158.

9. See *CCD*, 2:529 and 2:217n5.

10. John Grant Malcolmson to Darwin, 24 July 1839, 31 August 1839, 7 October 1839, [after 7 October 1839], and 30 November 1839, *CCD*, 2:207–12, 2:215–18, 2:223–27, and 2:244–46. See also Darwin, *Structure and Distribution of Coral Reefs*, 77–79, 134–37, 194.

11. On these working maps, see Camerini, "Darwin, Wallace, and Maps," 95–98.

12. Latour, "Give Me a Laboratory." Latour himself described the metropolitan map rooms of imperial powers as "centres of calculation" where knowledge was created through a "cycle of accumulation." See Latour, *Science in Action*, chap. 5.

13. On the composition and structure of the *Journal of Researches* see Tallmadge, "From Chronicle to Quest."

14. Darwin, *Journal of Researches* (1839), viii.

15. Charles Lyell to Gideon Mantell, 23 March 1829; quoted in Rudwick, *Worlds before Adam*, 288.

16. [Fitton], Review of *Elements of Geology* by Charles Lyell, 410. For the essay's attribution to Fitton, see Houghton, Slingerland, and Wellesley College, *Wellesley Index to Victorian Periodicals, 1824–1900*, vol. 1.

17. [Fitton], Review of *Elements of Geology* by Charles Lyell, 417. Fitton also cited (433) Darwin's contributions to Lyell's argument about the origin of the Chalk.

18. [Hall], Review of *Narrative of . . . "Adventure" and "Beagle."* For the tentative attribution of this review to Basil Hall, see Houghton, Slingerland, and Wellesley College, *Wellesley Index to Victorian Periodicals, 1824–1900*, vol. 1. I am personally convinced based on internal evidence that Hall was indeed the reviewer (for example, he wrote about the coast of Chile as if from firsthand experience, and he devoted considerable attention to the value of chronometers for navigation, a personal cause of his).

19. [Hall], Review of *Narrative of "Adventure" and "Beagle,"* 489–91.

20. Ibid., 492–93; emphasis added.

21. [Broderip], Review of *Narrative of the . . . "Adventure" and "Beagle" . . .* and *Journal of Researches*, 228–29.

22. Ibid., 232–33.

23. Darwin to Lyell, [19 February 1840], *CCD*, 2:253–54.

24. For earlier uses of atoll as a geographically specific term, see Horsburgh and Owen, "Some Remarks relative to the Geography of the Maldiva Islands"; Moresby, "Extracts from Commander Moresby's Report on the Northern Atolls of the Maldivas." In his 1842 book, Darwin followed Owen in quoting the seventeenth-century account of François Pyrard de Laval on the "atollons" of the Maldivas. See Horsburgh and Owen, "Some Remarks relative to the Geography of the Maldiva Islands," 88; Darwin, *Structure and Distribution of Coral Reefs*, 2.

25. Darwin to Lyell, [19 February 1840], *CCD*, 2:253–54.

26. Darwin to W. D. Fox, 24 October [1839], *CCD*, 2:234–35.

27. Darwin to Caroline Wedgwood, [27 October 1839], *CCD*, 2:235–37.

28. Darwin to Lyell, [19 February 1840], *CCD*, 2:253–54.

29. See Darwin's journal entry of 28 May 1841, *CCD*, 2:434.

30. Lyell, *Principles of Geology* (1840), 1:xii.

31. Ibid., 1:309–12.

32. Compare 1:315 with Darwin's "Connexion" paper, and 1:314 with Lyell to Darwin, 26 December 1836, *CCD*, 1:532–33.

33. Lyell, *Principles of Geology* (1840), 3:379.

34. Lyell also added the term atolls to his glossary for this edition of the *Principles*, defining them as "Coral islands of an annular form, or consisting of a circular strip or ring of coral surrounding a central lagoon." (Neither "atoll" nor "lagoon island" appeared in the glossary to the first edition.) Ibid., 3:409.

35. Ibid., 3:385. In the first edition Lyell had used "atoll" only in its geographically limited sense applying to the Maldives. See Lyell, *Principles of Geology* (1830), 2:286.

36. Lyell, *Principles of Geology* (1840), 3:385–90. The extracted sections may

be found (in the order presented in the sixth edition) in Lyell, *Principles of Geology* (1840), 2:296, 2:293, and 2:293–94.

37. Lyell, *Principles of Geology* (1840), 3:388.

38. For a similar example of an anachronistic, theory-laden self-reading, see my discussion of Darwin's account of the genesis of the coral reef theory in chapter 11 below. A good discussion of this phenomenon appearing in a similar context may be found in the closing chapters of Secord, *Controversy in Victorian Geology*.

39. Lyell, *Principles of Geology* (1840), 3:391.

40. In his March 1842 lecture on coral reefs at the Broadway Tabernacle in New York, Lyell offered a more graceful (if slightly self-contradictory) justification for pointing out his former comments: that they provided independent support for Darwin's theory. "I may as well mention," he explained, "that [the] theory of subsidence was not invented for the purpose of explaining these phenomena [the shape of coral islands]. Long before Darwin had made his examinations of these coral islands . . . I published my opinion upon this point, that the sinking down of the Pacific might be in excess: that its depression might be greater than its upheaval. . . . The theory [of Pacific subsidence], then, was not made for the purpose of fitting the facts [of reef shapes]—though it is a perfectly legitimate reason for adopting a theory that you find it will explain all the known phenomena which no other theory will explain.—Still, it is somewhat more satisfactory if the principle was not formed expressly to suit the facts of the case." [Lyell], "Mr. Lyell's Fourth Lecture on Geology." For a discussion of public science in New York in the first half of the nineteenth century, see Burnett, *Trying Leviathan*.

41. Lyell, *Principles of Geology* (1840), 3:391–92.

42. Ibid., 3:392n.

43. Ibid., 3:392–95.

44. Ibid., 3:398.

45. Ibid., 3:400.

46. Ibid., 3:384; CUL, DAR LIB.

47. Browne, *Charles Darwin: The Power of Place*, 219.

48. See entries in Darwin's journal, quoted in *CCD*, 2:434.

49. See Darwin to T. F. Jamieson, 6 September [1861], "Your arguments [on Glen Roy] seem to me conclusive. I give up the ghost. My paper is one long gigantic blunder." *CCD*, 9:255–56.

50. Darwin to Lyell, [14] September [1838], *CCD*, 2:104–8.

51. Secord, *Evolutionary Writings*, 399.

52. Undated diary entry immediately after 2 March 1841, *CCD*, 2:434.

53. Charles Lyell to J. S. Henslow, 10 April 1841, APS Scientists Collection, 509. L56.

54. My caution here is guided by Ian Hacking's argument that the naming of human behavioral disorders does more than just describe previously existing states, it also provides ways for individuals to feel. See Hacking, "Looping Effect of Human Kinds"; Hacking, "Making Up People".

55. Leader, *Writer's Block*, 1–3, 8.

56. Ibid., 252.

Chapter Ten

1. For an overview of Lyell's American lectures, see Dott, "Charles Lyell in America." On Lyell's travels in the United States and his life in the 1840s, see Wilson, *Lyell in America.*

2. Like the *Elements* and the *Principles*, this lecture was to discuss fossil and living coral reefs. In a field trip to Wales in June 1841, the month before his scheduled departure, Lyell saw Silurian beds containing fossil corals in their position of growth. He believed these formations, much older than the Chalk, had also been formed by subsidence similar to what Darwin had demonstrated. Darwin responded with constructive criticism for this view in his letter to Lyell of 6 [July 1841], *CCD*, 2:297–99. See also Wilson, *Charles Lyell: The Years to 1841*, 516.

3. Lyell informed Darwin that his clerk would be returning the "charts on Coral reefs you kindly lent me" in a letter written on the eve of his departure for America, ca. 16 July 1841. *CCD*, 2:299–300.

4. All the lectures were to be illustrated with large views and sectional diagrams, like the lectures he delivered during his short tenure in the geology chair at King's College. Some of the pictures used in the 1841–42 lectures were painted in the United States to illustrate examples from Lyell's recent fieldwork there. On the content of Lyell's lectures, see Rudwick, "Charles Lyell and His London Lectures"; Rudwick, "Charles Lyell Speaks." On Lyell's teaching illustrations see Dott, "Charles Lyell in America," 105–6; Rudwick, *Worlds before Adam*, 362–64. On geological illustrations of the period, see Rudwick, "Emergence of a Visual Language for Geological Science."

5. Darwin to Lyell, 6 [July 1841], *CCD*, 2:297–99.

6. Ibid.

7. For Darwin's former adulation at its most colorful, see his letter to Lyell of [14] September [1838]. On Lyell's stated "hope" that the *Principles* would stand the test of time, for example, Darwin responded, "Begin to hope,; why the possibility of a doubt has never crossed my mind for many a long day: this may be very unphilosophical, but my geological salvation is staked on it." *CCD*, 2:104–8.

8. Lyell to Benjamin Silliman, 6 February 1842, Sir Charles Lyell Papers, APS B L981, box 1.

9. William Lonsdale to Charles Lyell, [2?] December 1841, EUL Lyell correspondence GEN 113 Lyell 1/3502–3526 on fol. 3506.

10. Darwin to Leonard Jenyns, [9 May 1842], *CCD*, 2:319–20.

11. Darwin to Emma Darwin, [9 May 1842], *CCD*, 2:318–19. Darwin also bemoaned the expense of printing the volume, which contained a large colored map (see below). Given its estimated cost of £130 to £140, he feared that they would have to take £200 or £300 out of the family budget to supplement the printing of his other geological volumes.

12. Because of its closely set type, the appendix is even longer, compared with the rest of the book, than it appears when consulting the table of contents. I performed an electronic word count of the 1842 London first edition using the online text at www.darwin-online.org.uk. The appendix is about 25,000 words, while the

main text of the book, from the beginning of the introduction to the end of chapter 6, is roughly 54,000 words. Chapter 6 is 12,000 words, meaning the total proportion of the book devoted to discussion of the map is about 47 percent: 37,000 of 79,000 words.

13. The individual reefs were (in Darwin's spellings) North and South Keeling, Vliegen, Bow, Clermont Tonnere, Rimsky Korsacoff, Menchicoff, Caroline, Oulleay, Namonouito, Peros Banhos, Great Chagos Bank, Diego Garcia, Elizabeth, Egmont, Cardoo, Ducie's, the Australian barrier, the Red Sea reefs, Bermuda, Gambier, Vanikoro, Matilda, Whitsunday, Romanzoff, Milla dou-Madou, Suadiva, Mahlos Madou, North and South Nillandou, Male, Powell's, Horsburgh, and Heawandoo Pholo. The groups described collectively (of which some of the above are part) were the Low Archipelago, the Marshall Islands, the Carolines, the Maldives, and the Chagos group. The sources named in chapter 1 were Liesk (the resident at South Keeling), Beechey, Kotzebue, Chamisso, Nelson, Lütke, F. Bennett, Freycinet, Cook, Powell, Moresby, Prentice, Flinders, and Ehrenberg. He also mentioned the Cape Verde Islands and cited Lyell on the likelihood of reef channels' becoming obstructed. Darwin, *Structure and Distribution of Coral Reefs*, chap. 1.

14. Ibid., 60–63. In the text and a long footnote (p. 61) Darwin described his belief that cold-water currents determined the absence of reefs at the Galápagos, but he explained that this view could not account for the lack of coral reefs in many other districts.

15. Ibid., 65.

16. Ibid., 67.

17. Ibid., 76.

18. Ibid., 69.

19. Ibid., 67–71.

20. "It may be concluded, first, that considerable thicknesses of rock have certainly been formed within the present geological æra by the growth of coral and the accumulation of its detritus; and, secondly, that the increase of individual corals and of reefs, both outwards or horizontally and upwards or vertically, under the peculiar conditions favourable to such increase, is not slow, when referred either to the standard of the average oscillations of level in the earth's crust, or to the more precise but less important one of a cycle of years." Ibid., 79.

21. Couthouy, "Remarks upon Coral Formations."

22. Darwin, *Structure and Distribution of Coral Reefs*, 82.

23. Ibid.

24. Ibid., 89.

25. Ibid., 90.

26. Ibid., 92–94.

27. Ibid., 94.

28. Darwin, *Journal of Researches*, 558.

29. Darwin, *Structure and Distribution of Coral Reefs*, 98, 100.

30. Ibid., 102.

31. Ibid., 95–98.

32. Ibid., 103–13; quotations from 109 and 114.

33. Ibid., 114; emphasis added.

34. The map itself was bound into the front of the book.

35. The research map at Down House is English Heritage provenance number 88202837.

36. Darwin, *Structure and Distribution of Coral Reefs*, 124.

37. Ibid.

38. Camerini follows Michael Ghiselin in arguing that "his geological work evinces the same formal properties that characterize his later research," and she argues that one of the primary roles of maps in Darwin's thinking was to help him "make the transition from geology to biogeography." Thus she argues that the distributional and cartographical aspects of the coral reef theory "provided both practical and cognitive frameworks for his work on biogeography, which in turn formed the basis of his first theory of speciation." One important weakness of this argument is that it glosses over the simultaneity of these developments by ignoring the fact that Darwin worked on the coral reef book and created its map after he had begun his species project, although of course his (pre-species) 1835 coral islands essay contained prose embodying the kind of geographical perspective she includes under her category of visual thinking. Camerini, "Darwin, Wallace, and Maps," 105.

39. In 1842, the same year *Coral Reefs* was published, he commented on the value of maps in his referee's report on a manuscript for the Royal Geographical Society, noting, "The geographical details are well given, but might be greatly condensed, as inspection of the map gives nearly the same information." Charles Darwin, "Extract of a report on the Falkland Islands by Lieutenant Moody," RGS JMS/6/33 (1842).

40. Camerini was reacting to a prepublication version of Stoddart, "Darwin and the Seeing Eye."

41. Camerini, "Darwin, Wallace, and Maps"; see especially 102–3 and 194–95.

42. Ibid., 88, 96–97, 101.

43. Darwin, *Structure and Distribution of Coral Reefs*, 1.

44. Ibid., 41.

45. Darwin, *Structure and Distribution of Coral Reefs*, 57–58; emphasis added.

46. Ibid., 202; emphasis added. The reefs were Turneffe, Lighthouse, and Glover. Darwin made the same point about offshore reefs in the Red Sea (195), "which approach in structure to the true atolls of the Indian and Pacific Oceans; but they present only imperfect miniature likenesses of them."

47. Emphasis added. I believe the double negative in this sentence was unintended. Ibid., 89.

48. Emphasis added. Ibid. Note that the theory attributed to Aldelbert von Chamisso appeared in a piece that had in fact been written by his shipmate, J. F. Eschscholtz. See [Eschscholtz], "On the Coral Islands"; Sluiter, "Eine geschichtliche Berichtigung"; Sponsel, "Coral Reef Formation," chap. 1.

49. Darwin, *Structure and Distribution of Coral Reefs*, 122.

50. Ibid., 145.

51. Ibid.; emphasis added.

52. Ibid., 146.

53. Ibid., 147–48.

54. Ibid., 148; emphasis added.

55. James Keir, as quoted in Golinski, *Science as Public Culture*, 149.

56. I suspect that Darwin changed his reef classification in part for conscious strategic reasons, but also partly because the terminology and taxonomy simply were faithful to his frame of mind *at the time he was finishing the book*. Such was the power of a useful theory, that it could reshape memory to make a Darwin feel that he had always thought this way. I will make a similar argument in greater detail at the end of the book when I discuss Darwin's autobiographical "Recollections."

57. Darwin to Smith, Elder & Co., [17 May 1842]. *CCD*, 24:32–33.

58. Darwin's list of presentation copies is in DAR 69:A108.

59. A. B. Becher to Darwin, 1 September 1842, UKHO LB10 1841–42, p. 510. Curiously, no mention of Darwin's coral reef theory was given in the immediate post-1842 instructions to surveyors that I cited in chapter 1.

60. Darwin to Anne Susan Horner, [4 October 1842], *CCD*, 2:335–36.

61. Jackson, Review of *The Structure and Distribution of Coral Reefs . . .* by Charles Darwin, 120. It seems that Jackson was initially undecided whether to place the review in the journal of the Royal Geographical Society or to submit it to a popular periodical. Darwin saw more "dignity" in the journal but believed a magazine might publish the review more quickly. See Darwin to Julian Jackson, 13 October [1842], *CCD*, 2:338–39.

62. Driver, *Geography Militant*, 51–53.

63. Jackson, Review of *The Structure and Distribution of Coral Reefs . . .* by Charles Darwin, 115.

64. Maclaren, "On Coral Islands and Reefs," 39. This essay was an abridgment of a two-part review that appeared in the *Scotsman*, the periodical Maclaren edited, on 29 October and 9 November 1842. The full text from the *Scotsman*, along with the woodcuts that appeared as illustrations in both versions mentioned above, appeared as a separate pamphlet, an example of which may be seen at the NHM Earth Sciences Library, Murray Collection, section 15, 19.

65. Ibid., 40.

66. Ibid., 42.

67. Jackson, Review of *The Structure and Distribution of Coral Reefs . . .* by Charles Darwin, 118. Darwin interpreted the red-colored areas of the map as regions where the form of reefs suggested an absence of subsidence (as opposed to the presence of elevation), although he had pointed out that in most of these locations there was independent evidence of elevation in the form of upraised shells and corals.

68. Maclaren, "On Coral Islands and Reefs," 46.

69. Ibid.

70. Ibid., 47.

71. Darwin to Lyell, [September-December 1842], *CCD*, 2:328–30. On Lyell's having mentioned Maclaren's criticisms to Darwin, see Darwin, "Remarks on the Preceding Paper." Lonsdale commented to Lyell on Maclaren's objections in a letter of March 1843, EUL GEN 113 Lyell 1/3513–3514.

72. Ibid., 50; emphasis added.

73. Lyell to J. F. W. Herschel, 24 and 26 May 1837, RS HS 11.422 and 11.450; emphasis added.

74. Darwin, "Remarks on the Preceding Paper," 50.

75. On the arrangements see Darwin to Charles Lyell, [8 February 1845], and Darwin to John Murray, 17 March [1845], *CCD*, 3:136 and 3:158.

76. Darwin, *Journal of Researches* (1845), iii. Notwithstanding the dedication to Lyell, one of the most noteworthy additions to the revised edition of the journal was a fiery denunciation of slavery that appears to have been inspired at least in part by Lyell's impassive discussion of that subject in his recently published book (also published by John Murray) about the Lyells' 1841–42 geological tour of the United States. Lyell, *Travels in North America*; Desmond and Moore, *Darwin's Sacred Cause*, chap. 7.

77. Darwin, *Journal of Researches* (1845), v.

78. Camerini, "Darwin, Wallace, and Maps," 89.

79. Darwin to Hooker, 11 March [1844], and annotations on Hooker to Darwin, 28 October 1844; see also Darwin to Hooker, [6 March 1844], *CCD*, 3:16–17, 19–20, 68–73.

80. See, e.g., Darwin to Lyell, 5 July 1856 and [3 November 1869], *CCD*, 6:167–70. In 1856 this was the topic of an impassioned set of exchanges between Darwin, Lyell, and Hooker that ran concurrently with their discussions over when, and in what form, Darwin should publish his species theory (on which see chapter 11 below).

81. Wilson, *Sir Charles Lyell's Scientific Journals on the Species Question*, 61, 135. Lyell quoted Sedgwick in a letter to Darwin, 17 June 1856, *CCD*, 6:144–46.

82. Lyell to Darwin, 8 September 1860, *CCD*, 8:348–49.

83. Darwin to Hooker, 19 July 1856, *CCD*, 6:190–91.

84. Darwin to Lyell, [3 November 1869], *CCD*, 6:167–70.

85. Hooker to Darwin, 11 August 1881, Huxley, *Life and Letters of Sir Joseph Dalton Hooker*, 2:225.

86. Darwin to Ernst Dieffenbach, 4 July [1843], *CCD*, 7:476–78. Dieffenbach mentioned *Coral Reefs* and included two of the book's woodcuts in his translator's remarks at the end of the second volume. Darwin, *Naturwissenschaftliche Reisen*, 2:300.

87. Darwin to C. H. Smith, 26 January 1845, *CCD*, 3:131–32.

88. The next uses of the term theorist in Darwin's extant correspondence were 28 February 1858 (to Hooker) and, notably, 22 November 1860 (to Henry Walter Bates), in which he wrote, "I have an old belief that a good observer really means a good theorist." He did not use the term in any of his transmutation notebooks, in either edition of the *Journal of Researches*, in *Coral Reefs*, or in *Volcanic Islands*.

89. Darwin to H. T. De la Beche, 7 April [1848], *CCD*, 4:129–31.

90. Darwin to J. B. Jukes, 8 October [1847], *CCD*, 4:87–88. The Jukes quotation comes from Jukes, *Narrative of the Surveying Voyage of H.M.S. "Fly,"* 1:347.

91. Darwin to Charles Lyell, 4 December 1849, *CCD*, 4:284–85. On this exchange and on Dana's coral reef work more broadly see Sponsel, "Pacific Islands and

the Problem of Theorizing"; Stoddart, "'This Coral Episode.'" On Dana's career see Igler, "On Coral Reefs, Volcanoes, Gods, and Patriotic Geology"; Appleman, "James Dwight Dana and Pacific Geology"; Prendergast, "James Dwight Dana"; Gilman, *Life of James Dwight Dana.*

92. Darwin to Charles Lyell, [7? December 1849], *CCD*, 4:288–90.

93. For a valuable discussion of the distinction between utility and truth as possible aims of a theory in the earth sciences, see Oreskes, *Rejection of Continental Drift*, 316.

Chapter Eleven

1. This question was posed (in the restrictive sense described above) by John van Wyhe and answered in the negative. Although van Wyhe lays out ample evidence in support of that claim, (in my view) he adopts such a narrow definition of "delay" that he is defending uncontested ground. There is a broader sense in which Darwin chose, for a variety of reasons (one of which is the focus of this chapter), to continue working and thereby push off the moment of publication. Lyell captured this broader meaning when, after reading proofs of the *Origin*, he wrote to Darwin, "I have just finished your volume & right glad I am that I did my best with Hooker to persuade you to publish it without waiting for a time which probably could never have arrived tho' you lived till the age of 100, when you had prepared all your facts on which you ground so many grand generalizations." Lyell to Darwin, 30 September [1859], *CCD*, 7:338–39. On the "delay" question, see van Wyhe, "Mind the Gap," and the response in Ruse, "Origin of the *Origin*." The case that Darwin's evolutionary theory kept changing throughout the long period between 1837 and 1859 was laid out many years ago in Ospovat, *Development of Darwin's Theory.*

2. Charles Darwin to Emma Darwin, 5 July 1844, *CCD*, 3:43–45.

3. Most notably, Adrian Desmond interpreted Darwin's intent this way in the very last sentence of his 1989 book: "In 1844 he left his wife £400 with instructions to publish on the event of his death." Desmond, *Politics of Evolution*, 414.

4. Forbes, "Report on the Mollusca and Radiata of the Aegean Sea"; emphasis added. On the similarities between Darwin's work on submarine distribution and Forbes's later work, see Sponsel, "Amphibious Being."

5. Darwin also named the paleontologist Richard Owen in his letter, but at some point he crossed out the name. He also later emphasized Hooker's suitability for the role. Janet Browne notes that Darwin's list of potential editors included a combination of his mentors (Lyell and Henslow) and exceptionally competent young naturalists (Forbes and Hooker), but not anyone, such as Grant, noted for his openness to transmutation or general radicalism. Neither did he include a more conservative mentor, Sedgwick. Browne, *Charles Darwin: Voyaging*, 448–53.

6. Emphasis added.

7. Dov Ospovat has discussed Darwin's broad conception of what constituted "facts" in *Development of Darwin's Theory*, 95–97. These facts ranged from individual data to the broad generalizations that Darwin described elsewhere as "laws" or "classes of facts."

8. I draw my use of the term warrant from Stark, *Behind Closed Doors*, chap. 1.

9. The assumption that Darwin's publications *did* contain reliable accounts of his mode of thinking has underlain some scholarship by philosophers of science, most notably Michael Ghiselin in his *Triumph of the Darwinian Method.* Howard Gruber and Paul Barrett used evidence from Darwin's manuscripts to criticize Ghiselin's dependence on published texts and consequent "blur[ring of] the distinction between Darwin's early ideas about evolution and his ideas as expressed in the *Origin*." Gruber and Barrett, *Darwin on Man*, 131.

10. I deduce the contents of the letter from Horner, which to my knowledge does not survive, from Darwin's reply of 29 August [1844], *CCD*, 3:54–55.

11. Charles Darwin to Leonard Horner, 29 August [1844], *CCD*, 3:54–55; emphasis added.

12. Charles Darwin to Adolph von Morlot, 10 October [1844], *CCD*, 3:64–66; emphasis added.

13. Charles Darwin to C. H. L. Woodd, 4 March [1850], *CCD*, 4:316–17.

14. In their footnotes to the Woodd letter, the editors of *CCD* cite Glen Roy and craters of elevation as the topics he had speculated on, and they also refer to his 1840 letter to John Phillips (discussed in chapter 8 above). They do not mention coral reefs.

15. Lyell's journal is quoted in Wilson, *Sir Charles Lyell's Scientific Journals on the Species Question*, xliv.

16. Lyell, *Life, Letters and Journals of Sir Charles Lyell*, 2:211–13; Wilson, *Sir Charles Lyell's Scientific Journals on the Species Question*, xlvi.

17. Charles Lyell to Charles Darwin, 1–2 May 1856, *CCD*, 6:89–92. On Lyell's own travels to Madeira in 1853–54, see Wilson, "Geological Travels of Sir Charles Lyell in Madeira and the Canary Islands."

18. Charles Lyell to Charles Darwin, 1–2 May 1856, *CCD*, 6:89–92.

19. Charles Darwin to Charles Lyell, 3 May [1856], *CCD*, 6:99–101.

20. Lyell to Horner, 12 March 1838, quoted in Wilson, *Charles Lyell: The Years to 1841*, 456; emphasis added.

21. Charles Darwin to Charles Lyell, 3 May [1856], *CCD*, 6:99–101; emphasis added.

22. Ibid.

23. Herbert, "Place of Man," 173.

24. Charles Darwin to Joseph Dalton Hooker, 9 May [1856], *CCD*, 6:106–7. On Hooker's own ideals of scientific conduct see Bellon, "Joseph Dalton Hooker's Ideals for a Professional Man of Science"; Endersby, *Imperial Nature*, 1–29.

25. Charles Darwin to Joseph Dalton Hooker, 9 May [1856], *CCD*, 6:106–7.

26. Charles Darwin to Joseph Dalton Hooker, 11 May 1856, *CCD*, 6:108–10.

27. Ibid.; emphasis added.

28. George Howard Darwin was not yet engaged in the geophysics career for which he would become well known by the end of the century. He had already proved himself an adept mathematician by finishing second in the final honors examination at Cambridge known as the mathematical tripos, but in the early 1870s he was taking steps toward a career in law and letters.

29. Charles Darwin to George Howard Darwin, 21 October [1873], *CCD*, 21:460–62.

30. Charles Darwin to Anton Dohrn, 4 January 1870, *CCD*, 18:3–4.

31. Darwin's note is written on the letter from Anton Dohrn to Charles Darwin, 30 December 1869, *CCD*, 17:535–38.

32. Lyell's plan was recorded, as a draft of a letter to Darwin, in one of the notebooks that remains at the Lyell family's Kinnordy House. It was first published in Wilson, *Sir Charles Lyell's Scientific Journals on the Species Question*, xlviii–xlix. See also comments on the draft in its published form as Charles Lyell to Charles Darwin, [29? June 1856], *CCD*, 6:167–70.

33. Darwin's journal contains an entry on 14 June 1856 that reads, "Began by Lyells advice <u>writing</u> species sketch." By 13 October he had noted, "Finished [second] Chapt.(& before [finished] part of Geograph. Distr.)," indicating that he was already working on a large scale. These chapters and the successive ones he wrote in 1857–58 would have formed part of Darwin's intended large book (generally known by the title "Natural Selection") that was the work interrupted by the arrival of Wallace's letter (on which see below).

34. Charles Darwin to Charles Lyell, 5 July [1856], *CCD*, 6:167–70. The actual letter in which Lyell communicated this plan to Darwin has not been found, but this response by Darwin indicates that he did receive one.

35. Charles Darwin to Joseph Hooker, 12 July [1856], *CCD*, 6:178–79.

36. Charles Darwin to Charles Lyell, 18 [June 1858], *CCD*, 7:107–8.

37. Ibid.

38. Ibid.

39. Charles Darwin to Charles Lyell, [25 June 1858], *CCD*, 7:117–18.

40. Ibid.; emphasis added.

41. Charles Darwin to Charles Lyell, 26 [June 1858], *CCD*, 7:119.

42. Charles Darwin to J. D. Hooker, [29 June 1858], *CCD*, 7:121.

43. Charles Darwin to J. D. Hooker, [29 June 1858], *CCD*, 7:121. Note that this was the second letter Darwin wrote to Hooker on this day. The earlier letter was headed "Tuesday," and this letter was headed "Tuesday Night."

44. Charles Lyell and Joseph Hooker to J. J. Bennett, secretary of the Linnean Society, 30 June 1858, *CCD*, 7:122–24.

45. Ibid.

46. This is not to say that the act of putting the theory into print was necessarily more momentous than having it read before the Linnean Society. Print publication had not yet altogether replaced formal spoken communication as the medium of record for scientific claims (and particularly for establishing priority). See Secord, "How Scientific Conversation Became Shop Talk"; Csiszar, "Objectivities in Print."

47. Charles Lyell to Charles Darwin, 3 October 1859. *CCD*, 7:339–42.

48. Darwin to W. D. Fox, 7 July [1837], *CCD*, 2:29–30.

49. Alfred Russel Wallace to Mary Ann Wallace, 6 October 1858, and Alfred Russel Wallace to George Silk, November 1858, quoted in Berry, *Infinite Tropics*, 64.

50. "Proceedings at the Annual General Meeting, 18 February 1859," xxiii–xxiv.

51. Charles Darwin to Charles Lyell, 28 March [1859], *CCD*, 7:169–71.

52. Charles Darwin to Alfred Russel Wallace, 6 April 1859, *CCD*, 7:279–80.

53. Ibid.

54. Charles Darwin to Adolph von Morlot, 10 October [1844], *CCD*, 3:64–66. Charles Lyell to Leonard Horner, 12 March 1838, Lyell, *Life, Letters and Journals of Sir Charles Lyell*, 2:39–41.

55. Charles Darwin to Charles Lyell, 28 March [1859], *CCD*, 7:169–71.

56. Charles Darwin to Charles Lyell, 30 March [1859], *CCD*, 7:272–73.

57. Charles Darwin to Charles Lyell, 2 September [1859], *CCD*, 7:328–330; emphasis added.

58. Ibid.

59. Lyell, "On the Occurrence of Works of Human Art in Post-Pliocene Deposits," 95.

60. Charles Darwin to Charles Lyell, 20 September [1859], *CCD*, 7:333–35.

61. Ibid.

62. Darwin, *On the Origin of Species*, 310. For evidence that Lyell had coordinated the wording of this passage with Darwin, see the letter from Darwin to Lyell, 20 September [1859], in which Darwin wrote, "Thank you much for allowing me to put in the sentence about your 'grave doubt.'" 7:333–35.

63. Charles Darwin to Charles Lyell, 20 September [1859], *CCD*, 7:333–35.

64. Charles Darwin to Charles Lyell, 25 September [1859], *CCD*, 7:336–37.

65. Charles Darwin to Charles Lyell, 30 September [1859], *CCD*, 7:338–39.

66. The original quotation was from Whewell, *Astronomy and General Physics Considered with Reference to Natural Theology*, 356.

67. Darwin, *On the Origin of Species*, 488. When Darwin received a letter expressing the same sentiment from the author and Anglican divine Charles Kingsley, he inserted an unattributed quotation from the letter into the revised text of the *Origin*.

68. The naturalist Hewett Cottrell Watson pointed out to Darwin that first-person pronouns appeared forty-three times in the opening four paragraphs of the *Origin*. See Darwin to Joseph Hooker, 27 [March 1861], *CCD*, 9:70–71.

69. Darwin, *On the Origin of Species*, 1.

70. Charles Darwin to Thomas Henry Huxley, 15 October [1859], *CCD*, 7:350–51; emphasis in the original.

71. Darwin, *On the Origin of Species*, 1; emphasis added.

72. Janet Browne has noted that this may have been the first announcement of Darwin's ill health to the broader public. Browne, "I Could Have Retched All Night," 241–42.

73. Darwin, *On the Origin of Species*, 2.

74. Grafton, *Footnote*, 1–33.

75. Charles Darwin to Adolph von Morlot, 10 October [1844]. *CCD*, 3:64–66.

76. Darwin, *On the Origin of Species*, 2; emphasis added.

77. Ibid. Throughout the *Origin* Darwin reiterated his pledge to remedy shortcomings of the text in unspecified "future work." He made such promises at least eight times in the first edition. On the topic of variation between individual

organisms of the same species (within natural populations), for example, he wrote, "To treat this subject at all properly, a long catalogue of dry facts should be given; but these I shall reserve for my future work." This promise remained present with substantially the same wording through four further editions until, in the heavily revised sixth edition of 1872, Darwin changed the specific phrase "my future work" to the even vaguer "a future work." He made the same promise, and later the same change from "my" to "a," about facts relating to botanical varieties (p. 53 of the first edition), checks on the increase in animal populations (p. 67), and secondary sexual characteristics (p. 151). On the topic of the struggle for existence he promised to expand at "much greater length;" that is, until the 1869 edition, in which he promised simply "greater length." In none of these places did he subsequently introduce references to any of his other post-1859 publications, such as *The Variation of Animals and Plants under Domestication* (1868) or *The Descent of Man* (1871), although that body of work was considerable by 1871–72 when he revised the *Origin* for the final time. And in only one of those places (a discussion of the difficulty of explaining the evolution of complex organs) did he replace the promise with an expanded discussion within the revised text of the *Origin* itself (compare p. 192 of the first edition with pp. 149–50 of the sixth edition). Ibid., 2, 44, 53, 62, 67, 151, 192, 216. My discussion of changes made to subsequent volumes draws from variorum editions of the *Origin*: the online version at http://darwin-online.org.uk/Variorum/ and Peckham, *Origin of Species: A Variorum Text*. On the variety of ways Darwin worked to extend and reinforce his species theory after 1859 see Browne, *Charles Darwin: The Power of Place*.

78. In an important recent article, Richard Bellon argues, in effect, that these efforts at self-fashioning did not work. He points out that many of the *Origin*'s early critics condemned Darwin for (in Bellon's words) "betray[ing] the moral virtues of scientific labor" by basing the book so much on original investigations by others. The *Origin* was branded by many as mere speculation, Bellon argues, until Darwin illustrated his evolutionary theory's value by applying it to his own research problems, notably in his 1862 book on orchids. Bellon, "Inspiration in the Harness of Daily Labor," quotation on p. 395.

79. Good places to begin reviewing the large existing literature on the *Origin* and its reception are Ellegård, *Darwin and the General Reader*; Glick, *Comparative Reception of Darwinism*; Ruse and Richards, *Cambridge Companion to the "Origin of Species."*

80. For an analysis of how Darwin's authorship of the *Origin* was evaluated in comparison with the anonymous 1844 book *Vestiges of the Natural History of Creation*, which advanced a progressionist developmental theory of physical and living matter, see Secord, *Victorian Sensation*, 511–18.

81. [Richard William Church], Review of *On the Origin of Species* by Charles Darwin.

82. Texts of many of the published reviews of Darwin's work are available (along with all of Darwin's own publications) at van Wyhe, *Complete Work of Charles Darwin Online*.

83. My account of the Copley Medal episode draws chiefly on Bartholomew, "Award of the Copley Medal to Charles Darwin"; Burkhardt, "Darwin and the Copley Medal."

84. Mill, *System of Logic* (1862), 18. On Carpenter's use of the quotation see Erasmus Alvey Darwin to Charles Darwin, 9 November [1863], *CCD*, 11:662. In the footnote Mill argued that the *Origin* had demonstrated that natural selection was a *vera causa* that in principle *could* be proved. "The rules of Induction are concerned with the conditions of Proof. Mr. Darwin has never pretended that his doctrine was proved. He was not bound by the rules of Induction but by those of Hypothesis. And these last have seldom been more completely fulfilled. . . . And is it not a wonderful feat of scientific knowledge and ingenuity to have rendered so bold a suggestion, which the first impulse of everyone was to reject at once, admissible and discussable, even as a conjecture?" Mill was later to express dismay that, in his view, little progress had been made in moving past the hypothesis of natural selection. See Hull, *Darwin and His Critics*, 27; Hull, "Darwin's Science and Victorian Philosophy of Science." On the disparity between Whewell's and Mill's philosophical assessments of natural selection, see Snyder, *Reforming Philosophy*, chap. 3.

85. Adam Sedgwick to Charles Darwin, 24 November 1859, *CCD*, 7:396–98; [Sedgwick], "Objections to Mr. Darwin's Theory."

86. Significantly, Sabine listed "theoretical or speculative geology" as a separate endeavor of Darwin's, specifying under that heading several papers (but no books) including the "Connexion" paper and the Glen Roy essay.

87. On the alternative theory proposed by John Murray of the *Challenger*, and on the contentious subsequent history of Darwin's coral reef theory, see Sponsel, "Coral Reef Formation"; Deacon, *Scientists and the Sea*.

88. Darwin to Smith, Elder & Co., 17 December [1873], DAR 96:159–60. See also Charles Darwin to Horace Darwin, 9 January [1874], DAR 185. Amusingly, Darwin himself then faced the problem of finding a copy of his first edition. He wrote to Hooker to ask, "Did I give you a copy of my Coral reefs book? If so I wish you w[ould] give it me, & I will let you have a copy of a new edition; for I cannot buy one, & yet want a copy very much for correction." Within ten days he had two copies in hand, courtesy of Hooker and Lyell. Darwin to Hooker, 8 January 1874 and 18 January [1874], DAR 95:311–12 and 95:313–16. He decided to use Lyell's copy, which had arrived first, and immediately returned Hooker's copy.

89. Darwin, *Structure and Distribution of Coral Reefs* (1874), v. He referred to Dana, *Corals and Coral Islands*, which was a self-contained treatise drawing on Dana's earlier publications. If Dana's extraordinary researches into corals and reefs constituted the standard for new "important work" on the topic, then it certainly was in that class by itself. The bulk of the changes Darwin incorporated into the 1874 edition of *Coral Reefs* were responses to the work of Dana and a few others. Some of the new parts were silent revisions, but many were annotations clearly meant to preserve on record the exact wording of the original edition.

90. Darwin, *Evolutionary Writings*, 398.

91. Darwin's "Recollections" have been published in various forms since 1887 and are often referred to as his "autobiography." Throughout these notes I cite the

unabridged text found in Darwin, *Evolutionary Writings*, 355–425. The most widely cited edition is Barlow, *Autobiography of Charles Darwin*, which has been superseded by the edition I cite.

92. Secord, "Introduction," xxvii.

93. Critiques of specific claims made in the "Recollections" may be found in Gruber and Barrett, *Darwin on Man*, 172–73; Rudwick, "Darwin and Glen Roy," 101; Sulloway, "Darwin's Conversion," 327; Pearson and Nicholas, "'Marks of Extreme Violence.'"

94. Darwin, *Evolutionary Writings*, 371.

95. Ibid., 367.

96. Ibid., 422.

97. Ibid., 391.

98. Ibid., 400–401. Of Murchison Darwin wrote, "The services rendered to geology by Murchison by his classification of the older formations cannot be overestimated; but he was very far from possessing a philosophical mind."

99. Ibid., 398–99.

100. For an extended defense of this claim, see Sponsel, "Coral Reef Formation," chap. 2.

101. Sulloway, "Darwin and His Finches," 32; emphasis added.

102. Ibid., 47.

103. Darwin, *Evolutionary Writings*, 410; emphasis added.

104. Barrett et al., *Charles Darwin's Notebooks*. See also my discussion in chapter 7 above. Paul Barrett and Howard Gruber were among the first scholars to use evidence from Darwin's manuscripts to undermine claims he made in his recollections. See Gruber and Barrett, *Darwin on Man*, 172–73.

105. Charles Darwin to Leonard Horner, 29 August [1844], *CCD*, 3:54–55.

106. Ibid.; emphasis added.

107. Porter, *Making of Geology*.

108. Herbert, "Place of Man," 163–64.

Conclusion

1. I placed "reception" in scare quotes because it falsely implies transmission without alteration to a passive recipient. See Warwick, "Cambridge Mathematics and Cavendish Physics, Part I," and "Cambridge Mathematics and Cavendish Physics, Part II."

2. William Whewell to John Herschel, 4 December 1836, in Todhunter, "William Whewell," 250.

3. Among the most thought-provoking discussions of the attribution of credit, in science and beyond, is Michel Foucault's lecture "What Is an Author?" Although Foucault's stated objective was to set aside "analysis of the author as an individual [person]" and instead to analyze the function served for a community by the notion of "an author" (to indicate the common creator of several different works, for example), much of his discussion focused on the opposite theme: the reasons Karl Marx and Sigmund Freud should be credited for authoring entire discourses.

Throughout his discussion of the "author function" Foucault posited a distinction between the natural sciences and other domains including the human sciences. He argued that modern literary works, for example, require the name of an author to underpin their authenticity, while scientific works ostensibly derive their authority from fidelity to scientific methods and to nature itself. The guarantor of a scientific claim in the modern episteme is not "the individual who produced [it]" but the fact that a true scientific statement is "always redemonstrable" through a fresh investigation of nature. Finding a previously unknown text by Newton could modify our understanding of the history of science, but it could not make us reassess the authoritativeness of present-day physics. By contrast, finding a new text by Marx could potentially change the very interpretation of historical materialism. In the (modern) natural sciences, Foucault claimed, "Authentification no longer require[s] reference to the individual who had produced [a scientific text]; the role of the author [has] disappeared as an index of truthfulness."

It is important to note that I take Foucault, in explicating this notion that nature itself is the ultimate author of verified scientific claims, to have been describing an *idealized* notion of how modern science works. Of course, many science-studies scholars—and many practicing scientists—have pointed out that it in fact matters a great deal *who* makes scientific claims, and when, how, and to whom they do so. As Foucault himself observed, "the name of a biologist or the particular laboratory in which research was performed could serve to indicate [a work's] trustworthiness." Foucault, "Quest-ce qu'un auteur," translations my own drawing upon those in Foucault, "What Is an Author?" ed. Harari, and Foucault, "What Is an Author," ed. Bouchard. See also Chartier, "Foucault's Chiasmus"; Iliffe, "Butter for Parsnips."

Issues related to authorship in this sense have also been discussed in terms of "credit" and "credibility," notably by Steven Shapin and Mario Biagioli. See Shapin, "House of Experiment"; Shapin, *Social History of Truth*; Biagioli, *Galileo, Courtier*; Biagioli, *Galileo's Instruments of Credit*; Biagioli, "Rights or Rewards." I have had their formulations, and Biagioli's discussions of Galileo's "self-fashioning," in mind throughout.

4. I refer here to the acknowledged gradients of authority within British geology (discussed in chapters 7 and 8 above) and do not personally consider the epistemic status of observations and specimens to have been uncomplicated, or the identity of a collector to have had less importance for the value of a specimen than the identity of a theorist had for a theory. For a thoughtful discussion on this head see Rieppel, "Albert Koch's *Hydrarchos* Craze."

5. Robert FitzRoy to Darwin, 16 November 1837. *CCD*, 2:57–59.

6. For an analysis of FitzRoy's chapter see Herbert, *Charles Darwin, Geologist*, 190–92. On Lyell's disdain for "scriptural geologists," see Charles Lyell to John Murray II, 7 August 1827, and Charles Lyell to John Murray III, October 1832 (folios 83–84), "Letters to John Murray, publishers, of Charles Lyell," MS.40726, NLS.

7. Nor was Darwin listed as a formal contributor to any of Lyell's publications. FitzRoy and Darwin did cosign one publication: a letter "containing remarks on the moral state of Tahiti, New Zealand, &c" that appeared in the *South African Christian Recorder* 2 (September 1836): 221–38.

8. Shapin, "Invisible Technician," 560.

9. An extreme example of tension between captain and naturalist was that between Charles Wilkes and Joseph Pitty Couthouy of the United States Exploring Expedition (to the Pacific, 1838–42). Wilkes dismissed Couthouy outright while the squadron was in Honolulu. On Couthouy's complaints, see Sponsel, "Pacific Islands and the Problem of Theorizing."

10. Merton, "Matthew Effect in Science," quotation on 3.

11. Corynne McSherry and Peter Galison, among others, have written about the diversity of contributions that have been rewarded with formal credit as a co-author. In the case of work carried out primarily by students, but with supervisors or other senior colleagues named as coauthors, McSherry details various types of contributions (such as financial, institutional, and technical support) that can result in coauthorship for individuals who contributed nothing directly to the words written in a paper or even to the research activities it describes. McSherry, "Uncommon Controversies"; Galison, "Collective Author."

The question of how esteem is (and, for some, how it *should* be) divided among coauthors of different ranks animates latter-day scientometric research; a special case is posed by the problem of assessing the accomplishments of PhD students whose dissertations are based on published papers that were themselves coauthored. (See, for example, Hagen, "Deconstructing Doctoral Dissertations." Hagen is concerned with the phenomenon of nonmonographic PhD theses, that is, those that consist of a few published papers, and with whether the effort of producing a PhD of this style matches that historically required to produce a single-author monographic PhD thesis.) Some such research assumes a zero-sum model of intellectual contribution, in which each publication merits exactly one unit of credit that must be allocated among all contributors. In these cases, as in some historical discussions of priority disputes, analysts implicitly or explicitly envision credit as a fixed resource competed for in zero-sum fashion. In the example of Lyell and Darwin's mutual benefit from the latter's ascent in the scientific community we have an illustration that scientific credit does not necessarily have to be, and indeed was not, a zero-sum game.

12. Morrell, "Chemist Breeders"; Geison, *Michael Foster and the Cambridge School of Physiology*; Geison and Holmes, "Research Schools"; Fruton, *Contrasts in Scientific Style*.

13. Charles Darwin to Charles Lyell, 1 September [1860], *CCD*, 8:339–41.

14. Merton, "Matthew Effect in Science," quotation on 3.

15. In addition to works already cited in the introduction, I am thinking here of studies such as Pickering, *Constructing Quarks*; Lynch, *Art and Artifact in Laboratory Science*; Pinch, *Confronting Nature*; Galison, *How Experiments End*; Traweek, *Beamtimes and Lifetimes*; and the individual chapters in Gooding, Pinch, and Schaffer, *Uses of Experiment* and Pickering, *Science as Practice and Culture*.

16. Pickering, *Science as Practice and Culture*, 6–7.

17. Kohler, *Lords of the Fly*, 3. As Gerald Geison noted of Kohler's influential book, however, this view constitutes not an *absence* of epistemology but an epistemology of the material over the mental. Geison, "Review of Kohler," 331.

18. Endersby, *Imperial Nature*, 6–7.

19. Ibid., 312.

20. Suman Seth suggests that such an initiative, when pursued by historians of physics, should be termed a "new intellectual history" of physics. Seth, "History of Physics after the Cultural Turn," 113.

21. My thinking has been shaped by the authors of the following works: Galison and Warwick, "Introduction: Cultures of Theory"; Warwick, *Masters of Theory*; Kaiser, *Drawing Theories Apart*; Kaiser, *Pedagogy and the Practice of Science*; Seth, *Crafting the Quantum.*

22. To the extent that "theoretical" was synonymous with "mathematical," the Cambridge mathematician William Hopkins came the closest at this time to participating in British natural history exclusively as a theorist. Lyell and Darwin were both indebted to Hopkins's modeling of the mechanical effects of a subterraneous fluid on the overlying crust of the earth. See Herbert, *Charles Darwin, Geologist*, 210–15.

23. I draw this phrasing from Herbert, "Place of Man." Martin Rudwick argued in direct response to Herbert, "The concept of a theorist in the geology of the 1830s is still valuable, however, if it is used to denote a person who publicly attached relatively high-level theoretical inferences to his observational descriptions. In this sense all those who were tacitly acknowledged to belong to the elite . . . could be regarded as potential theorists." Rudwick, "Charles Darwin in London," 194–95.

24. Darwin to H. T. De la Beche, 7 April [1848], *CCD*, 4:129–31.

25. Sponsel, "Constructing a 'Revolution in Science.'"

Bibliography

Manuscript Collections Consulted

APS	American Philosophical Society, Philadelphia
BGS	British Geological Survey
BL	British Library
CUL	Cambridge University Library
DAR	Darwin manuscripts, CUL
DAR LIB	Darwin's personal library of books, CUL
DH	Down House collection, Downe, Kent
EUL	Edinburgh University Library
GSL	Geological Society of London
HRC	Harry Ransom Center, University of Texas
RS HS	Royal Society Archives, Herschel Papers
KHM	Kinnordy House manuscripts, Forfarshire, Scotland
LMS	London Missionary Society collection, SOAS, University of London
LUA	Lehigh University Archives
NAK	UK National Archives, Kew
NHM	Natural History Museum, London
NLS	National Library of Scotland

RGS   Royal Geographical Society, London

SLNSW   Mitchell Library, State Library of New South Wales

UKHO   UK Hydrographic Office, Taunton

Works Cited

Anthony, Patrick. "Underground Enlightenment: How German Mining Shaped 'Humboldtian Science.'" Paper delivered at History of Science Society annual meeting, Atlanta, 2016.

Appel, Toby A. "Jean-René-Constant Quoy." In *Dictionary of Scientific Biography*, edited by C. Gillispie. New York: Scribner's, 1974.

Appleman, Daniel E. "James Dwight Dana and Pacific Geology." In *Magnificent Voyagers*, edited by Herman J. Viola and Carolyn Margolis, 89–118. Washington, DC: Smithsonian Institution Press, 1985.

Armstrong, Patrick. "Charles Darwin's Geological Notes on Mauritius." *Indian Ocean Review* 1, no. 2 (1988): 1–20.

———. *Darwin's Other Islands*. New York: Continuum, 2004.

———. *Under the Blue Vault of Heaven: A Study of Charles Darwin's Sojourn in the Cocos (Keeling) Islands*. Nedlands, Western Australia: Indian Ocean Centre for Peace Studies, 1991.

Barford, Megan. "Fugitive Hydrography: The *Nautical Magazine* and the Hydrographic Office of the Admiralty, c. 1832–1850." *International Journal of Maritime History*, 27, no. 2 (2015): 208-26.

———. "Naval Hydrography, Charismatic Bureaucracy, and the British Military State, c. 1825–1855." PhD thesis, University of Cambridge, 2016.

Barlow, Nora, ed. *The Autobiography of Charles Darwin, 1809–1882*. London: Collins, 1958.

Barrett, Paul H., ed. *The Collected Papers of Charles Darwin*. Chicago: University of Chicago Press, 1977.

Barrett, Paul H., Peter J. Gautrey, Sandra Herbert, David Kohn, and Sydney Smith, eds. *Charles Darwin's Notebooks, 1836–1844: Geology, Transmutation of Species, Metaphysical Enquiries*. Ithaca, NY: Cornell University Press, 1987.

Bartholomew, M. J. "The Award of the Copley Medal to Charles Darwin." *Notes and Records of the Royal Society of London* 30, no. 2 (1976): 209–18.

Beaglehole, J. C. *The Life of Captain James Cook*. London: Hakluyt Society, 1974.

Beechey, Frederick William. *Narrative of a Voyage to the Pacific and Beering's Strait: To Co-operate with the Polar Expeditions, Performed in His Majesty's Ship "Blossom" . . . in the Years 1825, 26, 27, 28*. 2 vols. London: Henry Colburn and Richard Bentley, 1831.

———. *Narrative of a Voyage to the Pacific and Beering's Strait: To Co-operate with the Polar Expeditions , Performed in His Majesty's Ship "Blossom," under the Command of Captain F. W. Beechey, R.N. . . . in the Years 1825, 26, 27, 28. . . .* Philadelphia: Carey and Lea, 1832.

Bellon, Richard. "Inspiration in the Harness of Daily Labor: Darwin, Botany, and the Triumph of Evolution, 1859–1868." *Isis* 102, no. 3 (2011): 393–420.

———. "Joseph Dalton Hooker's Ideals for a Professional Man of Science." *Journal of the History of Biology* 34, no. 1 (2001): 51–82.

Berry, Andrew. *Infinite Tropics: An Alfred Russel Wallace Anthology*. London: Verso, 2002.

Biagioli, Mario. *Galileo, Courtier: The Practice of Science in the Culture of Absolutism*. Chicago: University of Chicago Press, 1993.

———. *Galileo's Instruments of Credit: Telescopes, Images, Secrecy*. Chicago: University of Chicago Press, 2006.

———. "Rights or Rewards? Changing Frameworks of Scientific Authorship." In *Scientific Authorship: Credit and Intellectual Property in Science*, edited by Mario Biagioli and Peter Galison, 253–79. New York: Routledge, 2002.

Blewitt, Mary. *Surveys of the Seas: A Brief History of British Hydrography*. London: Macgibbon and Kee, 1957.

Bloor, David. *Knowledge and Social Imagery*. Boston: Routledge and Kegan Paul, 1976.

Bory de Saint-Vincent, Jean Baptiste Geneviève Marcellin, ed. *Dictionnaire classique d'histoire naturelle*. 17 vols. Paris: Rey et Gravier, 1822.

Bougainville, Louis-Antoine de. *A Voyage round the World*. Translated by J. R. Forster. London: J. Nourse and T. Davies, 1772.

Bougainville, Louis-Antoine de, Michel Bideaux, and Sonia Faessel. *Voyage autour du monde par la frégate du roi la Boudeuse et la flûte l'Étoile; en 1766, 1767, 1768 & 1769*. Paris: Saillant et Nyon, 1771.

Bowlby, John. *Charles Darwin: A New Life*. New York: W. W. Norton, 1991.

Brannigan, Augustine. *The Social Basis of Scientific Discoveries*. New York: Cambridge University Press, 1981.

Brinkman, Paul D. "Charles Darwin's *Beagle* Voyage, Fossil Vertebrate Succession, and 'The Gradual Birth and Death of Species.'" *Journal of the History of Biology* 43, no. 2 (2010): 363–99.

[Broderip, W. J.]. Review of *Narrative of the . . . "Adventure" and "Beagle" . . . and "Journal of Researches." Quarterly Review* 65 (1839): 194–234.

Browne, Janet. *Charles Darwin: The Power of Place*. Princeton, NJ: Princeton University Press, 2002.

———. *Charles Darwin: Voyaging*. Princeton, NJ: Princeton University Press, 1996.

———. "I Could Have Retched All Night: Charles Darwin and His Body." In *Science Incarnate: Historical Embodiments of Natural Knowledge*, edited by Christopher Lawrence and Steven Shapin. Chicago: University of Chicago Press, 1998.

———. *The Secular Ark: Studies in the History of Biogeography*. New Haven, CT: Yale University Press, 1983.

Browne, Janet, and Michael Neve, eds. *Voyage of the "Beagle": Charles Darwin's "Journal of Researches."* London: Penguin, 1989.

Burkhardt, Frederick. "Appendix V: Darwin's Early Notes on Coral Reef Formation." In *The Correspondence of Charles Darwin*, edited by Frederick Burkhardt. Cambridge: Cambridge University Press, 1985.

———. "Darwin and the Copley Medal." *Proceedings of the American Philosophical Society* 145, no. 4 (2001): 510–18.

———. "Darwin's Early Notes on Coral Reef Formation." *Earth Sciences History* 3, no. 2 (1984): 160–63.

Burkhardt, Frederick H., et al, eds. *The Correspondence of Charles Darwin*. 24 vols. New York: Cambridge University Press, 1985–2017.

Burkhardt, Richard W., Jr. "Unpacking Baudin: Models of Scientific Practice in the Age of Lamarck." In *Jean-Baptiste Lamarck, 1744–1829*, edited by Goulven Laurent, 497–514. Paris: Éditions du CTHS, 1997.

Burnett, D. Graham. *Trying Leviathan: The Nineteenth-Century New York Court Case That Put the Whale on Trial and Challenged the Order of Nature*. Princeton, NJ: Princeton University Press, 2007.

Cain, P. J., and A. G. Hopkins. *British Imperialism, 1688–2000*. Harlow, UK: Longman, 2002.

———. "Gentlemanly Capitalism and British Expansion Overseas. I. The Old Colonial System, 1688–1850." *Economic History Review* 39, no. 4 (1986): 501–25.

———. "Gentlemanly Capitalism and British Expansion Overseas. II: New Imperialism, 1850–1945." *Economic History Review* 40, no. 1 (1987): 1–26.

Camerini, Jane R. "Darwin, Wallace, and Maps." PhD thesis, University of Wisconsin, 1987.

———. "Evolution, Biogeography, and Maps: An Early History of Wallace's Line." In *Victorian Science in Context*, edited by Bernard Lightman, 70–109. Chicago: University of Chicago Press, 1997.

Cannon, Susan Faye. *Science in Culture: The Early Victorian Period*. New York: Science History Publications, 1978.

Cannon, W. Faye [Susan Faye]. "Charles Lyell, Radical Actualism, and Theory." *British Journal for the History of Science* 9, no. 2 (1976): 104–20.

Cannon, Walter F. [Susan Faye]. "History in Depth: The Early Victorian Period." *History of Science* 3 (1964): 20–38.

Chamisso, Adelbert von. "On the Coral Islands of the Pacific Ocean." *Edinburgh Philosophical Journal* 6 (1822): 37–40.

Chancellor, Gordon. "Darwin's Geological Diary from the Voyage of the *Beagle*." In *The Complete Work of Charles Darwin Online*, edited by John van Wyhe, 2012. http://darwin-online.org.uk/EditorialIntroductions/Chancellor_Geological Diary.html.

———. "Humboldt's Personal Narrative and Its Influence on Darwin." In *The Complete Work of Charles Darwin Online*, edited by John van Wyhe, 2011. http://darwin-online.org.uk/EditorialIntroductions/Chancellor_Humboldt.html.

Chancellor, Gordon, and John van Wyhe, eds. *Charles Darwin's Notebooks from the Voyage of the "Beagle."* Cambridge: Cambridge University Press, 2009.

Chartier, Roger. "Foucault's Chiasmus: Authorship between Science and Literature in the Seventeenth and Eighteenth Centuries." In *Scientific Authorship: Credit and Intellectual Property in Science*, edited by Mario Biagioli and Peter Galison, 13–31. New York: Routledge, 2002.

[Church, Richard William]. Review of *On the Origin of Species*, by Charles Darwin. *Guardian*, February 8, 1860.

Clark, John Willis, and Thomas McKenny Hughes. *The Life and Letters of the Reverend Adam Sedgwick*. 2 vols. Cambridge: Cambridge University Press, 1890.

Cock, Randolph. "Rear-Admiral Sir Francis Beaufort, RN, FRS: 'The Authorized Organ of Scientific Communication in England,' 1829–55." In *Science and the French and British Navies*, edited by Pieter van der Merwe, 99–116. London: National Maritime Museum, 2003.

———. "Scientific Servicemen in the Royal Navy and the Professionalisation of Science, 1816–55." In *Science and Beliefs: From Natural Philosophy to Natural Science*, edited by David M. Knight and Matthew D. Eddy, 95–112. Aldershot, UK: Ashgate, 2004.

———. "Sir Francis Beaufort and the Co-ordination of British Scientific Activity, 1829–55." PhD thesis, University of Cambridge, 2003.

Coleman, William. "Joseph Paul Gaimard." In *Dictionary of Scientific Biography*, edited by C. C. Gillispie. New York: Scribner's, 1974.

Collins, H. M. *Changing Order: Replication and Induction in Scientific Practice*. London: Sage, 1985.

Colp, Ralph. *Darwin's Illness*. Gainesville: University Press of Florida, 2008.

———. *To Be an Invalid: The Illness of Charles Darwin*. Chicago: University of Chicago Press, 1977.

Cook, James. *The Journals of Captain James Cook on His Voyages of Discovery: The Voyage of the "Endeavour."* Vol. 1. Edited by J. C. Beaglehole. Cambridge: Cambridge University Press for the Hakluyt Society, 1955.

Couthouy, Joseph Pitty. "Remarks upon Coral Formations in the Pacific; with Suggestions as to the Causes of Their Absence in the Same Parallels of Latitude on the Coast of South America." *Boston Journal of Natural History* 4 (1842): 66–105, 137–62.

Cowles, Henry M. "On the Origin of Theories: Charles Darwin's Vocabulary of Method." *American Historical Review* 122, no. 3 (forthcoming).

Crosland, Maurice P. *The Society of Arcueil: A View of French Science at the Time of Napoleon I*. Cambridge, MA: Harvard University Press, 1967.

Csiszar, Alex. "Objectivities in Print." In *Objectivity in Science*, edited by Flavia Padovani, Alan Richardson, and Jonathan Y. Tsou, 145–69. Boston Studies in the Philosophy and History of Science 310. New York: Springer International, 2015.

Dana, James Dwight. *Corals and Coral Islands*. New York: Dodd and Mead, 1872.

Darwin, Charles. *The Autobiography of Charles Darwin, 1809–1882*. London: Collins, 1958.

———. *Evolutionary Writings*. Edited by James A. Secord. New York: Oxford University Press, 2008.

———. *For Private Distribution. The Following Pages Contain Extracts from Letters Addressed to Professor Henslow by C. Darwin Esq. They Are Printed for Distribu-*

tion among the Members of the Cambridge Philosophical Society in Consequence of the Interest Which Has Been Excited by Some of the Geological Notices Which They Contain, Etc. Cambridge, 1835.

———. "Geological Notes Made during a Survey of the East and West Coasts of South America, in the Years 1832, 1833, 1834, and 1835, with an Account of a Transverse Section of the Cordilleras of the Andes between Valparaiso and Mendoza. Communicated to the Geological Society by Adam Sedgwick." *Proceedings of the Geological Society of London* 2 (1838): 210–12.

———. "Geological Society [Report of Darwin's 'Areas of Elevation and Subsidence' Paper]." *Athenaeum*, no. 503 (June 17, 1837): 443.

———. "Geological Society [Report of Darwin's Paper 'On the Formation of Mould']." *Athenaeum* 526 (November 25, 1837): 218.

———. "Geology." In *A Manual of Scientific Enquiry: Prepared for the Use of Her Majesty's Navy, and Adapted for the Travellers in General*, edited by J. F. W. Herschel, 156–95. London: John Murray, 1849.

———. *Journal of Researches into the Geology and Natural History of the Various Countries Visited by H.M.S. "Beagle," under the Command of Captain FitzRoy, R.N., from 1832 to 1836*. London: Henry Colburn, 1839.

———. *Journal of Researches into the Natural History and Geology of the Countries Visited during the Voyage of H.M.S. "Beagle" round the World*. 2nd ed. London: John Murray, 1845.

———. *Naturwissenschaftliche Reisen*. Translated by Ernst Dieffenbach. Braunschweig: Vieweg, 1844.

———. "Observations of Proofs of Recent Elevation on the Coast of Chili, Made during the Survey of His Majesty's Ship *Beagle*, Commanded by Capt. FitzRoy, R.N." [Abstract of the Paper Read to the Geological Society of London on 4 January 1837.] *Proceedings of the Geological Society of London* 4, no. 2 (January 1838): 446–49.

———. "Observations on the Parallel Roads of Glen Roy, and of Other Parts of Lochaber in Scotland, with an Attempt to Prove That They Are of Marine Origin." *Philosophical Transactions of the Royal Society* 2 (1839): 39–81.

———. "On Certain Areas of Elevation and Subsidence in the Pacific and Indian Oceans, as Deduced from the Study of Coral Formations." [Abstract of the Paper Read to the Geological Society of London on 31 May 1837.] *Proceedings of the Geological Society of London* 2 (1838): 552–54.

———. "On the Connexion of Certain Volcanic Phaenomena, and on the Formation of Mountain-Chains and Volcanos, as the Effects of Continental Elevations." *Proceedings of the Geological Society of London* 2 (1838): 654–60.

———. "On the Formation of Mould." *Proceedings of the Geological Society of London* 2, (1838): 574–76.

———. "On the Formation of Mould." *Transactions of the Geological Society of London* 2, no. 5 (1840): 505–9.

———. *On the Origin of Species*. London: John Murray, 1859.

———. *The Red Notebook of Charles Darwin*. Edited by Sandra Herbert. Ithaca, NY: Cornell University Press, 1980.

———. "Remarks on the Preceding Paper, in a Letter from Charles Darwin, Esq., to Mr. Maclaren." *Edinburgh New Philosophical Journal* 34 (1843): 47–50.

———. "A Sketch of the Deposits Containing Extinct Mammalia in the Neighbourhood of the Plata." *Proceedings of the Geological Society of London* 2 (1838): 542–44.

———. *The Structure and Distribution of Coral Reefs.* London: Smith, Elder, 1842.

———. *The Structure and Distribution of Coral Reefs.* 2nd ed. London: Smith, Elder, 1874.

Dawson, L. S. *Memoirs of Hydrography: Including Brief Biographies of the Principal Officers Who Have Served in H.M. Naval Surveying Service between the Years 1750 and 1885.* London: Cornmarket, 1969.

Day, Archibald. *The Admiralty Hydrographic Service, 1795–1919.* London: Her Majesty's Stationery Office, 1967.

Deacon, Margaret. *Scientists and the Sea, 1650–1900: A Study of Marine Science.* Aldershot, UK: Ashgate, 1997.

De la Beche, H. T. *A Geological Manual.* London: Treuttel and Wuertz, 1831.

———. *Sections and Views Illustrative of Geological Phenomena.* London: Treuttel and Wuertz, 1830.

Desmond, Adrian. "Grant, Robert Edmond (1793–1874)." In *Oxford Dictionary of National Biography.* Oxford: Oxford University Press, 2004.

———. *The Politics of Evolution.* Chicago: University of Chicago Press, 1989.

Desmond, Adrian, and James Moore. *Darwin: The Life of a Tormented Evolutionist.* New York: Warner Books, 1991.

———. *Darwin's Sacred Cause: How a Hatred of Slavery Shaped Darwin's Views on Human Evolution.* London: Penguin, 2009.

Desmond, Adrian, and Sarah E. Parker. "The Bibliography of Robert Edmond Grant (1793–1874): Illustrated with a Previously Unpublished Photograph." *Archives of Natural History* 33, no. 2 (2006): 202–13.

Dettelbach, Michael. "Humboldtian Science." In *Cultures of Natural History*, edited by Nicholas Jardine, James A. Secord, and E. C. Spary, 286–304. Cambridge: Cambridge University Press, 1996.

Dobbs, David. *Reef Madness: Charles Darwin, Alexander Agassiz, and the Meaning of Coral.* New York: Pantheon, 2005.

Dott, Robert H., Jr. "Charles Lyell in America–His Lectures, Field Work and Mutual Influences, 1841–1853." *Earth Sciences History* 15 (1996): 101–40.

Driver, Felix. *Geography Militant: Cultures of Exploration and Empire.* Oxford: Blackwell, 2001.

Ebach, Malte Christian. *Origins of Biogeography: The Role of Biological Classification in Early Plant and Animal Geography.* Dordrecht: Springer, 2015.

Eddy, M. D. "Fallible or Inerrant? A Belated Review of the 'Constructivist's Bible,' Jan Golinski, *Making Natural Knowledge: Constructivism and the History of Science* (Cambridge History of Science, Cambridge: Cambridge University Press, 1999)." *British Journal for the History of Science* 37, no. 1 (March 2004): 93–98.

Egerton, Frank N. "History of the Ecological Sciences, Part 35: The Beginnings of British Marine Biology: Edward Forbes and Philip Gosse." *Bulletin of the Ecological Society of America* 91, no. 2 (April 1, 2010): 176–201.

———. "History of Ecological Sciences, Part 37: Charles Darwin's Voyage on the *Beagle*." *Bulletin of the Ecological Society of America* 91, no. 4 (October 1, 2010): 398–431.

———. "Humboldt, Darwin, and Population." *Journal of the History of Biology* 3, no. 2 (1970): 325–60.

Ehrenberg, Christian Gottfried. *Über die Natur und Bildung der Coralleninseln und Corallenbänke im Rothen Meere*. Berlin: Königlichen Akademie der Wissenschaften, 1834.

Ellegård, Alvar. *Darwin and the General Reader: The Reception of Darwin's Theory of Evolution in the British Periodical Press, 1859–1872*. Gothenburg: Almqvist och Wiksell, 1958.

Ellis, William. *Polynesian Researches during a Residence of Nearly Six Years in the South Sea Islands*. 2 vols. London: Fisher, Son, and Jackson, 1829.

Endersby, Jim. *Imperial Nature: Joseph Hooker and the Practices of Victorian Science*. Chicago: University of Chicago Press, 2008.

[Eschscholtz, J. F.]. "On the Coral Islands." In *A Voyage of Discovery into the South Sea and Bering's Straits for the Purpose of Exploring a North-East Passage, Undertaken in the Years 1815–1818*, edited by Otto von Kotzebue, 3:331–36. London: Longman, Hurst, Rees, Orme and Brown, 1821.

[Fitton, William Henry]. Review of *Elements of Geology* by Charles Lyell. *Edinburgh Review* 69, no. 140 (1839): 406–66.

FitzRoy, Robert. *Narrative of the Surveying Voyages of His Majesty's Ships "Adventure" and "Beagle" between the Years 1826 and 1836*. Vol. 2. London: Henry Colburn, 1839.

Forbes, Edward. "Report on the Mollusca and Radiata of the Aegean Sea, and on Their Distribution, Considered as Bearing on Geology." In *Report of the Thirteenth Meeting of the British Association for the Advancement of Science*, 130–93. London: John Murray, 1844.

Forster, J. R. *Observations Made during a Voyage round the World, on Physical Geography, Natural History and Ethic Philosophy*. London: G. Robinson, 1778.

Foucault, Michel. "What Is an Author?" In *Language, Counter-memory, Practice: Selected Essays and Interviews*, edited by Donald F. Bouchard, 113–38. Ithaca, NY: Cornell University Press, 1977 [1969].

———. "What Is an Author." In *Textual Strategies: Perspectives in Post-structural Criticism*, edited by Josué Harari, 141–60. Ithaca, NY: Cornell University Press, 1979.

Fruton, Joseph S. *Contrasts in Scientific Style: Research Groups in the Chemical and Biochemical Sciences*. Philadelphia: American Philosophical Society, 1990.

Galison, Peter. "The Collective Author." In *Scientific Authorship: Credit and Intellectual Property in Science*, edited by Mario Biagioli and Peter Galison, 325–53. New York: Routledge, 2002.

———. *How Experiments End*. Chicago: University of Chicago Press, 1987.

Galison, Peter, and Andrew Warwick. "Introduction: Cultures of Theory." *Studies in History and Philosophy of Modern Physics* 29, no. 3 (1998): 287-94.

Geison, Gerald L. *Michael Foster and the Cambridge School of Physiology: The Scientific Enterprise in Late Victorian Society*. Princeton, NJ: Princeton University Press, 1978.

———. "Review of Kohler, 'Lords of the Fly.'" *Isis* 87, no. 2 (1996): 328–31.

Geison, Gerald L., and Frederic Lawrence Holmes, eds. "Research Schools: Historical Reappraisals." *Osiris* 8 (1993).

"Geological Society." *Athenaeum*, no. 421 (December 21, 1835): 875–76.

Ghiselin, Michael T. "Introduction." In *The Structure and Distribution of Coral Reefs*, by Charles Darwin, vii–xii. Tucson: University of Arizona Press, 1984.

———. *The Triumph of the Darwinian Method*. Mineola, NY: Dover, 2003.

Gilman, Daniel Coit. *The Life of James Dwight Dana, Scientific Explorer, Mineralogist, Geologist, Zoologist, Professor in Yale University*. New York: Harper, 1899.

Glick, Thomas F., ed. *The Comparative Reception of Darwinism*. Chicago: University of Chicago Press, 1988.

Golinski, Jan. *Science as Public Culture: Chemistry and Enlightenment in Britain, 1760–1820*. Cambridge: Cambridge University Press, 1999.

Gooding, David, Trevor Pinch, and Simon Schaffer, eds. *The Uses of Experiment: Studies in the Natural Sciences*. Cambridge: Cambridge University Press, 1989.

Goodman, Jordan. *The "Rattlesnake": A Voyage of Discovery to the Coral Sea*. London: Faber and Faber, 2005.

Grafton, Anthony. *The Footnote: A Curious History*. Cambridge, MA: Harvard University Press, 1999.

Greene, Mott T. *Geology in the Nineteenth Century: Changing Views of a Changing World*. Ithaca, NY: Cornell University Press, 1982.

Gross, Alan G. *Starring the Text: The Place of Rhetoric in Science Studies*. Carbondale: Southern Illinois University Press, 2006.

Gruber, Howard E., and Paul H. Barrett. *Darwin on Man: A Psychological Study of Scientific Creativity; With Darwin's Early and Unpublished Notebooks Transcribed and Annotated by Paul H. Barrett*. New York: Dutton, 1974.

Gutch, John. *Beyond the Reefs: The Life of John Williams, Missionary*. London: Macdonald, 1974.

Hacking, Ian. "The Looping Effects of Human Kinds." In *Causal Cognition: A Multidisciplinary Debate*, edited by Dan Sperber, David Premack, and Ann James Premack, 351–94. Oxford: Oxford University Press, 1995.

———. "Making Up People." *London Review of Books*, August 17, 2006.

Hagen, Nils T. "Deconstructing Doctoral Dissertations: How Many Papers Does It Take to Make a PhD?" *Scientometrics* 85, no. 2 (November 2010): 567–79.

[Hall, Basil]. "*Narrative of . . . 'Adventure' and 'Beagle'* . . . by Captains King and FitzRoy, R.N., and Charles Darwin, Esq., Naturalist of the *Beagle*." *Edinburgh Review* 69, no. 140 (1839): 467–93.

Hannaway, Owen. Review of *The Society of Arcueil: A View of French Science at the Time of Napoleon I*, by Maurice Crosland. *Isis* 60, no. 4 (1969).

Harwood, Jonathan. *Styles of Scientific Thought: The German Genetics Community, 1900–1933*. Chicago: University of Chicago Press, 1993.

Herbert, Sandra. "Charles Darwin as a Prospective Geological Author." *British Journal for the History of Science* 24 (1991): 159–92.

———. *Charles Darwin, Geologist*. Ithaca, NY: Cornell University Press, 2005.

———. "Darwin the Young Geologist." In *The Darwinian Heritage*, edited by David Kohn, 483–510. Princeton, NJ: Princeton University Press, 1985.

———. "Doing and Knowing: Charles Darwin and Other Travellers." In *Four Centuries of Geological Travel: The Search for Knowledge on Foot, Bicycle, Sledge and Camel*, edited by Patrick Wyse Jackson, 311-23. Special Publication 287. London: Geological Society, 2007.

———. "From Charles Darwin's Portfolio: An Early Essay on South American Geology and Species." *Earth Sciences History* 14, no. 1 (1995): 23–36.

———. "The Place of Man in the Development of Darwin's Theory of Transmutation, Part 2." *Journal of the History of Biology* 10 (1977): 155–227.

———. "'A Universal Collector': Charles Darwin's Extraction of Meaning from His Galápagos Experience." In *Darwin and the Galápagos*, edited by Michael T. Ghiselin and Alan E. Leviton, 45–68. San Francisco: California Academy of Sciences; Pacific Division of the American Association for the Advancement of Science, 2010.

Herschel, John F. W. *Preliminary Discourse on the Study of Natural Philosophy*. London: Longman, Brown, Green and Longmans, 1830.

Hodge, M. J. S. "Darwin and the Laws of the Animate Part of the Terrestrial System (1835–1837): On the Lyellian Origins of His Zoonomical Explanatory Program." *Studies in the History of Biology* 6 (1982): 1–106.

———. "Darwin as a Lifelong Generation Theorist." In *The Darwinian Heritage*, edited by David Kohn, 207–44. Princeton, NJ: Princeton University Press, 1985.

———. "The Development of Darwin's General Biological Theorizing." In *Evolution from Molecules to Men*, edited by D. S Bendall, 43–62. New York: Cambridge University Press, 1983.

———. "The Notebook Programmes and Projects of Darwin's London Years." In *The Cambridge Companion to Darwin*, edited by M. J. S. Hodge and Gregory Radick, 40–68. Cambridge: Cambridge University Press, 2003.

———. "The Structure and Strategy of Darwin's 'Long Argument.'" *British Journal for the History of Science* 10, no. 3 (1977): 237–46.

Horsburgh, James, and W. F. W. Owen. "Some Remarks relative to the Geography of the Maldiva Islands, and the Navigable Channels (at Present Known to Europeans) Which Separate the Atolls from Each Other." *Journal of the Royal Geographical Society of London* 2 (1832): 72–92.

Houghton, Walter Edwards, Jean Harris Slingerland, and Wellesley College. *The Wellesley Index to Victorian Periodicals, 1824–1900*. Toronto: University of Toronto Press, 1966.

Hull, David L. *Darwin and His Critics: The Reception of Darwin's Theory of Evolution by the Scientific Community*. Cambridge, MA: Harvard University Press, 1973.

———. "Darwin's Science and Victorian Philosophy of Science." In *The Cambridge*

Companion to Darwin, edited by M. J. S. Hodge and Gregory Radick, 168–91. Cambridge: Cambridge University Press, 2003.

Humboldt, Alexander von. *Essai sur la géographie des plantes, accompagné d'un tableau physique des régions équinoxiales fondé, sur des mesures exécutées, depuis le 10e degré de latitude boréale jusqu'au 10e degré de latitude australe, pendant les années 1799, 1800, 1801, 1802 et 1803. Par Al. de Humboldt et A. Bonpland. Rédigé par Al. de Humboldt* . . . Paris: F. Schoell, 1805.

———. "New Inquiries into the Laws Which Are Observed in the Distribution of Vegetable Forms." Anonymous translation. *Edinburgh Philosophical Journal* 6 (1822): 273–79.

Humboldt, Alexander von, and Aimé Bonpland. *Personal Narrative of Travels to the Equinoctial Regions of the New Continent during the Years 1799–1804.* Translated by Helen Maria Williams. 7 vols. London: Longman, Hurst, Rees, Orme, and Brown, 1818-29.

Hutton, Charles. *A Philosophical and Mathematical Dictionary.* 2 vols. London, 1815.

Huxley, Leonard. *Life and Letters of Sir Joseph Dalton Hooker Based on Materials Collected and Arranged by Lady Hooker.* 2 vols. London: Murray, 1918.

Igler, David. "On Coral Reefs, Volcanoes, Gods, and Patriotic Geology, or James Dwight Dana Assembles the Pacific Basin." *Pacific Historical Review* 79 (2010): 23–49.

Iliffe, Robert. "Butter for Parsnips: Authorship, Audience and the Incomprehensibility of the *Principia.*" In *Scientific Authorship: Credit and Intellectual Property in Science,* edited by Mario Biagioli and Peter Galison, 33–66. Boston: Routledge, 2002.

Jackson, Julian. Review of *The Structure and Distribution of Coral Reefs* . . . , by Charles Darwin. *Journal of the Royal Geographical Society of London* 12 (1842): 115–20.

Jameson, Robert. "On the Growth of Coral Islands." In *Essay on the Theory of the Earth,* by Georges Cuvier, 379–98, translated by Robert Jameson. Edinburgh: William Blackwood, 1827.

Judd, J. W. "Darwin and Geology." In *Darwin and Modern Science,* edited by A. C. Seward, 337–84. Cambridge: Cambridge University Press, 1909.

Jukes, J. Beete. *Narrative of the Surveying Voyage of H.M.S. "Fly," Commanded by Captain F. P. Blackwood, R.N., in Torres Strait, New Guinea, and Other Islands of the Eastern Archipelago during the Years 1842–1846: Together with an Excursion into the Interior of the Eastern Part of Java.* London: Boone, 1847.

Kaiser, David. *Drawing Theories Apart: The Dispersion of Feynman Diagrams in Postwar Physics.* Chicago: University of Chicago Press, 2005.

———, ed. *Pedagogy and the Practice of Science: Historical and Contemporary Perspectives.* Cambridge, MA: MIT Press, 2005.

Keynes, Richard. *The "Beagle" Record: Selections from the Original Pictorial Records and Written Accounts of the Voyage of H.M.S. "Beagle."* New York: Cambridge University Press, 1979.

———, ed. *Charles Darwin's "Beagle" Diary.* Cambridge: Cambridge University Press, 1988.

————, ed. *Charles Darwin's Zoology Notes and Specimen Lists from H.M.S. "Beagle."* Cambridge: Cambridge University Press, 2000.

————. *Fossils, Finches, and Fuegians: Darwin's Adventures and Discoveries on the "Beagle."* New York: Oxford University Press, 2003.

Keynes, Simon. *Charles Darwin, Robert FitzRoy and the Voyage of HMS "Beagle."* London: Henry Sotheran Limited (forthcoming).

Knell, Simon J. *The Culture of English Geology, 1815–1851: A Science Revealed through Its Collecting.* Burlington, VT: Ashgate, 2000.

Knorr-Cetina, Karin. *The Manufacture of Knowledge: An Essay on the Constructivist and Contextual Nature of Science.* New York: Pergamon Press, 1981.

Kohler, Robert E. *Lords of the Fly: "Drosophila" Genetics and the Experimental Life.* Chicago: University of Chicago Press, 1994.

Kohn, David. "Theories to Work By: Rejected Theories, Reproduction, and Darwin's Path to Natural Selection." *Studies in History of Biology* 4 (1980): 67–170.

Kottler, Malcolm J. "Charles Darwin's Biological Species Concept and Theory of Geographic Speciation: The Transmutation Notebooks." *Annals of Science* 35, no. 3 (May 1, 1978): 275–97.

Kuhn, Thomas S. *The Essential Tension: Selected Studies in Scientific Tradition and Change.* Chicago: University of Chicago Press, 1977.

Lamouroux, Jean Vincent Félix. *Histoire des polypiers coralligènes flexibles, vulgairement nommés zoophytes.* Caen: Poisson, 1816.

Larkum, A. W. D. *A Natural Calling: Life, Letters and Diaries of Charles Darwin and William Darwin Fox.* New York: Springer, 2009.

Larsen, Anne Laurine. "Not since Noah: The English Scientific Zoologists and the Craft of Collecting, 1800–1840." PhD diss., Princeton University, 1993.

Latour, Bruno. "Give Me a Laboratory and I Will Raise the World." In *Science Observed*, edited by Karin Knorr-Cetina and Michael Mulkay, 141–70. Beverly Hills, CA: Sage, 1983.

————. *Science in Action: How to Follow Scientists and Engineers through Society.* Cambridge, MA: Harvard University Press, 1987.

Latour, Bruno, and Steve Woolgar. *Laboratory Life: The Social Construction of Scientific Facts.* Beverly Hills, CA: Sage, 1979.

Leader, Zachary. *Writer's Block.* Baltimore: Johns Hopkins University Press, 1991.

Leask, Nigel. "Darwin's Second Sun: Alexander von Humboldt and the Genesis of *The Voyage of the 'Beagle.'*" In *Literature, Science, Psychoanalysis, 1830–1970: Essays in Honour of Gillian Beer*, edited by Gillian Beer, Helen Small, and Trudi Tate, 13–36. Oxford: Oxford University Press, 2003.

"A List of Donations to the Library . . . Belonging to the Geological Society of London, from . . . June 1835, to . . . June 1839." *Transactions of the Geological Society of London* 5 (1840): Unpaginated appendix beginning on p. 755.

Litchfield, Henrietta Emma Darwin. *Emma Darwin, a Century of Family Letters, 1792-1896.* 2 vols. London: John Murray, 1915.

Love, Alan C. "Darwin and Cirripedia prior to 1846: Exploring the Origins of the Barnacle Research." *Journal of the History of Biology* 35 (2002): 251–89.

Lyell, Charles. "Address to the Geological Society, Delivered at the Anniversary, on the 19th of February, 1836." *Proceedings of the Geological Society of London* 2 (1837): 357–90.

———. "Address to the Geological Society, Delivered at the Anniversary, on the 17th of February, 1837." *Proceedings of the Geological Society of London* 2 (1838): 479–523.

———. *Elements of Geology*. London: John Murray, 1838.

[———]. "Mr. Lyell's Fourth Lecture on Geology." *New-York Tribune*, March 28, 1842.

———. "On the Occurrence of Works of Human Art in Post-Pliocene Deposits." Introductory address to the BAAS Section on Geology. In *Report of the Twenty-Ninth Meeting of the British Association for the Advancement of Science; Held at Aberdeen in September 1859*, 93–95. London: John Murray, 1860.

———. *Principles of Geology, or The Modern Changes of the Earth and Its Inhabitants*. 6th ed. 3 vols. London: John Murray, 1840.

———. *Principles of Geology, Being an Attempt to Explain the Former Changes of the Earth's Surface, by References to Causes Now in Operation*. 3 vols. London: John Murray, 1830-33.

———. *Travels in North America, with Geological Observations on the United States, Canada, and Nova Scotia*. 2 vols. London: John Murray, 1845.

Lyell, Katherine M., ed. *Life, Letters and Journals of Sir Charles Lyell, Bart*. 2 vols. London: John Murray, 1881.

Lynch, Michael. *Art and Artifact in Laboratory Science: A Study of Shop Work and Shop Talk in a Research Laboratory*. London; Boston: Routledge and Kegan Paul, 1985.

Maclaren, Charles. "On Coral Islands and Reefs, as Described by Mr. Darwin." *Edinburgh New Philosophical Journal* 34 (1843): 33–47.

MacLeod, Roy M., and Philip F. Rehbock, eds. *Darwin's Laboratory: Evolutionary Theory and Natural History in the Pacific*. Honolulu: University of Hawaii Press, 1994.

Martinez, Oscar A., Jorge Rabassa, and Andrea Coronato. "Charles Darwin and the First Scientific Observations on the Patagonian Shingle Formation (Rodados Patagónicos)." *Revista de La Asociación Geológica Argentina* 64, no. 1 (2009): 90–100.

McCalman, Iain. *Darwin's Armada: Four Voyages and the Battle for the Theory of Evolution*. New York: W. W. Norton, 2009.

McCartney, Paul J., and Douglas Anthony Bassett. *Henry De la Beche: Observations on an Observer*. Cardiff: Friends of the National Museum of Wales, 1977.

McSherry, Corynne. "Uncommon Controversies: Legal Mediations of Gift and Market Models of Authorship." In *Scientific Authorship: Credit and Intellectual Property in Science*, edited by Mario Biagioli and Peter Galison, 253–79. New York: Routledge, 2002.

Merton, Robert. "The Matthew Effect in Science." *Science* 159, no. 3810 (January 5, 1968): 56–63.

Mill, John Stuart. *A System of Logic, Ratiocinative and Inductive: Being a Connected View of the Principles of Evidence, and the Methods of Scientific Investigation.* 5th ed. London: Parker, 1862.

Montgomery, William. "Charles Darwin's Theory of Coral Reefs and the Problem of the Chalk." *Earth Sciences History* 7 (1988): 111-20.

Moresby, Robert. "Extracts from Commander Moresby's Report on the Northern Atolls of the Maldivas." *Journal of the Royal Geographical Society of London* 5 (1835): 398-404.

Morrell, J. B. "The Chemist Breeders: The Research Schools of Liebig and Thomas Thomson." *Ambix* 19, no. 1 (1972): 1-46.

———. *John Phillips and the Business of Victorian Science.* Aldershot, UK: Ashgate, 2005.

———. "London Institutions and Lyell's Career: 1820-41." *British Journal for the History of Science* 9 (1975): 132-46.

Morrell, Jack, and Arnold Thackray. *Gentlemen of Science: Early Years of the British Association for the Advancement of Science.* Oxford: Oxford University Press, 1981.

Murchison, Roderick Impey. "Address to the Geological Society, Delivered at the Anniversary, on the 15th of February, 1833." *Proceedings of the Geological Society of London* 1 (1834): 438-64.

O'Connor, Ralph. *The Earth on Show: Fossils and the Poetics of Popular Science, 1802-1856.* Chicago: University of Chicago Press, 2008.

Oreskes, Naomi. *The Rejection of Continental Drift: Theory and Method in American Earth Science.* New York: Oxford University Press, 1999.

Ospovat, Dov. *The Development of Darwin's Theory.* Cambridge: Cambridge University Press, 1981.

Page, Leroy E. "The Rivalry between Charles Lyell and Roderick Murchison." *British Journal for the History of Science* 9, no. 2 (1976): 156-65.

Pearn, Alison M., ed. *A Voyage round the World: Charles Darwin and the "Beagle" Collections in the University of Cambridge.* Cambridge: Cambridge University Press, 2009.

Pearson, P. N., and C. J. Nicholas. "'Marks of Extreme Violence': Charles Darwin's Geological Observations at St Jago (São Tiago), Cape Verde Islands." In *Four Centuries of Geological Travel: The Search for Knowledge on Foot, Bicycle, Sledge and Camel*, edited by Patrick Wyse Jackson, 239-54. Special Publication 287. London: Geological Society, 2007.

Peckham, Morse, ed. *The Origin of Species: A Variorum Text.* Philadelphia: University of Pennsylvania Press, 1959.

Pence, Charles H. "Sir John F. W. Herschel and Charles Darwin: Nineteenth-Century Science and Its Methodology." Under review (consulted November 9, 2015).

Phillips, John. *Treatise on Geology.* Vol. 1. *Lardner's Cabinet Cyclopedia.* London: Longman, 1837.

Pickering, Andrew. *Constructing Quarks: A Sociological History of Particle Physics.* Chicago: University of Chicago Press, 1984.

————, ed. *Science as Practice and Culture*. Chicago: University of Chicago Press, 1992.

Pinch, Trevor. *Confronting Nature: The Sociology of Solar-Neutrino Detection*. Boston: Reidel, 1986.

Porter, Duncan. "The *Beagle* Collector and His Collections." In *The Darwinian Heritage*, edited by David Kohn, 973–1020. Princeton, NJ: Princeton University Press, 1985.

Porter, Roy. "Charles Lyell and the Principles of the History of Geology." *British Journal for the History of Science* 9, no. 2 (1976): 91–103.

————. *The Making of Geology: Earth Science in Britain, 1660–1815*. New York: Cambridge University Press, 1977.

Prendergast, Michael Laurent. "James Dwight Dana: The Life and Thought of an American Scientist." PhD diss., UCLA, 1978.

"Proceedings at the Annual General Meeting, 18 February 1859." *Quarterly Journal of the Geological Society of London* 16 (1860): xxii–lv.

Quoy, J. R. C., and Paul Gaimard. "Mémoire sur l'accroissement des polypes lithophytes considéré géologiquement." *Annales des Sciences Naturelles* 6 (1825): 273–90.

Radin, Joanna. *Life on Ice: A History of New Uses for Cold Blood*. Chicago: University of Chicago Press, 2017.

Rehbock, Philip F. "The Early Dredgers: Naturalizing in British Seas, 1830–1850." *Journal of the History of Biology* 12, no. 2 (1979): 293–368.

Reidy, Michael S. *Tides of History: Ocean Science and Her Majesty's Navy*. Chicago: University of Chicago Press, 2008.

Report of the First and Second Meetings of the British Association for the Advancement of Science. London: John Murray, 1833.

Review of *A Narrative of Missionary Enterprises*, by John Williams. *Athenaeum*, no. 502 (June 10, 1837): 413–14.

Rhodes, Frank H. T. "Darwin's Search for a Theory of the Earth: Symmetry, Simplicity, and Speculation." *British Journal for the History of Science* 24 (1991): 193–229.

Richards, Robert J. *The Romantic Conception of Life: Science and Philosophy in the Age of Goethe*. Chicago: University of Chicago Press, 2002.

Richardson, R. Alan. "Biogeography and the Genesis of Darwin's Ideas on Transmutation." *Journal of the History of Biology* 14, no. 1 (1981): 1–41.

Rieppel, Lukas. "Albert Koch's *Hydrarchos* Craze: Credibility, Identity, and Authenticity in Nineteenth-Century Natural History." In *Science Museums in Transition: Cultures of Display in Nineteenth-Century Britain and America*, edited by Carin Berkowitz and Bernard Lightman, 139-61. Pittsburgh: University of Pittsburgh Press, 2017.

Ritchie, G. S. *The Admiralty Chart: British Naval Hydrography in the Nineteenth Century*. New York: American Elsevier, 1967.

Robinson, Arthur H. "The Genealogy of the Isopleth." *Cartographic Journal* 8, no. 1 (June 1, 1971): 49–53.

Rosen, Brian Roy, and Darrell, Jill. "A Generalized Historical Trajectory for Charles Darwin's Specimen Collections, with a Case Study of His Coral Reef Specimen List in the Natural History Museum, London." In *Darwin tra scienza, storia e società: 150o anniversario della pubblicazione di Origine delle specie*, edited by Francesco Stoppa and Roberto Veraldi, 133–98. Rome: Edizioni Universitarie Romane, 2010.

Rozwadowski, Helen M. *Fathoming the Ocean: The Discovery and Exploration of the Deep Sea*. Cambridge, MA: Harvard University Press, 2005.

Rudwick, Martin J. S. "Charles Darwin in London: The Integration of Public and Private Science." *Isis* 73 (1982): 186–206.

———. "Charles Lyell, F.R.S. (1797–1875) and His London Lectures on Geology, 1832–33." *Notes and Records of the Royal Society of London* 29, no. 2 (1975): 231–63.

———. "Charles Lyell Speaks in the Lecture Theatre." *British Journal for the History of Science* 9, no. 2 (1976): 147–55.

———. "Charles Lyell's Dream of a Statistical Palaeontology." *Palaeontology* 21 (1978): 225–44.

———. "Darwin and Glen Roy: A 'Great Failure' in Scientific Method?" *Studies in the History and Philosophy of Science* 5 (1974): 97–185.

———. "Darwin and the World of Geology." In *The Darwinian Heritage*, edited by David Kohn, 511–18. Princeton, NJ: Princeton University Press, 1985.

———. "The Emergence of a Visual Language for Geological Science, 1760–1840." *History of Science* 14 (1976): 149–95.

———. *The Great Devonian Controversy: The Shaping of Scientific Knowledge among Gentlemanly Specialists*. Chicago: University of Chicago Press, 1985.

———. "Uniformity and Progression: Reflections on the Structure of Geological Theory in the Age of Lyell." In *Perspectives in the History of Science and Technology*, edited by Duane H. D. Roller, 209–27. Norman: University of Oklahoma Press, 1971.

———. *Worlds before Adam: The Reconstruction of Geohistory in the Age of Reform*. Chicago: University of Chicago Press, 2008.

Ruse, Michael. "Charles Lyell and the Philosophers of Science." *British Journal for the History of Science* 9, no. 2 (1976): 121–31.

———. "Darwin's Debt to Philosophy: An Examination of the Influence of the Philosophical Ideas of John F. W. Herschel and William Whewell on the Development of Charles Darwin's Theory of Evolution." *Studies in the History and Philosophy of Science* 6 (1975): 159–81.

———. "The Origin of the *Origin*." In *The Cambridge Companion to the "Origin of Species,"* edited by Michael Ruse and Robert J Richards, 1–13. New York: Cambridge University Press, 2009.

Ruse, Michael, and Robert J. Richards. *The Cambridge Companion to the "Origin of Species."* New York: Cambridge University Press, 2009.

Saint-Pierre, Jacques Henri Bernardin de. *Paul et Virginie*. Lausanne: J. Mourer, 1788.

———. *Paul and Virginia. Translated from the French . . . by Helen Maria Williams.* London: G. G. and J. Robinson, 1795.

[———]. *Voyage à l'Isle de France, à l'Isle de Bourbon, au Cap de Bonne Espérance, &c. Avec des observations nouvelles sur la nature et sur les hommes, par un officier du roi.* 2 vols. Amsterdam: Merlin, 1773.

"Scientific Voyage." *Athenaeum*, no. 217 (December 24, 1831): 834–35.

Secord, James A. *Controversy in Victorian Geology: The Cambrian-Silurian Dispute.* Princeton, NJ: Princeton University Press, 1986.

———. "The Discovery of a Vocation: Darwin's Early Geology." *British Journal for the History of Science* 24 (1991): 133–57.

———. "Edinburgh Lamarckians: Robert Jameson and Robert E. Grant." *Journal of the History of Biology* 24 (1991): 1–18.

———. "How Scientific Conversation Became Shop Talk." *Transactions of the Royal Historical Society*, ser. 6, 17 (2007): 129–56.

———. "Introduction." In *Evolutionary Writings*, by Charles Darwin, vii–xxxvii. New York: Oxford University Press, 2008.

———. "Introduction." In *Principles of Geology*, by Charles Lyell, ix–xliii. London: Penguin, 1997.

———. *Victorian Sensation: The Extraordinary Publication, Reception, and Secret Authorship of "Vestiges of the Natural History of Creation."* Chicago: University of Chicago Press, 2000.

[Sedgwick, Adam]. "Objections to Mr. Darwin's Theory of the Origin of Species." *Spectator*, March 24, 1860.

Seth, Suman. *Crafting the Quantum: Arnold Sommerfeld and the Practice of Theory, 1890–1926.* Cambridge, MA: MIT Press, 2010.

———. "The History of Physics after the Cultural Turn." *Historical Studies in the Natural Sciences* 41, no. 1 (2011): 112–22.

Shapin, Steven. "History of Science and Its Sociological Reconstructions." *History of Science* 20 (1982): 157–211.

———. "The House of Experiment in Seventeenth-Century England." *Isis* 79, no. 3 (1988): 373–404.

———. "The Invisible Technician." *American Scientist* 77, no. 6 (1989): 554–63.

———. *A Social History of Truth: Civility and Science in Seventeenth-Century England.* Chicago: University of Chicago Press, 1994.

Shapin, Steven, and Simon Schaffer. *Leviathan and the Air-Pump: Hobbes, Boyle, and the Experimental Life.* Princeton, NJ: Princeton University Press, 1985.

Sivasundaram, Sujit. *Nature and the Godly Empire: Science and Evangelical Mission in the Pacific, 1795–1850.* New York: Cambridge University Press, 2005.

———. "Science." In *Pacific Histories: Ocean, Land, People*, edited by David Armitage and Alison Bashford, 237–60. New York: Palgrave Macmillan, 2014.

Sloan, Phillip. "Darwin's Invertebrate Program, 1826–1836: Preconditions for Transformism." In *The Darwinian Heritage*, edited by David Kohn, 71–120. Princeton, NJ: Princeton University Press, 1985.

———. "The Making of a Philosophical Naturalist." In *The Cambridge Companion to Darwin*, edited by M. J. S. Hodge and Gregory Radick, 17–39. Cambridge: Cambridge University Press, 2003.

Sluiter, C. Ph. "Eine geschichtliche Berichtigung: Die Korallentheorie von Eschscholtz." *Zoologischer Anzeiger* 15 (1892): 326–27.

Snyder, Laura J. *Reforming Philosophy: A Victorian Debate on Science and Society.* Chicago: University of Chicago Press, 2006.

Sponsel, Alistair. "An Amphibious Being: How Maritime Surveying Reshaped Darwin's Approach to Natural History." *Isis* 107, no. 2 (June 1, 2016): 254–81.

———. "Charles Darwin's Notes on the Geology and Corals of the Keeling Islands." In *The Complete Work of Charles Darwin Online*, edited by John van Wyhe, 2010. http://darwin-online.org.uk/content/frameset?viewtype=side&itemID=CUL-DAR41.40-57&pageseq=1.

———. "Constructing a 'Revolution in Science': The Campaign to Promote a Favourable Reception for the 1919 Solar Eclipse Experiments." *British Journal for the History of Science* 35, no. 4 (2002): 439–67.

———. "Coral Reef Formation and the Sciences of Earth, Life, and Sea, c. 1770–1952." PhD diss., Princeton University, 2009.

———. "Darwin and Humboldt." In *A Voyage round the World: Charles Darwin and the "Beagle" Collections in the University of Cambridge*, edited by Alison M. Pearn, 13–15. Cambridge: Cambridge University Press, 2009.

———. "From Cook to Cousteau: The Many Lives of Coral Reefs." In *Fluid Frontiers: New Currents in Marine Environmental History*, edited by John Gillis and Franziska Torma, 137–61. Cambridge: White Horse Press, 2015.

———. "Pacific Islands and the Problem of Theorizing: The U.S. Exploring Expedition from Fieldwork to Publication." In *Soundings and Crossings: Doing Science at Sea, 1800–1970*, edited by Katharine Anderson and Helen M Rozwadowski, 79–112. Sagamore Beach, MA: Science History Publications, 2016.

Stark, Laura Jeanine Morris. *Behind Closed Doors: IRBs and the Making of Ethical Research.* Chicago: University of Chicago Press, 2012.

Stoddart, David R., ed. "*Coral Islands* by Charles Darwin." *Atoll Research Bulletin* 88 (1962): 1–20.

———. "Darwin and the Seeing Eye: Iconography and Meaning in the *Beagle* Years." *Earth Sciences History* 14 (1995): 3–22.

———. "Darwin, Lyell, and the Geological Significance of Coral Reefs." *British Journal for the History of Science* 9 (1976): 199–218.

———. "Grandeur in This View of Life: Darwin and the Ocean World." *Bulletin of Marine Science* 33 (1983): 521–27.

———. "Theory and Reality: The Success and Failure of the Deductive Method in Coral Reef Studies—Darwin to Davis." *Earth Sciences History* 13 (1994): 21–34.

———. "'This Coral Episode': Darwin, Dana, and the Coral Reefs of the Pacific." In *Darwin's Laboratory : Evolutionary Theory and Natural History in the Pacific*, edited by Roy M. MacLeod and Philip F. Rehbock, 24–48. Honolulu: University of Hawaii Press, 1994.

Stott, Rebecca. *Darwin and the Barnacle*. New York: W. W. Norton, 2003.

Sulloway, Frank J. "Darwin and His Finches: The Evolution of a Legend." *Journal of the History of Biology* 15, no. 1 (1982): 1–53.

———. "Darwin's Conversion: The *Beagle* Voyage and Its Aftermath." *Journal of the History of Biology* 15, no. 3 (1982): 325–96.

———. "Further Remarks on Darwin's Spelling Habits and Dating the *Beagle* Voyage Manuscripts." *Journal of the History of Biology* 16, no. 3 (1983): 361–90.

Tallmadge, John. "From Chronicle to Quest: The Shaping of Darwin's *Voyage of the 'Beagle.'*" *Victorian Studies* 23, no. 3 (1980): 325–45.

Thackray, John C. *To See the Fellows Fight: Eye Witness Accounts of Meetings of the Geological Society of London and Its Club, 1822–1868*. Stanford in the Vale, Faringdon, Oxfordshire: British Society for the History of Science, 2003.

Todhunter, Isaac. "William Whewell: An Account of His Writings with Selections from His Literary and Scientific Correspondence." In *Collected Works of William Whewell*, edited by Richard Yeo, vol. 16. Bristol, UK: Thoemmes, 2001.

Traweek, Sharon. *Beamtimes and Lifetimes: The World of High Energy Physicists*. Cambridge, MA: Harvard University Press, 1988.

Turnbull, David. "Cook and Tupaia: A Tale of Cartographic Méconnaissance?" In *Science and Exploration in the Pacific: European Voyages to the Southern Oceans in the Eighteenth Century*, edited by Margarette Lincoln, 117–31. Rochester, NY: Boydell and Brewer, 1998.

———. *Masons, Tricksters and Cartographers: Comparative Studies in the Sociology of Scientific and Indigenous Knowledge*. New York: Routledge, 2000.

van Wyhe, John, ed. *The Complete Work of Charles Darwin Online*, 2002. http://darwin-online.org.uk.

———, ed. *Charles Darwin's Shorter Publications, 1829–1883*. Cambridge: Cambridge University Press, 2009.

———. "Mind the Gap: Did Darwin Avoid Publishing His Theory for Many Years?" *Notes and Records of the Royal Society* 61 (2007): 177–205.

———. "'My Appointment Received the Sanction of the Admiralty': Why Charles Darwin Really Was the Naturalist on HMS *Beagle*." *Studies in History and Philosophy of Science Part C: Studies in History and Philosophy of Biological and Biomedical Sciences* 44, no. 3 (2013): 316–26.

Vorzimmer, Peter J. "The Darwin Reading Notebooks (1838–1860)." *Journal of the History of Biology* 10, no. 1 (March 1, 1977): 107–53.

[Walsh, John Henry]. *Manual of British Rural Sports: Comprising Shooting, Hunting, Coursing, Fishing, Hawking, Racing, Boating, Pedestrianism, and the Various Rural Games and Amusements of Great Britain*. 2nd ed. London: G. Routledge, 1857.

Walters, S. M., and E. A Stow. *Darwin's Mentor: John Stevens Henslow, 1796–1861*. New York: Cambridge University Press, 2001.

Warwick, Andrew C. "Cambridge Mathematics and Cavendish Physics: Cunningham, Campbell, and Einstein's Relativity, 1905-1911. Part I: The Uses of Theory." *Studies in History and Philosophy of Science* 23 (1992): 625–56.

————. "Cambridge Mathematics and Cavendish Physics: Cunningham, Campbell, and Einstein's Relativity, 1905-1911. Part II: Comparing Traditions in Cambridge Physics." *Studies in History and Philosophy of Science* 24 (1993): 1–25.

————. *Masters of Theory: Cambridge and the Rise of Mathematical Physics*. Chicago: University of Chicago Press, 2003.

Weisgall, Jonathan M. *Operation Crossroads: The Atomic Tests at Bikini Atoll*. Annapolis, MD: Naval Institute Press, 1994.

Whewell, William. "Address to the Geological Society, Delivered at the Anniversary, on the 16th of February, 1838." *Proceedings of the Geological Society of London* 2 (1838): 624–49.

————. *Astronomy and General Physics Considered with Reference to Natural Theology*. Bridgewater Treatises on the Power, Wisdom and Goodness of God as Manifested in the Creation, Treatise 3. London: W. Pickering, 1833.

————. *History of the Inductive Sciences, from the Earliest to the Present Times*. London: J. W. Parker, 1837.

————. "The Philosophy of the Inductive Sciences, Founded upon Their History," 2nd ed., 1847. In *Collected Works of William Whewell*, edited by Richard Yeo, vols. 3–5. Bristol, UK: Thoemmes, 2001.

————. Review of *Principles of Geology*, vol. 2 (1832), by C. Lyell. *Quarterly Review* 47, no. 93 (March 1832): 103–32.

White, Paul. "Darwin, Concepción, and the Geological Sublime." *Science in Context* 25, no. 1 (2012): 49–71.

Williams, John. *A Narrative of Missionary Enterprises in the South Sea Islands*. London, 1837.

Wilson, Leonard G. *Charles Lyell: The Years to 1841, the Revolution in Geology*. New Haven, CT: Yale University Press, 1972.

————. *Lyell in America: Transatlantic Geology, 1841–1853*. Baltimore: Johns Hopkins University Press, 1998.

————. "The Geological Travels of Sir Charles Lyell in Madeira and the Canary Islands, 1853–1854." In *Four Centuries of Geological Travel: The Search for Knowledge on Foot, Bicycle, Sledge and Camel*, edited by Patrick Wyse Jackson, 207–28. Special Publication 287. London: Geological Society, 2007.

————. *Sir Charles Lyell's Scientific Journals on the Species Question*. New Haven, CT: Yale University Press, 1970.

Young, J. L. "Names of the Paumotu Islands, with the Old Names so Far as They Are Known." *Journal of the Polynesian Society* 8 (1899): 264–68.

Index

Note: Page numbers followed by "f" refer to figures. Endnotes are listed only when they contain information beyond simple citations.

Aconcagua River (Chile), 53–54, plate 2
Agassiz, Louis, 181–82, 225, 243
Alison, Robert, 54–55, 60, 112
Allan, J., 172, 203
amphibious being: CD as, 8, 33–34, 48, 52, 102, 266; Lyell's concept of, 15–16, 33–34, 144
Anthony, Patrick, 288n22
Armstrong, Patrick, 294n2
Athenaeum (periodical), 26, 110, 203, 304n1, 308n15
atoll, as new term for a "lagoon island," 177, 179, 189, 198, 208, 292
atolls: CD's theory-laden definition of, 199–200; form of, 19–29, 74, 171, 195f, plate 4; problem of their formation, 112, 127–29, 192–93, 194f, 205–6; as remnants of former land, 132, 136, 139, 209–11
authorship: CD and, 11, 148–50, 256, 259–62; scientific, 143, 261–63, 325–27; of theories, 106

Autobiography of Charles Darwin, The (Darwin), 250. *See also* Darwin, Charles (CD): autobiographical recollections of

Babbage, Charles, 119, 135, 138, 155, 174
Banks, Joseph, 111
Baudin, Nicolas, 21
Beagle (ship). *See* HMS *Beagle*
Beaufort, Francis: attitude toward CD, 101, 111, 171, 196, 203; and CD's *Structure and Distribution of Coral Reefs*, 203; instructions to FitzRoy, 24–27, 30–32, 38, 44, 61, 73, 75, 82, 91–92, 98–99; Lyell and, 115, 171; philosophy on science, 27, 79
Beaumont, Léonce Élie de, 109, 119, 161–62, 179, 203, 251, 297
Beche, Henry De la. *See* De la Beche, Henry

Becher, A. B., 317n59
Beechey, Frederick William, 22–26, 32, 73–74, 96, 111, 115, 171, 191, 203, 306n47, 315n13
Bennett, F. D., 306n47, 315n13
Bennett, J. J., 235
Bermuda, 158, 186, 315n13
Biagioli, Mario, 326n3
Blackwood, Francis Price, 27
Bligh, William, 19, 306n47
Bonpland, Aimé, 30
Bougainville, Louis-Antoine de, 19–20, 65, 68, 288, 306n47
Bowlby, John, 150, 278n7, 279n9, 298n3
Boyle, Robert, 262
British Association for the Advancement of Science, 44–45, 242–43
Broderip, William, 114, 175–76
Brown, Robert, 252
Browne, Janet, 181, 311, 319n5
Brubaker, Rogers, 277–78n3
Buch, Leopold von, 23, 109, 203, 222
Buckland, William, 110, 126, 157, 163
Bunbury, Charles, 224
Bynoe, Benjamin (*Beagle* assistant surgeon), 153

Camerini, Jane, 197–98, 209, 316n38
Cannon, Susan Faye, 34–35
Cape Verde Islands, 36–37, 50
Carpenter, William, 249
Cavahi (atoll). *See* Kauehi Atoll (Noon Island)
Chalk (geological formation), 121, 157–58, 161, 179, 181
Chamisso, Adelbert von, 21, 22, 31, 199, 203, 282, 286, 315n13, 316n48
Christian, Fletcher, 19
Clunies-Ross, John, 82
Cocos (Keeling) Islands, 81, 294n2.
    See also Keeling reefs
Coldstream, John, 28, 32

Coleridge, Samuel Taylor, 184
Colp, Ralph, 4, 278n7, 279n9, 311n1
"consilience of inductions," Whewell's notion of, 134
Conybeare, William, 293n43
Cook, James, 19–20, 72, 111, 306n47, 315n13; chart of Tahiti and Eimeo, 69f
Cooper, Frederick, 277–78n3
coral reef formation: CD's, and history of life, 136–40; CD's autobiographical recollections of, 253–54; CD's theory of, 3, 7–8, 55–70, 253; claims made about how CD developed, 105–7; as compelling issue, for Europeans between 1770 and 1830, 19–24; contrast between different iterations of CD's theory about, 126–28, 138–39, 170–72, 196, 200, 249–57; development of, 8, 11; Lyell's appropriation of CD's, 178–82; shift by CD to expand, 131–36; status relative to CD's species theory, 248–50; studying, as objective of *Beagle* voyage, 24–27; whether CD's theory thereof was developed in South America, 48–49, 55–61, 77, 80
Corbin, Alain, 126
Couthouy, Joseph Pitty, 191; depiction of Eimeo (Moorea), 70f
Covington, Syms, 111, 262
crater theory (of reef formation), 22–23, 45, 106, 112, 118, 130, 138–39, 203; CD's reasons for rejecting, 70, 74–75, 78, 83, 92, 97, 130, 199–200, 211; Lyell's embrace of, 23, 139; Lyell's eventual rejection of, 118–19, 122–23, 140–43, 179–80, 221
credit, attribution of, 6, 12, 117, 141, 143, 146, 180, 238, 259, 262–64, 325–26n3, 327n11

Cuvier, Georges, 28, 31, 42–44, 112, 243, 256

Dampier, William, 306n47
Dana, James Dwight, 211, 249–50, 324n89
Dangerous Archipelago. *See* Tuamotu Archipelago (Dangerous or Low Archipelago)
Darwin, Annie (CD's daughter), 232
Darwin, Caroline (CD's sister), 52, 106, 108, 112, 121
Darwin, Catherine (Emily Catherine) (CD's sister), 55, 61
Darwin, Charles (CD): anxiety of, and Lyell, 169–70; authorship and, 11, 106, 148–50, 256, 259–62; autobiographical recollections of, 12, 31, 55–57, 60, 77, 80, 126, 248–57, 291, 302, 324–25; awarded Wollaston Medal, 238–40; beginning of anxiety of, about speculation, 160–66; debt to Lyell, 262–63; diary of, 4–5; entry into scientific community of, 5–6; eureka moment of, 253; factors of success, 263–64; fear of speculation and, 3–4; finches of Galápagos Archipelago and, 253–54; geological career of, 2–3; glacial theory, acceptance of, 181–82; growing anxiety of, 149–51; homelife of, 234; honors awarded to, 248–49; illness, 155–56, 169–70, 178, 182–84; Lyell and, 241–44, 259–60; Lyell's influence on, 255–56; Lyell's patronage of, 261–63; marriage to Emma Wedgwood, 170; obligations of, as student of Lyell, 156–60; persona of, after coral reef paper to Geological Society, 151–56; and the "pleasure of gambling," 2, 54, 155; as practitioner of Lyellian geological speculation, 144–48; publishing

advice to younger authors, 222–23, 228–30, 251; relationship with Lyell, 11; reliance on local informants, 72; restrained approach to publishing of, 1–2; scholarly conflict and, 149–50; seeking of causal explanations and, 256–57; theories and, 266–68; theorizing and, 256; theory of coral reef formation of, 3, 7; as tormented theorist, 4. *See also* theories
Darwin, Charles (CD), and *Beagle* voyage: ambitious plan for studying zoophytes of, 41–45; as "amphibious being," 8, 33–34, 48, 52, 102, 266; contexts of ambitions of, 18–19; convergence of areas of interest of, 61; coral reef studies of, 210–11; in dangerous reefs of Low Archipelago, 65–68; "1835 Coral Islands" essay, 73–80; enthusiasm for South Sea Islands, 30–32; eureka moment of, 63–80; geology of South America and, 50–61, plate 3; hydrographic initiative at Mauritius, 98–102; hydrographic surveying and, 37–41; in Keeling Islands, 81–82; maritime perspective of, for theories of South American landscape, 47–49; Patagonian "pebbles" and, 50–52, 51f, 55; Red Notebook of, 109; return from voyage, 108–12, 115–17; Santiago Book of, 55–60; scientific approach of, at beginning of, 36–37; scientific training of, 27–30; sea-level study of South Keeling reef, 82–91; significance of sounding for, 33–34; in Tahiti, 63–64, 68–73; use of leaping poles, 85, 86f, 91; view on elevation of South America, 52–54; vigor of, during *Beagle voyage*, 18

Darwin, Charles (CD), and coral reef theory, 7–8, 136–40, 224; as aspiring geologist, 108–12; authoring theories and, 106; B Notebook of, 137; coral reef map, 209–10, plate 10; "1835 Coral Islands" essay, 111–12, 128; 1837 paper to Geological Society, 125–26, 138–44, 148; going public on his coral reef theory at Geological Society, 126–31; primacy of geology in life of, 121–23; Red Notebook of, 121; student relationship with Lyell, 114–21; voted into Geological Society, 117–18; whether the theory was more "Lyellian" than Lyell's, 106–7, 260

Darwin, Charles (CD), species theory of: attitude of "facts" vs. "theory" for, 220–21, 222–24; debate on publishing of, 216; instructions to Emma Darwin on publishing, 217–21; Lyell's choreographing publishing of, 231–32; Lyell's prompting to publish, 230; main text of, 245–47; preface of, 244–45; promotion of, 221–22; publishing of, as "abstract, 240–48; rejection of Lyell's suggestion to publish sketch of, 224–28; release of, 244–45; reluctant approach to publishing, 215–16; reviewers of, 247–48; selection of ideal editor for, 219–20; Wallace's letter and, 232–34

Darwin, Charles Waring (CD's son), 232, 234

Darwin, Emma (Wedgwood), 155, 170, 178, 188, 224, 232, 233, 280n11, 310n57, 311n62; CD's publishing instructions to, 217–22, 225, 242; grief over death of son Charles, 234

Darwin, Erasmus (CD's grandfather), 251

Darwin, Erasmus Alvey (CD's brother), 23, 110–11, 118, 297n61

Darwin, Francis (CD's son), 251

Darwin, George (CD's son), 85, 228–30, 251

Darwin, Henrietta (CD's daughter). *See* Litchfield, Henrietta (Darwin)

De la Beche, Henry, 73, 113, 116f, 142f, 152, 163, 211, 256, 283n17, 295n7, 306n47

Demosthenes, 230

descriptive geology, 146–47

Desmond, Adrian, 4, 115, 184, 280n14, 307n71, 319n3

Dillon (navigator), 306n47

Dohrn, Anton, 230

ecology, 7, 34, 41, 281n17

Eddington, Arthur Stanley, 278n5

Edinburgh Review (periodical), 173–74

Ehrenberg, Christian Gottfried, 118, 130, 186, 191, 306n47, 315n13

Eimeo (Moorea): CD's eureka moment while viewing, 70–74, 126, 132, 201, 294n48; Cook's chart of Tahiti and, 69f; Couthouy's depiction of, 70f

Elements of Geology (Lyell), 114, 156–60, 174, 188

Endersby, Jim, 265–66

Eschscholtz, Johann Friedrich, 21–22, 31, 92, 199, 316n48

Essay on the Principle of Population (Malthus), 137, 254, 281n15

finches, of Galápagos Archipelago, 253–54

Fitton, William, 174

FitzRoy, Robert, 18–19, 150, 203, 306n47; Admiralty's orders for, 18, 24–27; on CD's insufficient acknowledgment of *Beagle* officers' assistance, 150–54, 262; charts by, 49f, 66f, 84f; commentary on CD's draft essay on coral reefs, 111; decision to call at the Keeling Islands, 81–82; favorable reports on CD's

work, 111; gift of Lyell's *Principles*
(vol. 1) to CD, 36; *Narrative* of the
Beagle voyage, 98, 150, 160, 173;
navigation of Low Archipelago,
65–68, 66f; obligation to study coral
reef formation, 61; scientific authority compared to CD's, 262; scientific
observations by, 53, 68, 78, 88,
91–93, 112; surveying work of (*see*
hydrography); zeal for Alexander
von Humboldt's work, 30
Flinders, Matthew, 21, 31, 306n47,
315n13
Forbes, Edward, 28, 210, 220, 225, 243
Forster, Georg (George), 20
Forster, Johann Reinhold, 20–21, 22, 31,
73–76, 190, 306n47
Fox, William Darwin, 54, 143, 153–55,
233, 237
Franklin, John, 22–23
Freycinet, Louis-Claude de, 22, 315n13

Gaimard, Joseph Paul, 22–23, 31, 45, 60,
73, 75–76, 83, 84, 89, 93, 100, 130,
138, 190–91, 199, 203
Galápagos Islands, 47, 65, 253–54,
315n14; craters analyzed retrospectively by CD in light of Tahiti coral
reef theory, 78–79, 293n48
Galison, Peter, 327n11
Gärtner, Karl Friedrich von, 230
Geoffroy Saint-Hilaire, Étienne, 28
*Geological Observations on Volcanic
Islands* (Darwin), 3, 155, 173, 208,
222
Geological Society, 10, 108, 110; CD
awarded Wollaston Medal by, 238–
40; CD going public on his coral
reef theory at, 126–31; CD made
fellow of, 117–18; Lyell's 1836 presidential address, 112–13
geology: descriptive and dynamic approaches described by Whewell,
146–47; social makeup of British

geological community, in 1830s,
152–53
glaciers, theories of, 120, 161, 181–82,
223, 239, 240
Glen Roy, "parallel roads" of, 155, 164–
65, 170, 181–82, 223, 239, 255
Gould, John, 117
Grant, Robert, 28–29, 32, 36–38, 42, 45,
106, 115, 147, 160, 263
Gray, Asa, 234–35
Great Barrier Reef, 20, 27, 74, 129, 177,
211, 315n13
Greenough, George, 112, 117
Gressier, C. L., 196

Hacking, Ian, 313n54
Hall, Basil, 174–75, 230
Henslow, John Stevens, 29–30, 110,
115–18, 141–43, 147, 155, 183, 220,
237, 252, 303n77; CD's *Beagle* letters to, 36–37, 42–45, 51, 54, 61,
107, 110, 112, 126
Herbert, Sandra, 106, 163, 227, 256,
271, 289n39, 293n48, 294n2, 296n45
Herschel, John, 101, 109, 117–18, 122–
23, 135, 136, 140–41, 143, 144, 152,
162, 163, 202, 207, 225, 245, 256,
260, 299n17, 303n86
Hewett, William, 26
HMS *Beagle*: chart showing rate of
travel, 1831–36, 10f; chart showing
track of, 1831–36, 9f; key destinations visited in South America by,
49f; library of, 73; in longitudinal
section and in overhead view, 35f;
studying coral reef formation as
objective of voyage, 24–27
Hodge, M. J. S. (Jonathan), 28, 282n20,
284n28, 289n41
Honden Island (Puka-Puka), 65
Hooker, Joseph, 111, 209–10, 220–22,
224, 225, 227–28, 230; Wallace's letter and, 234–36, 261, 265
Hopkins, William, 135, 328n22

Horner, Leonard, 163, 204, 222, 255
Humboldt, Alexander von, 18, 19, 29–30, 34, 36, 38, 40, 64, 69, 71, 73, 203, 280n15, plate 5; his books as sources of data and theory for CD, 64, 77, 79–80, 92, 97, 109, 162, 190; as CD's hero, 29, 34; as surveyor-style naturalist, 41, 288n22
Humboldtian science, 34–35, 38, 79, 96
Huxley, Thomas Henry, 224, 225, 241, 246, 249
hydrographers, 26, 80, 144, 191, 203; as underacknowledged contributors, by CD, 153–54, 262
hydrography, 24–27, 334; as resource for CD, 7–8, 16–17, 33–34, 37–41, 39f, 48, 50–52, 61, 79–80, 82, 92–102, 266

identity, as an analytical term, 277n3; of theories, 266–67
intellectual property, 261

Jackson, Julian, 204–5, 317n61
Jameson, Robert, 31, 283n17
Jenyns, Leonard, 165, 170, 188, 221
John Murray (publishing house), 113–14, 207–8, 241
Journal of Researches (Darwin), 150, 154, 155, 160, 248; German translation of, 210; reviews of, 173–76; second edition of, 207–9
Jukes, Joseph Beete, 211

Kauehi Atoll (Noon Island), 67–68
Keeling reefs, 101, 118, 127, 129–30, 133, 189–91; CD's sea-level study of, 82–91; hydrographic survey of, 91–98
Keynes, Simon, 92
King, Phillip Parker, 26, 191
King's College (London), 152, 220, 314n20
Kohler, Robert, 264–65

Kotzebue, Otto von, 315n13
Krusenstern, Johann von, 65, 67, 74, 203

Labillardière, Jacques, 306n47
Lagoon islands. *See* atolls
Lamarck, Jean-Baptiste, 28, 43–44, 110, 121, 224
Lamarckism, 225
Lamouroux, J. V. F., 21, 28, 43, 100, 112
Lapérouse, comte de (Jean-François de Galaup), 306n47
Lardner, Dionysius, 144
Latour, Bruno, 172, 278n5, 280n14, 311n12
Lauder, Thomas Dick, 164, 181
Lavoisier, Antoine-Laurent de, 202
Leader, Zachary, 184
leaping poles, plate 8; CD's use of, 85, 86f, 172, 189
Lesson, René, 136, 305n37, 306n47
Liesk, William, 315n13
life, history of, and CD's coral reef theory, 136–40
Linnean Society, 110, 227, 235–40
Litchfield, Henrietta (Darwin), 232, 234, 280n11
Lonsdale, William, 188, 220, 304n2
Low Archipelago. *See* Tuamotu Archipelago (Dangerous or Low Archipelago)
Lütke, Friedrich (Fyodor Litke), 171, 315n13
Lyell, Charles, 22, 145f, 147, 306n47, 315n13; "amphibious being" metaphor, 15–16, 33–34, 144; anxiety of CD and, 169–70; appropriation of CD's coral reef theory, 178–82; as author, 112–14; CD's commemoration of, 251–52; CD's coral reef map and, 209; CD's reef formation theory and, 118–19, 122–23; as cherished reader of CD's work, 241–42; choreographing of CD's

public presentation as an author, 140–44, 230–38; criticized by De la Beche for being too quick to theorize, 113, 116f, 142f; *Elements of Geology*, 114, 156–60, 174, 188; embrace of crater theory by, 23; "heretical doctrines," 119, 163, 225, 241; as mentor to CD, 11, 107–8, 113–21, 156–60, 259–63; *Principles of Geology*, 11, 15–16, 17f, 22–24, 26–27, 36, 43, 73, 77, 108, 109, 112, 114, 156, 157, 174, 178–79; rejection of crater theory by, 122–23; suggestion to publish a "sketch" of his species theory, 224–28; Wallace's letter and, 232–36

MacCulloch, John, 164, 181
Maclaren, Charles, 204–8
Malcolmson, John, 171–72
Malthus, Thomas Robert, 137, 254, 281n15
Mantell, Gideon, 26, 174
Manual of Scientific Enquiry (Herschel), 101–2
Matthew Effect (Merton), 263, 264
Mauritius, 82, 106–9, 123, 127–28, 192; CD's hydrographic study of, 98–102
McSherry, Corynne, 327n11
Merton, Robert, 263–64, 279n10
Mill, John Stuart, 229, 249
Moore, James, 4, 115, 184, 280n14, 307n71
Moorea. *See* Eimeo (Moorea)
Moresby, Robert, 186, 191, 203, 297n68, 312n24, 315n13
Morlot, Adolph von, 222–23, 229, 233, 240, 247, 251
Murchison, Roderick Impey, 108, 140–42, 152, 163, 243, 252, 308n10, 310n56, 325n98
Murray, John (marine chemist), 324n87
Murray, John, II (publisher). *See* John Murray (publishing house)

Narrative of a Voyage to the Pacific and Beering's Strait (Beechey), 22–23, 32, 73–74, 284n35
natural selection. *See* species theory
Nelson, Richard, 158, 315n13
Neptunism. *See* Wernerian geognosy/geology
Noon Island. *See* Kauehi Atoll (Noon Island)

On the Origin of Species (Darwin) (1859). *See* species theory
Ospovat, Dov, 281–82n19, 319n7
Owen, Richard, 115, 117, 119–20, 147, 238–39, 243, 319n5
Owen, William, 306n47

"parallel roads" of Glen Roy. *See* Glen Roy, "parallel roads" of
Peacock, George, 28
pebbles, of Patagonia (Rodados Patagónicos), 50–52, 51f, 55, 57–59, 58f
Phillips, John, 144–46, 163–64, 239
Philosophy of the Inductive Sciences (Whewell), 134
Plinian Society (Edinburgh), 29, 42
Powell, F. T., 315n13
practices, as a topic of study for historians of science, 6, 259, 264–65
Preliminary Discourse on the Study of Natural Philosophy (Herschel), 122–23, 144, 310n52. See also *vera causa* (as desideratum in theorizing)
Principles of Geology (Lyell), 11, 15–16, 17f, 22–24, 26–27, 36, 43, 73, 77, 108, 109, 112, 114, 156, 157, 174; CD's contributions to, 178–79
procrastination, 183
Pye-Smith, John, 308n5

Quarterly Review (periodical), 113–14, 173, 175–76

Quoy, Jean René Constant, 22–23, 31, 45, 60, 73, 75–76, 83, 84, 89, 93, 100, 130, 138, 190–91, 199, 203, 306n47

Red Notebook (Darwin), 109, 121–22
Rodados Patagónicos. *See* pebbles, of Patagonia (Rodados Patagónicos)
Royal Society: CD's election to, 170; CD's Glen Roy paper presented to, 149; Copley Medal awarded to CD, 249. *See also* Glen Roy, "parallel roads" of
Rudwick, Martin, 107, 152, 198, 279n8, 279n11, 328n23

Sabine, Edward, 249
Santiago Book (Darwin), 55–60, 70, 109
Schaffer, Simon, 303n86
Scrope, George Poulett, 310n56
Secord, James, 28, 109, 250
Sedgwick, Adam, 29–32, 36–37, 106, 108, 110, 114–15, 147, 152, 153, 163, 203, 209, 237, 243, 249, 263, 310n56, 319n5
Serapis. *See* Temple of Serapis (Pozzuoli, Italy)
Seth, Suman, 328n20
Shapin, Steven, 262, 325–26n3
Silliman, Benjamin, 187
Sivasundaram, Sujit, 153
Sloan, Phillip, 28, 285n47, 289n41
Smith, Charles Hamilton, 210
Smith and Elder (publishing house), 155, 182, 203, 250
Smyth, William Henry, 171
sociology of scientific knowledge, 5–6, 34, 172, 259, 262–63, 279n10, 279n11, 281n16
Solander, Daniel, 111
sounding lead, 20, 22, 33–34, 48, 50, 72, 93, plate 1; CD's acquisition of spec-

imens from, 40–42, 56f, 95f, 99–102, 189, 266. *See also* hydrography
soundings. *See* hydrography
South Keeling reef, plate 8, plate 9; CD's collection of zoological and geological specimens on, 87, 88f; CD's cross-sectional diagrams of structure, 89–91, 90f, 91f, 97f; CD's sea-level study of, 84f; difficulty of taking soundings on, 93–96; hydrographic survey of, 91–98
species theory, 1, 169, 185, 215; CD's instructions to Emma Darwin on posthumous publication, 217–21; CD's reluctant approach to publishing, 215–16; CD's study of, as diversion, 4, 182–84; debate on publishing of, 216; Lyell's choreographing publishing of, 231–32; Lyell's prompting to publish, 230; and *Origin of Species*, 240–48; as preoccupation of Darwin scholars, 12, 182–84, 259–60; promotion of, 221–22; rejection of Lyell's suggestion to publish sketch of, 224–28; selection of ideal editor for, in case of posthumous publication, 219–20; Wallace's letter and, 232–34. *See also* theories
speculation. *See* theories
Spencer, Herbert, 252
Stanley, Owen, 27
Stark, Laura, 320n8
Stewart, Peter Benson, 153
St. Jago. *See* Cape Verde Islands
Stoddart, David, 106, 198, 296n44
Stokes, John Lort, 153, 191
Structure and Distribution of Coral Reefs, The (Darwin), 170–72, 185; CD's intended audience for, 203–4; CD's objectives for, 188–89; CD's study of Keeling reefs and, 83; chapter 5's theory, 192–203; chapters 1 to

4 of, 189–92; facts in, 188; Lyell's pressure on CD to finish, 186–88; pressure of public expectations and, 173–78; reviews of, 204–9; subsidence theory, 33, 118, 130, 131

Sulivan, Bartholomew J., 92, 94f, 95f, 95–96, 99, 126, 142–43, 153

Sulloway, Frank, 253–54, 294n48

Tahiti: CD in, 63–64, 68–73; Cook's chart of, 69f; vegetation of, 68–69, plate 6

Takapoto Atoll, plate 4

Temple of Serapis (Pozzuoli, Italy), 15–16, 17f, 36, 122

theories, ix, 2–4, 5–6, 18–19, 204, 224–25, 248, 249; CD and, 106, 149–50, 154–55, 165, 170, 185, 198–99, 215–24, 251–52, 266–68, 281–82; CD apologizing for being overly speculative, 163, 210–12, 245–47, 254–55; CD criticized for being overly speculative, 173–75, 205–6; CD describing theories in terms of their "truth," 131, 134, 137, 162, 212, 217, 243–44, 266–67; CD describing theories in terms of their utility, 207, 212; CD's "pleasure of gambling," 2, 54, 155; distributed by CD in the same fashion as physical specimens, 117; facts and, 170–71, 220–24, 233, 245; imposing retrospective coherence on theorist's activities, 253–54; in natural history as opposed to physics, 265–66; and memory, 5, 252–57; as possessing an "identity," 8, 266–68; as presenting special challenges for authorship, 3, 106, 182–84, 186–87, 215, 219, 221, 226–27, 240, 256–57, 261–64; as the preserve of a select few within the geological community, 140–41, 152–53

"theories of the earth," 256; CD's, 3, 121–22, 155, 178

Tuamotu Archipelago (Dangerous or Low Archipelago), 20, 65–68, 66f, 74, 193, plate 4

Tupaia, 72

United States Exploring Expedition (1838–42), 70f, 191, 211, 249

Usborne, Alexander Burns, 153

van Wyhe, John, 300n28, 319n1

vera causa (as desideratum in theorizing), 123, 163, 324n84

Voyage of the Beagle (Darwin). See Journal of Researches (Darwin)

Wallace, Alfred Russel, 246; CD cultivating scientific allegiance from, 240–41; 1858 letter to CD and enclosed species paper, 12, 230–37, 238; reaction to Lyell and Hooker's presentation of his paper to Linnean Society, 237–38; stature raised by Lyell's actions as CD's had previously been, 238, 261

Warwick, Andrew, 279n11, 287n3, 325n1, 328n21

Watson, Hewett Cottrell, 322n68

Wedgwood, Emma. See Darwin, Emma (Wedgwood)

Wedgwood, Hensleigh, 118, 217

Werner, Abraham Gottlob, 29, 286n62, 299n12

Wernerian geognosy/geology, 28, 30, 256, 286n62, 309n35

Wernerian Society (Edinburgh), 29

Whewell, William, 152, 162, 163, 200, 202, 244–45, 256, 260; coining of term "uniformitarianism" to refer to Lyell's geology, 132; commentary on CD's "Lyellist" geology, 118, 146–48, 239, 261, 304n20;

Whewell, William (*continued*)
    notion of consilience, 134; *Phi-
    losophy of the Inductive Sciences,*
    134; as president of the Geological
    Society, 117–18, 123, 132, 134, 143,
    144, 146–48
Wickham, John Clements, 81–82
Wilkes, Charles, 191, 327n9
Williams, Helen Maria, 297n74
Williams, John, 151–53, 196, 262
Wollaston, Thomas, 224, 225
Wollaston Medal (Geological Society):

awarded to CD, 238–40; awarded to
    Richard Owen, 147
Woodd, C. H. L., 223, 229, 251
writer's block, 183–84

Zoological Society, 110, 116
zoology, as the initial frame of reference
    for CD's work on corals, 35–36, 37,
    39f, 40, 44–45, 61, 93, 105–6, 112
zoophytes, 18, 24, 28–29, 31, 32, 35–
    36, 60, 71; CD's ambitious plan for
    studying, 41–45, 61